PHARMACOLOGY
OF THE HYPOTHALAMUS

PHARMACOLOGY OF THE HYPOTHALAMUS

Proceedings of a British Pharmacological Society International Symposium on the Hypothalamus held on Thursday, September 8th, 1977 at the University of Manchester, U.K.

Edited by

B. COX, I. D. MORRIS
and
A. H. WESTON

Department of Pharmacology, Materia Medica and Therapeutics
University of Manchester, U.K.

UNIVERSITY PARK PRESS
Baltimore

First published 1978 by
The Macmillan Press Ltd
London and Basingstoke

Published in North America by
UNIVERSITY PARK PRESS
233 East Redwood Street
Baltimore, Maryland 21202

Printed in Great Britain

Library of Congress Cataloging in Publication Data

British Pharmacological Society International
Symposium on the Hypothalamus, University of
Manchester, 1977.
Pharmacology of the hypothalamus.

Bibliography: p.
Includes index.
1. Hypothalamus—Congresses. 2. Hypothalamic hormones—Congresses. I. Cox, Barry, 1937– II. Morris, Ian D., 1948– III. Weston, Arthur Henry, 1944– IV. British Pharmacological Society. V. Title.
QP383.7.B74 1977 612'.826 78–15946

ISBN 0–8391–1299–8

The generous financial contributions of the following organisations are gratefully acknowledged:

Allen and Hanburys
Astra Chemicals
Ayerst Laboratories
Boehringer Ingelheim
Ciba Laboratories
Geigy Pharmaceuticals
Hoechst U.K.
ICI Pharmaceuticals Division
Janssen Pharmaceuticals
Leo Laboratories
Organon Scientific Development Group
Sandoz Products
Schering AG
Smith Kline and French Laboratories
Stuart Pharmaceuticals
Wellcome Trust
Winthrop Laboratories

Symposium speakers

W. Feldberg, National Institute for Medical Research, Mill Hill, London, NW7 1AA, UK

B. Halász, Second Department of Anatomy, Histology and Embryology, Semmelweis University Medical School, Tüzoltó u. 58, 1094 Budapest, Hungary

K. Fuxe, Department of Histology, Karolinska Institutet, Stockholm, Sweden

J. S. Kelly, MRC Neurochemical Pharmacology Unit, Department of Pharmacology, Medical School, Hills Road, Cambridge, CB2 2QD, UK

A. S. Milton, Department of Pharmacology, University Medical Buildings, Foresterhill, Aberdeen, AB9 2ZD, UK

E. Flückiger, Biological and Medical Research Division, Sandoz Ltd, Ch-4002 Basle, Switzerland

A. V. Schally, Endocrine and Polypeptide Laboratory, Veterans' Administration Hospital, 1601 Perdido Street, New Orleans, Louisiana 70146, USA

G. M. Besser, Department of Endocrinology, St. Bartholomew's Hospital Medical College, London, EC1A 7BE, UK

L. Martini, Universita di Milano, Instituto di Endocrinologia, Via A. del Sarto 21, 20129 Milano, Italy

Contents

Preface

When Ian Morris proposed that a symposium on the pharmacology of the hypothalamus should be held during the September, 1977 meeting of the British Pharmacological Society in Manchester the idea was immediately accepted, since recent advances in this field suggested that the theme would be of widespread interest.

In the time available it was impossible to include all aspects of hypothalamic function and so a programme was selected which attempted to stress new developments and to relate experimental findings to potential therapeutic applications. Interest in this research field and its importance was emphasised by the fact that some 500 people attended the symposium. Moreover the editors note with pleasure the award of the Nobel Prize in Medicine to one of the contributors, A. V. Schally, shortly after the meeting. No causal relationship is claimed!

We gratefully thank all the organisations (listed earlier) who provided generous financial support and also the academic, secretarial and technical staff of the Department of Pharmacology at Manchester University without whose help neither the symposium itself nor this publication would have been possible. We are grateful to L. Martini, who chaired the afternoon session, for providing an additional chapter and we also thank the speakers themselves whose stimulating contributions played such an important role in the success of the symposium.

Manchester, 1978

B. C.
I. D. M.
A. H. W.

Introduction

W. Feldberg*

A good way to explain present-day politics is to go back to Adam and Eve, to show how they looked upon their problems and then how the problems have developed over the centuries. Although the hypothalamus is one of the most ancient parts of the vertebrate forebrain, speaking phylogenetically, we need not go quite as far back as that in order to explain its functioning. Physiologists can stay in the twentieth century. For instance, in the 1900 edition of Schaefer's famous textbook of physiology, the hypothalamus is not mentioned at all, either in the index or in the chapter which deals with the parts of the brain below the cerebral cortex. This chapter was written by no less a person than Sherrington. The anatomist has to go a little further back in history because in 1865 a brain atlas was published by Luys in which the hypothalamus is described as the nucleus of Luys.

The first evidence for a physiological function of the hypothalamus was obtained in 1909 by Karplus and Kreidl (1909). They showed that its electrical stimulation caused a discharge in the cervical sympathetic nervous system which led to pupillary dilatation, opening of the eyes and withdrawal of the nictitating membranes. Later, it was shown that practically every effect mediated by the sympathetic nervous system can be reproduced by its electrical stimulation. The sympathetic nervous system is, as we say today, represented in the hypothalamus.

Recently, Wei and I (Feldberg and Wei, 1977; 1978), analysing the cardiovascular effects of morphine, have been able to influence selectively the sympathetic discharge from the hypothalamus to one organ, the heart. Injected into the third cerebral ventricle of anaesthetised cats, morphine produces pronounced cardio-acceleration without signs of a sympathetic discharge to other organs, for instance, to the muscles of the eye, to the pupils or nictitating membranes.

The upper record of figure 0.1 shows the morphine tachycardia. On injection at the arrow of 400 μg of morphine into the third ventricle, heart rate rises from 200 to 330 per minute. The lower record shows the associated pressor response resulting from the cardiac effect because after removal of both stellate ganglia, both tachycardia and pressor response are abolished. Morphine thus induces, probably through an action on the paraventricular nuclei, a sympathetic discharge to the heart alone without even affecting vasomotor tone.

But to return to the early history. We go back to 1914 because in that year Isenschmid and Schnitzler (1914) found that the chief mechanism for the control of body temperature was localised in the hypothalamus. But it was not until 1932 that Keller and Hare (1932) demonstrated that without its hypo-

*National Institute for Medical Research, Mill Hill, London, NW7 1AA

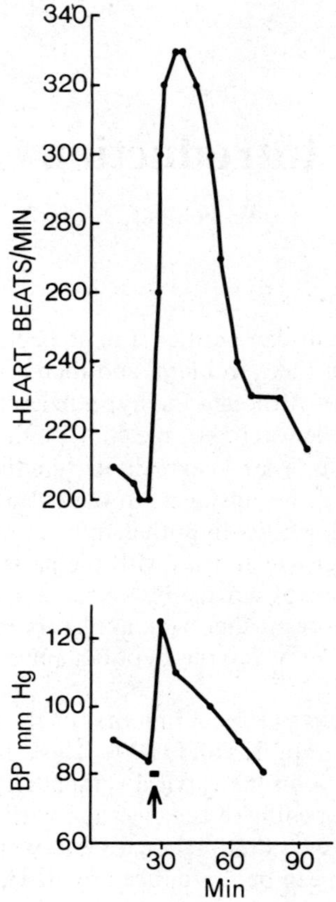

Figure 0.1 Pulse rate per minute (top record) and mean arterial blood pressure (bottom record) from a cat anaesthetised with chloralose. A Collison cannula was implanted into the third cerebral ventricle and the aqueduct cannulated. At the arrow, injection of 400 μg of morphine sulphate into the third ventricle. (After Feldberg and Wei, unpublished observations)

thalamus the rabbit had lost completely its ability to maintain constant temperature, and that it was unable to shiver. It exhibited, however a remarkable release of the heat-loss mechanism: merely pressing the tail elicited typical panting.

This early anatomical dissociation between the sites for heat-producing and heat-eliminating mechanisms is of particular interest to those who study the pharmacology of body temperature and use the method of injecting drugs into the cerebral ventricles. Many drugs applied in this way appear to be unable to discriminate between the synapses: they produce strong shivering with a rise in body temperature, but at the same time panting.

In the 1920s and 1930s our knowledge of the hypothalamus advanced rapidly. In 1940 a symposium was held in the USA entitled *The Anatomy, Physiology and Clinical Significance of the Hypothalamus*. From the title it is evident that the pharmacology of the hypothalamus was not yet born. Yet the proceedings included over a thousand references. Since then, progress has accelerated from year to year, and there is certainly no indication of its slowing down.

This book commences, as it should, with the anatomy, but in a modern sense because the first contributor, Bela Halász, stresses the functional anatomy of the hypothalamus which, in the hardened human brain, weighs approximately 4 g – that is, about 0.3 per cent of the whole brain.

Today two kinds of anatomist exist: those who work on myelinated nerve fibres and those who work on non-myelinated nerve-fibres. We owe our present knowledge of non-myelinated nerve fibres in the central nervous system to a great extent to our Swedish colleagues, and we are fortunate that K. Fuxe has contributed a review of the transmitters of the hypothalamus. The contributions of Fuxe and his colleagues to this newly opened and continuously expanding field is enormous. Today it includes not only the classical transmitters, the monoamines and acetylcholine, and the inhibitory transmitters, the amino acids, but also a wide range of peptides (like the enkephalins, substance P, angiotensin, the thyrotrophin releasing factor and so on) for which transmitter functions have been postulated but not yet always convincingly proved.

The following four chapters are devoted to the pharmacology of the hypothalamus. J. S. Kelly and L. P. Renaud stress the general difficulties we have in understanding how the electrophysiological activity in the hypothalamus is transformed into hormone release, release of vasopressin and oxytocin for instance, and in interpreting some of the results which are obtained with microinjections of substances into small regions of the hypothalamus and which show individual neurones consistently to be either excited or inhibited by the same putative neurotransmitter or neurohormones.

If the reader gets disheartened by these difficulties the next chapter by A. S. Milton on the pharmacology of temperature regulation should cheer him up by illustrating the value of pharmacological methods in elucidating hypothalamic functions. Milton himself made the fundamental discovery that prostaglandins of the E series produce fever when acting on the hypothalamus. A pharmacological approach to the thermoregulating function of the hypothalamus was urgently needed, otherwise we would still look upon thermoregulation in the hypothalamus as the physicist looks upon a thermostat, forgetting that the central synapse is a pharmacologically sensitive structure.

From the historical point of view the next chapter by E. Flückiger on ergot alkaloids might well have been given earlier, because as long ago as 1935 Taylor and Weld observed 'sham rage' on intravenous injection of ergotoxin and attributed the effect to an action on the hypothalamus. Looking at the 1937 edition of the textbook of Best and Taylor, we see that Taylor found it difficult at that time to reconcile his observation with that of Hess– that electrical stimulation of the hypothalamus induces sleep. It has been a long way from this initial observation to the modern aspects of the actions of ergot alkaloids on the endocrine effects, which have been brought to light by the fascinating work of Flückiger and his colleagues on the properties of these alkaloids to inhibit prolactin and to facilitate gonadotrophin secretion.

The final chapter is perhaps the most topical of all. Not only the ergot alkaloids exert endocrine effects. Numerous peptides, the regulatory peptides of the hypothalamus as they are called, have this property as well. A. V. Schally and his twelve associates review hypothalamic peptide hormones and their analogues.

In the Proceedings of the first symposium on the hypothalamus in 1940, already mentioned, over a thousand references were given. This is nothing compared to more than five hundred references given in a review in 1977 by Vale, Rivier and Brown (1977) dealing solely with these regulatory peptides. And, by the way, Schally himself and his associates are named as authors in sixty five of these references.

In our ambition to unravel the complexities of the hypothalamus we get so specialised that we may be in danger of finding ourselves in the same position as the people of antiquity on building the Tower of Babel. The danger is that we may soon all speak a different language which a person not working on precisely the same subject will be unable to understand. For instance, in the review mentioned, twenty abbreviations were used: CRF, GRF, TRF, HRP, MSH, SLA, VIP, and so on. However, the danger is greatly reduced by holding symposia from time to time and asking specialists to review their subject. We could not have a better specialist for the hypothalamic peptides than Schally, because since 1955, when he was one of the discoverers of the existence of the corticotrophin releasing factor, he has contributed so much to our knowledge in this field. His chapter discusses the efforts he and his associates are making to identify the structure of these peptides to establish their functions as well as those of the synthetic analogues. He has tried to cover in a few pages nine hormones and their analogues, obviously a very difficult task.

As it was appropriate to begin with the anatomy, so it is appropriate to finish with the clinical aspects of hypothalamic disorders. This difficult task has been undertaken by G. M. Besser whose contribution is on the evaluation of clinical endocrine disturbances of the hypothalamus. He illustrates how rapidly a physiological or pharmacological advance is taken up by the clinician. His is the most difficult task because he cannot simply take a rat or a cat, but has to wait until a suitable patient turns up. Then his first and foremost responsibility is not to research but to the patient. In addition, his methods of research are naturally limited. His contributions thus deserve our greatest admiration.

REFERENCES

Feldberg, W. and Wei, E. (1977). The central origin and mechanism of cardiovascular effects of morphine as revealed by naloxone in cats. *J. Physiol. (Lond.)*, **272**, 99–100P

Feldberg, W. and Wei, E. (1978). Central sites at which morphine acts when producing cardiovascular effects. *J. Physiol. (Lond.)*, **275**, 75P

Isenschmid, R. and Schnitzler, W. (1914). Beitrag zur Lokalisation des der Wärmeregulation vorstehenden Zentralapparates im Zwischenhirn. *Arch.exp.Path.Pharmak.*, **76**, 202–223

Karplus, J. P. and Kreidl, A. (1909). Gehirn und Sympathicus. I. Zwischenhirnbasis und Halssympathicus. *Pflüg.Arch.ges.Physiol.*, **129**, 138–144

Keller, A. D. and Hare, W. K. (1932). Hypothalamus and heat regulation. *Proc.Soc.exp.Biol. N.Y.*, **29**, 1069–1070

Vale, W., Rivier, C. and Brown, M. (1977). Regulatory peptides of the hypothalamus. *Ann.Rev.Physiol.*, **39**, 473–527

1
Functional anatomy of the hypothalamus

B. Halász*

INTRODUCTION

The hypothalamus, part of the diencephalon, is a rather small region of the brain. The human hypothalamus weighs about 4 grammes and thus represents only 0.3 per cent of the whole brain. In spite of its small size it plays a fundamental role in homeostasis as well as in reproduction. The structural organisation of this diencephalic region is most informative about the complex and manifold interrelated functions which have largely been discovered in this century, some significant contributions having been made in the last decades.

Knowledge concerning the morphology of the hypothalamus has accumulated considerably in recent years. This progress may first of all be ascribed to the introduction of new experimental techniques such as immunohistochemistry, fluorescence histochemistry for the demonstration of biogenic amines (Eränkö, 1955; Falck *et al.*, 1962), new tract-tracing methods like retrograde transport of horseradish peroxidase histochemistry, autoradiography (local injection of labelled amino acids which are synthesised to proteins in the nerve cell bodies and transported along the axons to the terminals) and hypothalamic deafferentation.

From the morphological point of view the hypothalamus can be subdivided into three longitudinal zones: periventricular, medial and lateral. Both the periventricular and medial zones are cell-rich, whereas the lateral one is dominated by the longitudinal fibre system of the medial forebrain bundle. For a detailed description of the anatomy of the hypothalamus the reader is referred to the excellent reviews of Diepen (1962) and Haymaker *et al.* (1969). By a combined double filling technique, using blue and red ink, the arterial supply and venous drainage of the rat hypothalamus have recently been studied in great detail by Ambach and Palkovits (1974*a*, *b*; 1975).

*Second Department of Anatomy, Histology and Embryology, Semmelweis University Medical School, Tüzoltó u. 58, 1094 Budapest, Hungary

MORPHOLOGICAL CHARACTERISTICS OF THE HYPOTHALAMIC CELL GROUPS

It is interesting to consider the number of nerve cells which constitute a nucleus. In the rat, for example, the arcuate nucleus contains 30,000 neurones on one side, and the suprachiasmatic nucleus, a fairly small cell group, consists of about 12,000 to 13,000 nerve cells (unpublished observations of our department).

A crude quantitative analysis, based on the cell density, shape and size of the cell nuclei, reveals that the known hypothalamic cell groups are not homogeneous and several subgroups can be distinguished.

A detailed Golgi analysis of the neuronal cell types which, together with their dendritic and axonal terminal arborisation, comprise the various hypothalamic nuclei, has been presented by Szentágothai (1968). To give an example, the ventromedial nucleus (figure 1.1) contains two main types of nerve cell: larger

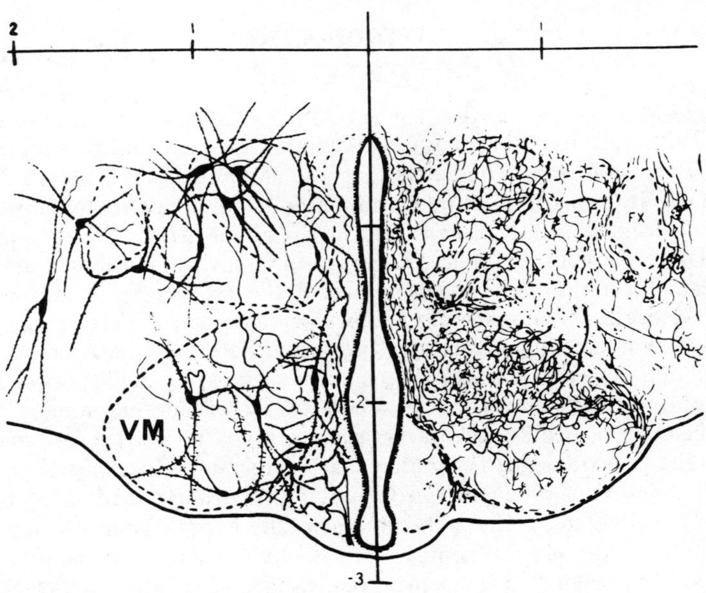

Figure 1.1 Cellular elements (left) and terminal axon ramifications (right) in the ventromedial nucleus (VM) of the rat hypothalamus. (After Szentágothai, 1968)

ones in the centre and smaller ones near the medial border. The larger cells generally have three main dendrites without any apparent orientation of the dendritic arborisation. The dendrites of the smaller border cells are orientated mainly parallel with or perpendicular to the borders of the nucleus. The axons of the large nerve cells leave the nucleus mostly in a dorsal direction where, having reached the capsule of the nucleus, they turn into an antero-posterior direction.

Each axon gives rise to several initial collaterals inside the ventromedial nucleus. These collaterals are very thin and ramify immediately into still finer branches which establish synaptic contact mostly with nerve cells in the same nucleus and in the vicinity of those from which they originate. Terminating axons enter the ventromedial nucleus from all sides and arborise freely with successive divisions or sometimes in an almost brush-like manner. These axons may show grape-like end arborisations. The arborisation area of a single axon may incorporate a space containing hundreds, if not more, of nerve cells in the ventromedial nucleus. Because a large number of terminal axon ramifications overlap in the nucleus there is not only a considerable divergence of the impulses conducted by the same presynaptic afferent fibre, but also a very large convergence of presynaptic influences impinging upon the same nerve cell from different channels (Szentágothai, 1968).

In the hypothalamus axosomatic synapses are rather rare. The great majority of synapses are axodendritic, the axons terminating on both dendritic shafts and dendritic spines. Quantitative analysis of the synaptic organisation of the rat supraoptic nucleus (Léránth *et al.*, 1975) revealed that there are over five million boutons on each side of this nucleus, an average of 596 per neurone. Raisman and Field (1973) reported sexual dimorphism in the neuropil of the rat preoptic area and its dependence on neonatal androgen. Thus in the normal female the number of non-amygdaloid synapses on dendritic spines in the preoptic area is higher than in the male. Interestingly, castration of the male within twelve hours after birth (but not after seven days) causes an increase in the number of spine synapses compared with females of similar age. Conversely, females treated on day 4 (but not on day 16) with testosterone propionate have a low number of spine synapses within the male range. In accordance with these findings neonatal oestradiol treatment results in an increased number of axodendritic synapses in the arcuate nucleus (Matsumoto and Arai, 1976).

The neurones of a nucleus display rich intrinsic connections. Two-thirds of the axon terminals in the supraoptic nucleus appear to be of intranuclear or otherwise of local origin (Léránth *et al.*, 1975). This could be explained by assuming either numerous intranuclear axon collaterals or interneurones with richly arborising axons, or possibly both. According to our (unpublished) findings, a significant number of nerve terminals in the arcuate nucleus remain unaltered after the complete isolation of the medial basal hypothalamus from the rest of the brain, indicating that these boutons belong to neurones within the isolated area.

INTRAHYPOTHALAMIC CONNECTIONS

Information on the intrahypothalamic connections is rather scanty, although some progress has recently been made in this area. Studies with lesions or using autoradiography (Szentágothai, 1968; Saper *et al.*, 1976*a*) indicate that the ventromedial nucleus, for example, has abundant intrahypothalamic connections (figure 1.2). The same also seems to hold true for the arcuate nucleus. It is well corroborated by both neurohistological (Szentágothai, 1964) and electrophysiological data (Yagi and Sawaki, 1970; Harris *et al.*, 1971; Makara *et al.*, 1972; Renaud, 1976*a*) that several axons of the arcuate neurones project on to

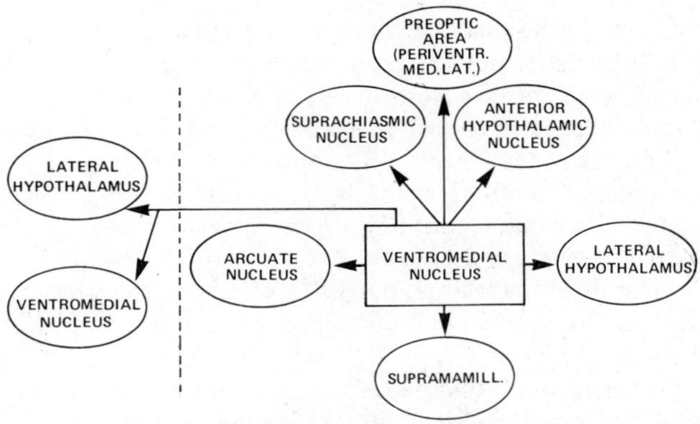

Figure 1.2 Intrahypothalamic efferent connections of the ventromedial nucleus

the surface zone of the median eminence and thus constitute a significant part of the tubero-infundibular tract (see later). In addition, there is morphological as well as electrophysiological evidence that arcuate neurones also project to several regions, including the ventromedial nucleus, lateral hypothalamic area, anterior hypothalamic and preoptic area (Harris and Sanghera, 1974; Makara and Hodács, 1975; Bodoky and Réthelyi, 1977).

Renaud (1976*a*) provided experimental evidence for the presence of axon collaterals within the tubero-infundibular system; some appear to terminate locally within the hypothalamus (anterior hypothalamic, preoptic region), whilst others extend rostrally and caudally into extrahypothalamic areas such as the thalamic medial dorsal nucleus. Arcuate neurones receive afferents from, among others, the medial preoptic area (Poulain and Partouche, 1967; Dyer and Cross, 1972; Dyer, 1973; Moss, 1975; Conrad and Pfaff, 1976*a*; Köves and Réthelyi, 1976), the ventromedial nucleus (see above), and presumably from several other hypothalamic regions.

The abundant intranuclear connections and in addition the rich connections between the various hypothalamic cell groups support the general impression that the hypothalamus should be considered a neuronal network of quasi-random internal connections. In this network, in which the impulses leave the hypothalamus through the main axons, excitation can spread from a given focus in any direction and can establish an infinite number of closed, self-re-exciting chains.

HYPOTHALAMIC OUTPUT AND INPUT CHANNELS

Both output and input channels of the hypothalamus are of two kinds, i.e. humoural and neural.

Output channels

These are shown in figure 1.3.

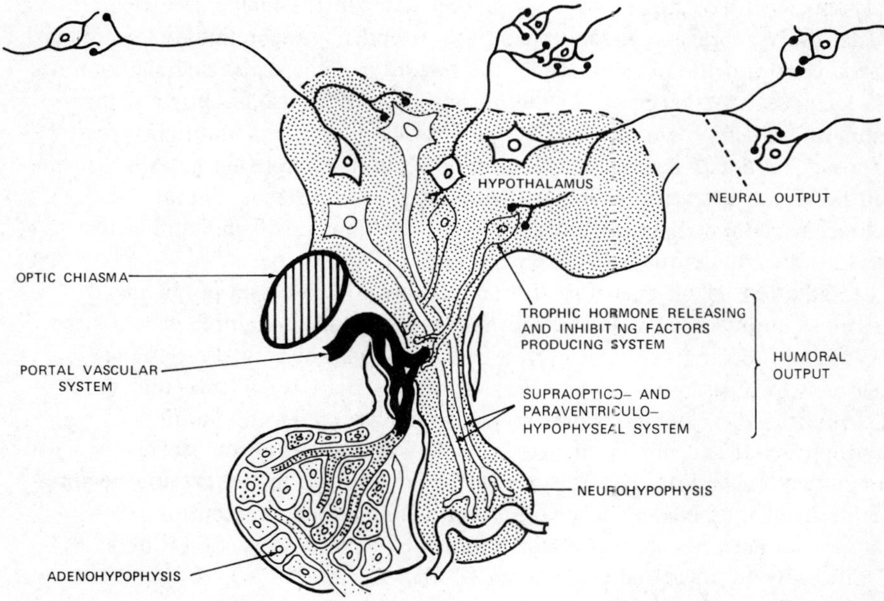

Figure 1.3 Schematic representation of the two kinds of output channels of the hypothalamus

Humoural output channels

Humoural output channels are represented by the classical magnocellular neurosecretory supraoptico- and paraventriculo-hypophyseal system, and the parvicellular neurosecretory system producing trophic hormone releasing and inhibiting factors or hormones.

Supraoptico- and paraventriculo-hypophyseal system It is well known that vasopressin and oxytocin are synthesised together with neurophysins, the carrier peptides in the supraoptic and paraventricular nuclei. In these magnocellular neuroendocrine cells the neuro-secretory material is packed in granules and transported along the neurotubules of the axons to the posterior pituitary. There is no need to deal with this in detail and the reader is referred to the excellent monographs of Scharrer and Scharrer (1954), Bargmann (1954) and Diepen (1962).

However, the immunohistochemical studies which attempt to localise the oxytocin- and vasopressin-producing elements should be mentioned. In the rat (Vandesande and Dierickx, 1975; Choy and Watkins, 1977) as well as in the bovine and pig hypothalamus (Vandesande *et al.*, 1975*a*; Watkins and Choy, 1977), oxytocin and vasopressin are synthesised in different neurones and are transported along different axons. Both the supraoptic and the paraventricular nuclei contain vasopressinergic and oxytocinergic neurones. In the supraoptic nuclei the vasopressin neurones are more prevalent, whilst in the paraventricular nuclei the oxytocin neurones predominate. In the rat paraventricular nuclei the vasopressinergic neurones show a tendency to preferential grouping in the centre of the nucleus, whilst the majority of the oxytocinergic neurones are located at the periphery of the nucleus. In the pars supraoptica of the supraoptic nuclei there also exists a preferential location of the two types of neurone: its dorsal regions consist mainly of oxytocinergic neurones but in the ventral region the vasopressinergic neurones predominate. In the rat the suprachiasmatic nuclei also contain a moderate number of immunoreactive vasopressinergic neurones (Vandesande *et al.*, 1975*b*).

In addition, it has been found that the bovine vasopressin-producing neurones are identical with the neurophysin II-producing neurones, while the oxytocin-producing neurones correspond to the neurophysin I-producing neurones (Vandesande *et al.*, 1975*a*). In the pig there is evidence that the distribution of oxytocin and vasopressin is similar to the distribution of neurophysin II and porcine neurophysin I (Watkins and Choy, 1977). Immunocytochemical investigations further reveal that in the external region of the median eminence, nerve fibres containing either oxytocin or vasopressin are present, and that the number of the positive fibres increases significantly following adrenalectomy (Dierickx *et al.*, 1976).

Oxytocin and vasopressin are contained within neurosecretory granules which are found interspersed in the axoplasm among tubular or vesicular structures and mitochondria. Regarding the mechanism of hormone liberation, it has been shown that the secretion of intragranular contents takes place after the fusion of the granule-limiting membrane with the plasma membrane (exocytosis) and that this is followed in turn by a process of membrane retrieval or endocytosis (for details see Dreifuss *et al.*, 1976).

The hypothalamic system producing trophic hormone releasing and inhibiting factors (hormones). It is well established that the mechanism by which the central nervous system controls anterior pituitary function is neurovascular. The so-called trophic hormone-releasing and inhibiting factors (hormones) are synthesised by hypothalamic neurones and transported along their axons in the median eminence where they are released into the portal capillaries and carried to the anterior pituitary.

This view is consistent with the structure of the median eminence (figure 1.4). Interestingly, the surface of contact between the nervous tissue and the portal capillary system is significantly larger in female rats than in males, as demonstrated by Réthelyi (unpublished observations). In recent years three of the hypothalamic neurohormones acting on the anterior pituitary have been isolated, their chemical structures determined and the hormones synthesised (for references see Blackwell and Guillemin, 1973). These are thyrotrophin-releasing hormone (TRH), luteinising hormone-releasing hormone (LH-RH), and growth-hormone release-inhibiting hormone or somatostatin. The discovery of the chemical structure of these hormones and their subsequent synthesis has enabled the production of antisera against these

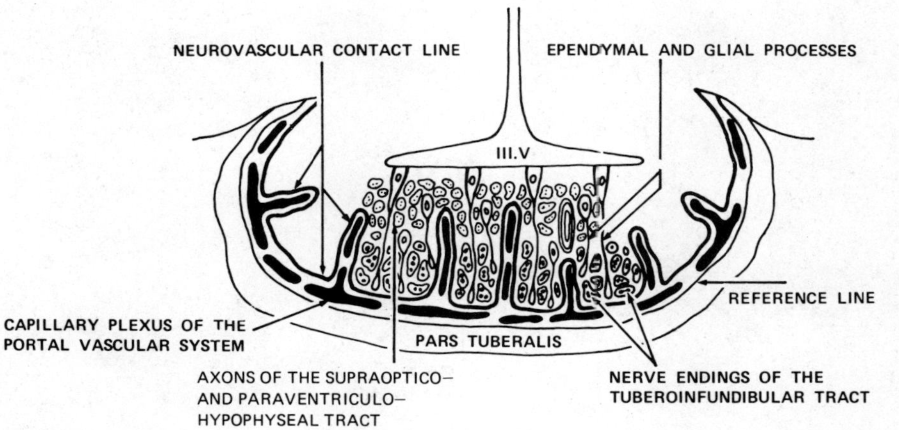

Figure 1.4 Schematic representation of the structural organisation of the median eminence as seen in coronal section

neurohormones which in turn has promoted the use of immunohistochemistry in order to obtain information on the site of production of these substances.

Guinea-pig LH-RH was the first releasing hormone to be localised by immunohistological methods (Leonardelli *et al.*, 1973). Although marked quantitative differences seem to exist in the different species regarding the location of LH-RH cell bodies, the observations to date indicate that the vast majority of the LH-RH neurones reside in the tuberal-premamillary and the preoptic-suprachiasmatic regions. In the human and the monkey the LH-RH neurones are concentrated mainly in the arcuate and premamillary nuclei of the mediobasal hypothalamus (Barry and Carette, 1975; Barry, 1977). In the dog, about 40 per cent of the cells are found in the tuberal-premamillary region. Preoptic-anterior hypothalamic localisation of the LH-RH cells is a characteristic of the rat (Sétáló *et al.*, 1976), guinea-pig and cat (Barry and Dubois, 1975). In the rabbit, perikarya containing LH-RH are detectable in both the supraoptic and tuberal regions (Sétáló, 1977).

The majority of the LH-RH-positive axons terminate in the median eminence, mainly in its lateral part (figure 1.5) and in the vascular organ of the lamina terminalis in close proximity to the capillary loops. Electron microscopic immunohistochemistry shows that LH-RH is contained in the neurosecretory granules (75–95 nm in diameter) of some nerve endings or axon profils located in the palisade layer of the median eminence (Pelletier *et al.*, 1974; Goldsmith and Ganong, 1975). Recently Silverman and Desmoyers (1976) have reported that the hormone is present within granules of diameter 90–120 nm in axons of the palisade zone and in nerve terminals abutting on to the portal plexus (granule diameter 40–80 nm). In man, primates, dog and cat the LH-RH cell bodies are easily detected, whereas in the rat and the guinea-pig, stimulation of LH-RH synthesis and/or the blockade of its axoplasmic transport or release is required to locate the cells which synthesise the hormone. LH-RH cell bodies as well as their fibres and tracts are not restricted to the above-mentioned hypothalamic regions but are also evident in several extra-

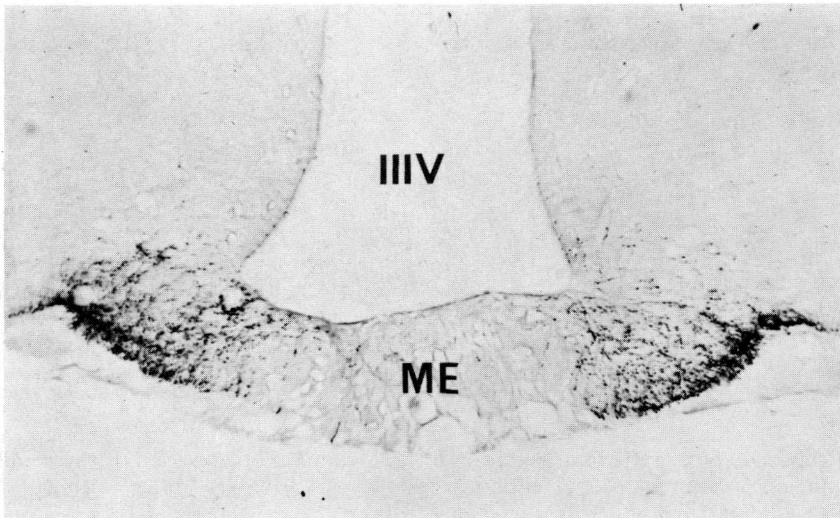

Figure 1.5 Coronal section of the rat medial basal hypothalamus showing immunohistologically stained LH-RH containing nerve fibres and terminals (black) in the rat median eminence (ME). Note that LH-RH axons accumulate in two bundles at both edges of the ME and that axon terminals are preferentially located in the lateral thirds of the ME. IIIV: third ventricle, x160 magnification. (Courtesy G. Sétáló and S. Vigh, Department of Anatomy, University Medical School, Pécs, Hungary)

hypothalamic regions, particularly in various limbic structures (Barry and Carette, 1975; Barry, 1977; Sétáló, 1977).

TRH-containing cell bodies have been demonstrated in the dorsomedial nucleus, perifornical area and in the paraventricular nucleus and TRH fibres have been detected in the medial part of the external layer of the median eminence. Hypothalamic regions rich in TRH fibres include the dorsomedial hypothalamic nucleus, perifornical regions and the parvicellular part of the hypothalamic paraventricular nucleus. TRH fibres have also been found in the periventricular suprachiasmatic and preoptic regions. In addition, such elements could also be detected in extrahypothalamic structures such as the septum pellucidum, brainstem and spinal cord (Hökfelt *et al*., 1975*b*).

Somatostatinergic perikarya have been identified mainly in the periventricular region of the hypothalamus extending from the anterior commissure to the rostral margin of the median eminence (Elde and Parsons, 1975; Hökfelt *et al*., 1975*a*; Alpert *et al*., 1976). Immunoreactive somatostatin-positive fibres form a dense plexus in the suprachiasmatic, arcuate and ventromedial nuclei, and in the median eminence where they mainly occupy the medial part of the internal and external zones (figure 1.6) (Hökfelt *et al*., 1974; King *et al*., 1975; Sétáló *et al*., 1975; Dubé *et al*., 1975). A dense plexus of somatostatin has also been detected immunohistologically in the organum vasculosum laminae terminalis (Dubé *et al*., 1975). Somatostatin-positive cell bodies are also evident in the spinal ganglia, in

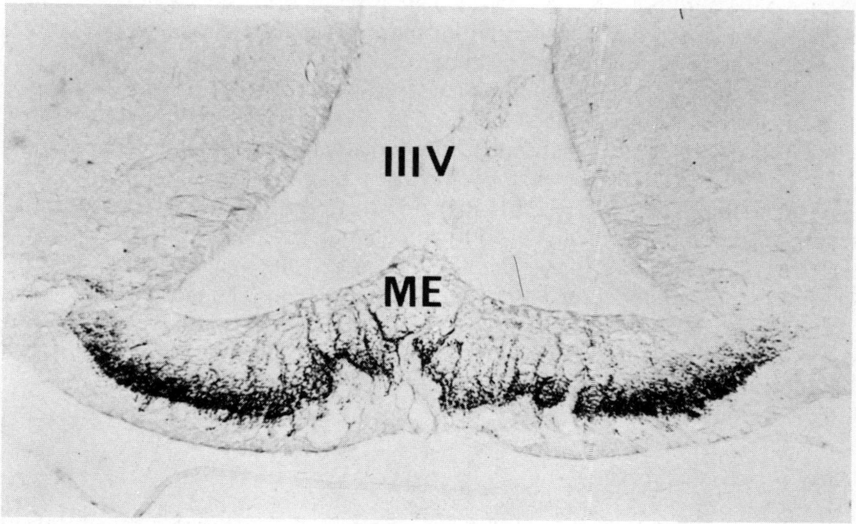

Figure 1.6 Immunohistological demonstration of the somatostatin-containing nerve fibres and terminals (black) in the rat median eminence (ME). These fibres and terminals are distributed mainly in the medial part of the ME, but in a more extended region than the LH-RH-containing elements. Otherwise similar to figure 1.5. (Courtesy G. Sétáló and S. Vigh)

the islets of the pancreas, and positive fibres have been detected in the dorsal horn of the spinal cord and also in the intestinal wall (Hökfelt *et al.*, 1975*a;* 1976).

There is no direct information about the site of production of other releasing and inhibiting hormones (for a summary of the indirect data see Mess, 1969). If one accepts that the hypothalamic releasing and inhibiting factors are released at the nerve terminals in the superficial layer of the median eminence and the proximal part of the pituitary stalk, useful information as to the site of production of these substances may be obtained by searching for the origin of these nerve endings. Studies made in this area indicate that the majority of the axon terminals in the superficial layer of the median eminence and proximal part of the pituitary stalk belong to neurones of the medial basal hypothalamus which extend from the mid-level of the suprachiasmatic region to the mamillary area (Réthelyi and Halász, 1970; Koritsánszky and Köves, 1976). Several neurones of this area project to the median eminence, forming the tubero-infundibular tract (Szentagothai, 1964). Thus it seems reasonable to assume that the majority of the neural elements producing the neurohormones which act on the anterior lobe are located in the medial basal hypothalamus, including the suprachiasmatic region, and that this system may be represented, at least partly, by the tubero-infundibular tract. The assumption that the medial basal hypothalamus may be the primary location for the cells which produce the trophic hormone releasing and inhibiting factors is supported by the finding that pituitary trophic function is fairly well maintained following the interruption of all neural connections of the medial basal hypothalamus (Halász, 1969; Butler *et al.*, 1975*a,b*; Krey *et al.*, 1975*a,b*). In the absence of any hypothalamic influence, hormone secretion of the anterior pituitary is markedly reduced.

One part of the tubero-infundibular system terminating in the surface zone of the median eminence is dopaminergic. The dopamine released by these neurones may represent at least one of the prolactin release inhibiting factors (PIF). For further details see chapter 2.

As already mentioned, TRH, LH-RH and somatostatinergic nervous elements are not restricted to the medial basal hypothalamus but are spread over a wide area. This fact does not disprove the assumption that the hypothalamic neurohormones acting on the anterior pituitary reside in the medial basal hypothalamus. Rather it suggests that these substances are not only neurohormones but also neurotransmitters or modulators. Thus their action is not limited to the pituitary but may also influence nervous elements or even non-neural structures, especially in the case of somatostatin.

Neural Output Channels

Anatomically, the connections between the hypothalamus and the rest of the brain are dominated by certain limbic structures: the hippocampus, septum, amygdala, piriform cortex and mesencephalon.

It will be apparent when dealing with the afferent connections of the hypothalamus that the efferent pathways of the hypothalamus appear to reciprocate several of the major afferent hypothalamic connections. Many such reciprocating connections are contained in the medial forebrain bundle, the dorsal longitudinal fasciculus, the stria medullaris and the stria terminalis. These pathways appear to close neural circuits between the hypothalamus on the one hand and the mesencephalon, thalamus and several of the limbic forebrain structures on the other.

The major brain structures receiving hypothalamic efferents are the amygdala, hippocampus, septum, thalamus and the lower brainstem, mainly the mesencephalon and pons.

Outputs to the amygdala, hippocampus and septum. According to the recent studies of Wakefield and Hall (1974), the anterior hypothalamic nucleus and the lateral preoptic area send projections to the central, lateral, medial and basal nuclei of the amygdala. These authors did not note any degeneration in the amygdala following lesions in the medial preoptic area, ventromedial nuclei or in the lateral hypothalamic area caudal to the anterior hypothalamic area. In contrast, electrophysiological and neurohistological studies (Conrad and Pfaff, 1976*a*; Saper *et al.*, 1976*a*; Renaud and Hopkins, 1977) indicate that the mediobasal hypothalamus, particularly the medial preoptic area and the ventromedial nucleus, do send efferents to the amygdala.

By retrograde transport of horseradish peroxidase it has been demonstrated that axons originating from the hypothalamic supramamillary nucleus and from the cell group located below the mamillo-thalamic tract project to the dorsal posterior hippocampus and more sparsely to the ventral hippocampus (Pasquier and Reinoso-Suarez, 1976).

Periventricular, medial as well as lateral hypothalamic neurones send axons to the septum, primarily to its lateral parts (Nauta and Haymaker, 1969; Saper *et al.*, 1976*a*; Conrad and Pfaff, 1976*a,b*).

Outputs to the thalamus and epithalamus. Among these efferent paths the mamillo-thalamic tract, arising largely from the medial mamillary nucleus, is the best known. This tract distributes its fibres to all three components of the anterior thalamic nucleus from which, in turn, originate massive projections to the gyrus cinguli. There is experimental evidence for the existence of a mamillo-subthalamic tract which diverges laterally from the former tract and terminates in the subthalamic region. Neurones of the preoptic, anterior hypothalamic and ventromedial areas project to the periventricular and mediodorsal thalamic nucleus, and, via the stria medullaris, to the lateral habenula (Nauta and Haymaker, 1969; Conrad and Pfaff, 1976*a, b*; Saper *et al.*, 1976*a*). In turn the lateral habenula has efferent connections with the nucleus interpeduncularis, tegmentum and central grey of the midbrain.

Outputs to the lower brainstem and spinal cord. Periventricular, medial and lateral hypothalamic regions have descending efferents. They project mainly to the tectal area, central grey, ventral and central tegmental area of the midbrain and pons. The fibres are incorporated into the dorsal longitudinal fasciculus and into the medial forebrain bundle.

Recently, Saper *et al.* (1976*b*) reported that certain neurones in the hypothalamus can be retrogradely labelled by injections into the spinal cord of the enzyme marker horseradish peroxidase. Labelled neurones were evident in the paraventricular nucleus, the lateral and dorsal hypothalamic areas, the dorsomedial nucleus and in the posterior hypothalamic area dorsal to the mamillary body. The majority of the labelled cells were on the ipsilateral side of the injection but a few were also present on the contralateral side. These observations indicate that hypothalamic neurones project directly to the spinal cord. Earlier studies could not unequivocally demonstrate such fibres and their terminations are still obscure.

Recent autoradiographic studies clearly indicate the very rich efferent connections of a hypothalamic nucleus or area suggesting the great divergency of the system. The following efferent connections of the hypothalamic ventromedial nucleus were, for example, observed (Saper *et al.*, 1976*a*): preoptic (periventricular, medial and lateral areas), bed nucleus of the stria terminalis, mainly ipsilateral partly contralateral side, amygdala (central and dorsal nucleus, part of medial amygdaloid nucleus), ventral part of the lateral septum, thalamic periventricular nuclei, ventral and central tegmental area, central grey, locus coeruleus and zona incerta.

Medial and periventricular preoptic area efferents were traced into the septum, midbrain central grey, raphe nuclei, ventral tegmental area, reticular formation, amygdala, lateral habenula and into the periventricular thalamus (Conrad and Pfaff, 1976*a*).

According to the autoradiographic studies of Conrad and Pfaff (1976*b*), area hypothalamica anterior neurones project to the lateral septum, midbrain central

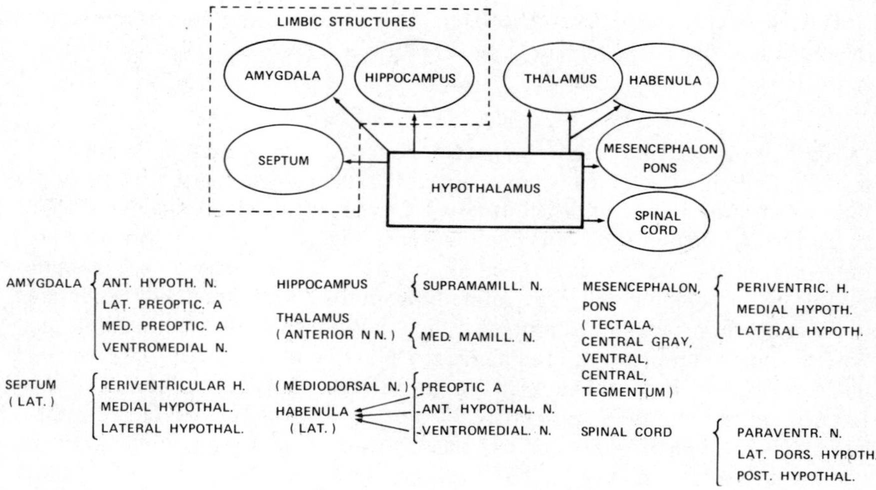

Figure 1.7 Neural efferent pathways from the hypothalamus. A: area; H: hypothalamus; N: nucleus

grey, ventral tegmental area, raphe nuclei, midbrain reticular formation, periventricular thalamus, medial amygdala and lateral habenula.

Hypothalamic efferent pathways are summarised in figure 1.7.

Input Channels

Since the hypothalamus is deeply involved in homeostatic adjustments, it needs abundant information from both the external and the internal environment and a rich afferent input is indeed evident. The information which flows continuously into the hypothalamus is carried in two ways: along humoural and neural pathways.

Humoural Input Channels

Several parameters of the blood influence hypothalamic functions. One of these is the blood hormone level and hormone receptors exist within the hypothalamus. It has been demonstrated that certain hypothalamic regions show a similar uptake and release pattern of sex steroids as do the uterus and vagina (McGuire and Lisk, 1969). By means of light microscopic autoradiography using ^{3}H-oestradiol or ^{3}H-testosterone it has been possible to map the steroid-concentrating cells. Such studies revealed that the oestrogen and androgen binding cells are concentrated in the medial preoptic area, medial anterior hypothalamus, ventromedial nucleus, arcuate nucleus and ventral premamillary nucleus (Pfaff and Keiner, 1973; Keefer and Stumpf, 1975; Sar and Stumpf, 1975; Pfaff *et al.*, 1976).

The sex steroid feedback is partly involved in pituitary gonadotrophic hormone secretion, exerting both a negative and a positive feedback action, and partly in sexual behaviour. With regard to the latter function, the preoptic area and the anterior hypothalamus appear to be primarily involved. Direct implantation of oestrogen into these regions induced lordosis in ovariectomised rats (Lisk, 1962). Natural glucocorticosteroids do not appear to be concentrated in the cells of the hypothalamus in contrast to certain limbic structures.

Several data suggest that pituitary trophic hormones may have a direct feedback action on the basal region of the hypothalamus. Follicle-stimulating hormone, luteinising hormone, prolactin, growth hormone or adrenocorticotrophic hormone, implanted into the medial basal hypothalamus, all influence the hormone secretion of the adeno-hypophysis (for details see Szentágothai *et al.*, 1968). It has been proposed (Hyyppä *et al.*, 1971) that hypothalamic releasing hormones, too, may have a direct feedback action on the hypothalamus itself.

Primarily electrophysiological data suggest that in addition to hormone receptors there are also thermoreceptors in the preoptic-anterior hypothalamic region, osmoreceptors in the anterior hypothalamus and glucoreceptors in the ventromedial nucleus (for details see Cross, 1964).

Neural Input Channels

These are shown in figure 1.8.

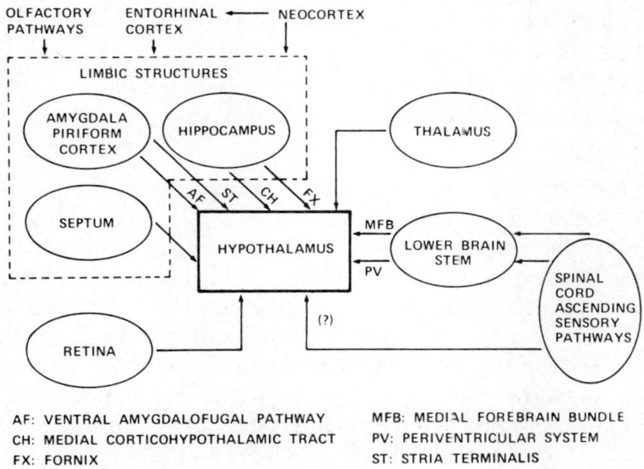

Figure 1.8 Regions projecting to the hypothalamus

Inputs from the amygdaloid complex. The amygdala consists of two main nuclear groups: the phylogenetically older corticomedial group and the more recent basal and lateral group. Both main divisions are under the influence of the olfactory bulbs but both groups receive afferents from the hypothalamus (see the section on hypothalamic efferent connections).

Two pathways connect the amygdala with the hypothalamus, the stria terminalis and the ventral amygdalo-fugal system. Both pathways contain afferent and efferent fibres.

The efferent fibres of the stria terminalis have their origin mainly in the corticomedial and basal nuclei of the amygdala but the lateral nucleus also contributes to such fibres. Depending on their relationship with the anterior commissure, a distinction is made between a supra- or precommissural and a postcommissural component of the stria terminalis. The precommissural part consists of fibres which run forward over the anterior commissure and turn backwards and ventrally to pass through the preoptic area and terminate in a region which completely surrounds the hypothalamic ventromedial nucleus and extends as far back as the ventral premamillary nucleus (Heimer and Nauta, 1969). In its caudal course this component traverses the postcommissural component which terminates in the medial preoptic and adjacent medial parts of the anterior hypothalamic areas (Heimer and Nauta, 1969). In the preoptic area the stria terminalis fibres are distributed more or less equally on dendritic shafts and dendritic spines, whereas in the ventromedial nucleus they terminate preferentially on dendritic spines (Field, 1972).

The ventral amygdalo-fugal pathway contains fibres from the piriform cortex and from the basolateral group of the amygdala, forwarding among others the olfactory input. In the hypothalamus it terminates in the lateral parts of the preoptic, anterior and midhypothalamic areas. In the lateral preoptic and anterior hypothalamic area these fibres join the medial forebrain bundle (for details and references see Lammers, 1972). According to Szentágothai (1968), some of the fibres of the ventral amygdalo-fugal pathway also terminate in the ventro- and dorsomedial nuclei. Electrophysiological studies (Renaud, 1976*b*) suggest that the amygdala exerts a prominent monosynaptic influence on the activity of many hypothalamic ventromedial neurones, coupled with polysynaptic activation of powerful local post-synaptic inhibitory mechanisms.

Inputs from the hippocampus. The hippocampal complex contains four basic components which differ in their intrinsic and extrinsic connections: Field CA1 or regio superior; Field CA3 or regio inferior; dentate gyrus; and prosubiculum (for details see Raisman *et al*., 1965, 1966). The intrinsic connections appear to form four levels of organisation (Raisman and Field, 1971). Principal afferents arise in the entorhinal area; CA3 receives afferents from the ipsilateral medial septal nuclei (for details and references see Raisman and Field, 1971).

The fornix forms the main efferent pathway from the hippocampus with its two distinct precommissural and postcommissural fibre groups. The precommissural fornix terminates mainly in the septal region but some fibres are distributed to the lateral preoptic region. The postcommissural fornix terminates in the mamillary body and issues additional fibre groups to hypothalamic regions rostral to its termination in the mamillary body. Such hypothalamic fornix fibres have been followed to the rostral part of the dorsal hypothalamic area and to the lateral hypothalamic region. Furthermore, in rodents a particularly striking hypothalamic offset of the postcommissural fornix, the so-called tractus corticohypothalamicus medialis arising from the subiculum, distributes itself to the rostral half of the periventricular hypothalamic zone including the rostral pole of the arcuate nucleus (for references see Raisman and Field, 1971).

Inputs from the septum. The septum acts as a nodal point relating the hippocampus, amygdala and piriform cortex with the hypothalamus (Nauta, 1960; Raisman, 1966). Analyses utilising retrograde horseradish peroxidase histochemistry and anterograde ^{3}H-leucine autoradiography transport methods (Meibach and Siegel, 1977) indicate that the dorsal septum projects to the lateral preoptic area, lateral hypothalamus and periventricular hypothalamus, whilst the ventral septum projects mainly to the medial mamillary nucleus. According to Conrad and Pfaff (1976*c*), the medial septum projects to the preoptic area, anterior and dorsomedial hypothalamus, and to the arcuate nucleus.

Inputs from the thalamus. In the monkey thalamofugal fibres in the inferior thalamic peduncle have been traced from the mediodorsal thalamic nucleus to the lateral preoptico-hypothalamic region. There are some findings indicating that thalamic midline nuclei project to the dorsal periventricular and medial hypothalamic areas (for references see Nauta and Haymaker, 1969).

Inputs from the lower brainstem. The connections between the lower brainstem and the hypothalamus are very rich. Axons arising from the rostral part of the midbrain central grey substance and from the dorsal tegmental nucleus, constituting partly the dorsal longitudinal fasciculus, appear to be distributed in the supramamillary, premamillary nuclei, lateral hypothalamic area, dorso- and ventromedial nucleus and in the suprachiasmatic nucleus (Szentágothai, 1968; Nauta and Haymaker, 1969). Earlier data and recent studies indicate that besides the mentioned fascicle, several pathways may be available for impulse conduction to the hypothalamus from the brainstem reticular formation. These terminate mainly in the lateral hypothalamic area, dorso- and ventromedial nucleus and the retrochiasmatic area (Szentágothai, 1968; Edwards and de Olmos, 1976). Axons of locus coeruleus and adjacent pontine tegmentum neurones also appear to terminate in the lateral hypothalamic area (McBride and Sutin, 1976). The nucleus raphe pontis projects to the paraventricular nucleus of the hypothalamus (Bobiller *et al.*, 1976).

Inputs from the basal ganglia. Pathways from the basal ganglia have been postulated but the existence of direct projections from the basal ganglia to the hypothalamus requires more evidence than is currently available. The pallido-hypothalamic tract really consists of an aberrant group of pallidofugal fibres which loop back into the major system of pallidal efferents, without apparent termination in the hypothalamus (for details see Nauta and Haymaker, 1969).

Inputs from the cortex. Efferents from the prefrontal cortex to the lateral hypothalamic region appear to exist (Nauta and Haymaker, 1969) but direct neocorticohypothalamic connections have not been convincingly demonstrated. Neural codes emanating from the primary sensory areas (visual, auditory, and somataesthetic) have no direct access to the hypothalamus. There is convincing experimental evidence that numerous fibres from the piriform cortex and olfactory tubercle join the medial forebrain bundle and terminate in the lateral preopticohypothalamic zone.

Sensory input channels

Although the hypothalamus receives a large amount of sensory input through the above-mentioned connections, there are some more or less direct pathways which should be mentioned.

Input from the retina. A direct projection from the retina to the hypothalamus, although suspected for several years, has only recently been demonstrated experimentally. After injection of ^{3}H-leucine or ^{3}H-proline into the posterior chamber of the eye, labelled protein has been shown autoradiographically in the suprachiasmatic nuclei of the medial hypothalamus, both ipsilateral and contralateral to the injected eye, in the rat, guinea-pig, rabbit, cat and monkey. Electron microscopy of the suprachiasmatic nucleus following orbital enucleation disclosed degenerating endings making synaptic contacts with small dendrites of the suprachiasmatic nucleus (Hendrickson *et al.*, 1972; Moore and Lenn, 1972; Wenisch, 1976). In addition to the suprachiasmatic nucleus, some retinal fibres also terminate on arcuate cells (Sousa-Pinto, 1970; Mason and Lincoln, 1976). The effect of light on the hypothalamus, particularly on its control of the adenohypophysis, may be mediated by this pathway.

Inputs from the olfactory bulb. Such projections also have a relatively free access to the hypothalamus, although direct connections from the olfactory bulb to the hypothalamus are not known. The piriform cortex which receives fibres from the olfactory bulbs, sends projections directly to the hypothalamus. Other olfactory pathways reach the hypothalamus mainly via the amygdala, primarily via its corticomedial nuclei (Heimer, 1975).

Inputs from the spinal cord and lower brainstem. In mammals none of the classical lemniscal pathways appears to establish a direct connection with the hypothalamus. This suggests that secondary sensory cell groups of the spinal cord and rhombencephalon have no direct access to the hypothalamus and can affect the latter only via intermediary processing mechanisms. Such intermediary neural organisations are suspected to exist at a brainstem level, in the thalamus, the limbic forebrain structures and in the neocortex. Numerous fibres of the spinothalamic tract are known to terminate in the central grey substance of the midbrain and also in the dorsal tegmentum. The mediodorsal thalamic nucleus receives afferents from the midbrain tegmentum, amygdalo-piriform complex and the septal region.

An increasing body of evidence suggests that there are neural connections between the gonads and the hypothalamus as well as the adrenal gland and the hypothalamus. More than fifteen years ago Halász and Szentágothai (1959) reported that removal of one adrenal gland resulted in different changes in the hypothalamic ventromedial neurones of both sides. On the side contralateral to the removed adrenal the cell nuclei of the neurones were enlarged, whilst on the ipsilateral side they were shrunken. Using light microscopic autoradiography, Gerendai and Halász (1976) found that the hypothalamic arcuate nuclei of the two sides displayed different rates of protein and RNA synthesis after unilateral gonadectomy: contralateral to the removed gonad the radioactive concentration was significantly higher than on the ipsilateral side. The effect was restricted to the arcuate neurones and did not occur in the hypothalamic dorsomedial nucleus, indicating that the hemigonadectomy-induced changes did not affect all cell groups on one side of the hypothalamus but that they were more limited. A similar effect was observed in the hypothalamic ventromedial nucleus following unilateral adrenalectomy (Gerendai *et al.*, 1974). It appears that the postulated pathways are both afferent as well as efferent. Anatomical details of these connections are obscure. Recent findings indicate that the functional significance of the neural pathways between the gonads and the hypothalamus as well as between the adrenals and the hypothalamus may be in the mechanism of the compensatory hypertrophy of the mentioned endocrine glands following unilateral adrenalectomy (Engeland and Dallman, 1975, 1976; Gerendai *et al.*, 1977).

Before summarising the structural features of the hypothalamus, mention should be made of the overlap in the regions involved in the various hypothalamic functions—such as the regulation of body temperature, food and water intake, control of the adenohypophysis and the role of the hypothalamus in sexual behaviour. Presumably there is a link between several functions such as temperature regulation and the control of pituitary TSH secretion, pituitary gonadotrophic function and sexual behaviour, water intake and the magnocellular neurosecretory system and so on. Apart from this, it appears that regions of dissimilar function are housed in the same area which may be an argument, among others, against a rigid, monolithic centre hypothesis of functions within the hypothalamus.

CONCLUSIONS

Figure 1.9 is an oversimplified scheme of the functional morphology of the hypothalamus. The following conclusions may be drawn on the basis of this scheme.

(1) The hypothalamus functions partly as nervous tissue and partly as an endocrine organ producing oxytocin, vasopressin and trophic hormone releasing and inhibiting factors (hormones).

(2) It receives abundant information streaming in continuously along neural and humoural channels.

(3) The sensory inputs of light and smell have a relatively free access to the hypothalamus, whereas other sensory inputs appear to be relayed by the lower brainstem, the thalamus or via a multisynaptic pathway through the neocortex.

(4) The brain regions directly connected with the hypothalamus are certain limbic structures (amygdala, hippocampus, septum), the thalamus and the midbrain.

(5) Most of the hypothalamic connections are reciprocal.

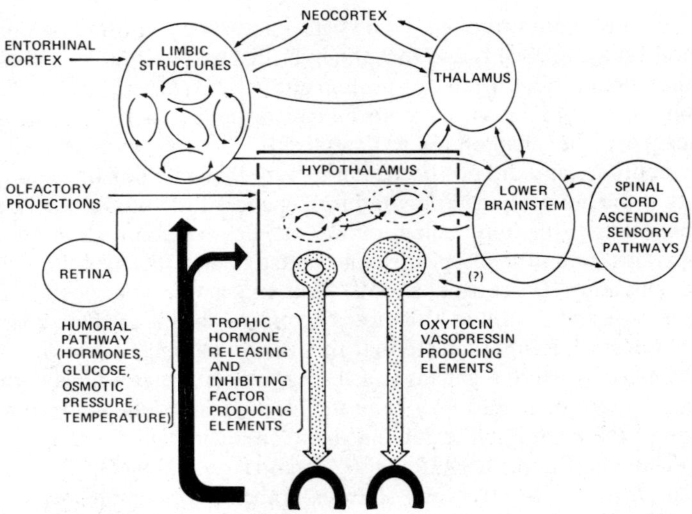

Figure 1.9 An oversimplified scheme of the hypothalamic output and input channels and of the connections between the related structures

(6) Neuronal networks, presumably forming circuits, may exist at all structural levels: within the hypothalamic cell groups, between various hypothalamic nuclei and areas as well as between hypothalamic and different extra-hypothalamic regions. Among others, limbic-hypothalamic-mesencephalic, mesencephalic-hypothalamic-thalamic (epithalamic), limbic-hypothalamic-thalamic and mesencephalic-hypothalamic circuits can be assumed. The circuits appear to be closely interconnected and are partly excitatory and partly inhibitory. This may be true also for intranuclear circuits (Yagi and Sawaki, 1975). This arrangement suggests that the hypothalamus is an open-loop system.

(7) The reciprocal connections of the hypothalamus with the limbic forebrain structures and the mesencephalon are of such magnitude that it appears possible to interpret the hypothalamus, at least partly, as a way-station in both the ascending and descending limbs of a polysynaptic neural circuit which extends between the limbic forebrain on the one hand and the paramedian mesencephalic region on the other (Nauta and Haymaker, 1969). It may be assumed that the functional state of the hypothalamus is determined to a significant extent by the neural events which take place in the limbic structures and the mesencephalon, both having a very integrated structural organisation with several reciprocal interconnections and neural circuits. In addition, they receive a vast amount of information from both the internal and the external environment flowing in along neural and humoural pathways (hormone-sensitive structures exist in the hippocampus, amygdala, etc).

(8) As neurones of several different regions have terminals in the same area and a cell group projects to many different areas, convergency and divergency are characteristic features of the hypothalamic connections.

(9) Among the humoural afferents that are partly hormonal feedback loops, sex steroids should be emphasised. They act on both the limbic structures and

the hypothalamus, influencing both sexual behaviour and pituitary gonadotrophic function. At an early age of life, at least in rodents, they appear to have an action on the structural organisation of the hypothalamus.

(10) Localisation of hypothalamic functions is overlapping as several are linked with others. When taking into account the extremely complex neuronal network of the hypothalamus and its rich reciprocal connections, it seems likely that, apart from a few exceptions such as the supraoptico- and paraventriculo-hypophyseal system, there are not well-defined regions (centres) and pathways that are specifically and exclusively concerned with a discrete hypothalamic function. Of course, this does not exclude the predominance of one or the other hypothalamic area in the involvement of a particular hypothalamic function. Instead of a mosaic type pattern the hypothalamus can rather be envisaged as some kind of a computer. This computer has a number of built-in programmes and its elements are involved in several processes. It elaborates the solution for each actual situation on the basis of a wealth of information that is partly stored and partly streaming in continuously by way of neural and humoural channels. The results are then distributed over a number of neural and humoural output channels.

(11) Presumably the hypothalamus cannot be regarded as the integrative centre of homeostatic adjustments and of reproductive processes but rather as a key-member of the system.

REFERENCES

Alpert, L.C., Brawer, J. R., Patel, Y. C. and Reichlin, S. (1976). Somatostatinergic neurones in anterior hypothalamus: immunohistochemical localisation. *Endocrinology*, 98, 255–258

Ambach, G. and Palkovits, M. (1974*a*). Blood supply of the rat hypothalamus. I: Nucleus supraopticus. *Acta morph. Acad. Sci. Hung.*, **22**, 291–310

Ambach, G. and Palkovits, M. (1974*b*). Blood supply of the rat hypothalamus. II: Nucleus paraventricularis. *Acta morph. Acad. Sci. Hung.*, **22**, 311–320

Ambach, G. and Palkovits, M. (1975). Blood supply of the rat hypothalamus. III: Anterior region of the hypothalamus. *Acta morph. Acad. Sci. Hung.*, **23**, 21–49

Bargmann, W. (1954). *Das Zwischenhirn-Hypophysensystem,* Springer, Berlin

Barry, J. (1977). Immunofluorescence study of LRF neurones in man. *Cell Tiss. Res.*, **181**, 1-14

Barry, J. and Carette, B. (1975). Immunofluorescence study of LRF neurones in primates. *Cell Tiss. Res.*, **164**, 163–178

Barry, J. and Dubois, M. P. (1975). Immunofluorescence study of LRF-producing neurones in the cat and the dog. *Neuroendocrinology*, **18**, 290–298

Blackwell, R. E. and Guillemin, R. (1973). Hypothalamic control of adenohypophyseal secretions. *Ann. Rev. Physiol.*, **35**, 357–390

Bobiller, P., Seguin, S., Petitjean, F., Salvert, D., Touret, M. and Jouvet, M. (1976). The raphe nuclei of the cat brainstem: a topographical atlas of their efferent projections as revealed by autoradiography. *Brain Res.*, **113**, 449–486

Bodoky, M. and Réthelyi, M. (1977). Dendritic arborization and axon trajectory of neurones in the hypothalamic arcuate nucleus of the rat. *Exp. Brain Res.*, **28**, 543–555

Butler, W. R., Krey,L. C., Espinosa-Campos, J. and Knobil, E. (1975*a*). Surgical disconnection of the medial basal hypothalamus and pituitary function in the rhesus monkey. III: Thyroxine secretion. *Endocrinology*, **96**, 1094–1098

Butler, W. R., Krey, L. C., Lu, K. H., Peckham, W. D. and Knobil, E. (1975*b*). Surgical disconnection of the medial basal hypothalamus and pituitary function in the rhesus monkey. IV: Prolactin secretion. *Endocrinology*, **96**, 1099–1105

Choy, V. J. and Watkins, W. B. (1977). Immunocytochemical study of the hypothalamo-neurohypophyseal system. *Cell Tiss. Res.*, **180**, 467–490

Conrad, L. C. A. and Pfaff, D. W. (1976*a*). Efferents from medial basal forebrain and hypothalamus in the rat. I: An autoradiographic study of the medial preoptic area. *J. comp. Neurol.*, **169**, 185–220

Conrad, L. C. A. and Pfaff, D. W. (1976*b*). Efferents from medial basal forebrain and hypothalamus in the rat. II: An autoradiographic study of the anterior hypothalamus. *J. comp. Neurol.*, **169**, 221–262

Conrad, L. C. A. and Pfaff, D. W. (1976*c*). Autoradiographic tracing of nucleus accumbens efferents in the rat. *Brain Res.*, **113**, 589–596

Cross, B. A. (1964). The hypothalamus in mammalian homeostasis. *Symp. Soc. exp. Biol.*, **18**, 157–193

Diepen, R. (1962). Der Hypothalamus. In von Möllendorff's *Hdb.d. Mikr. Anat. d. Menschen Nervensystem,* Springer, Berlin, IV, 7

Dierickx, K., Vandesande, F. and de Mey, J. (1976). Indentification in the external region of the rat median eminence of separate neurophysin-vasopressin and neurophysin-oxytocin containing nerve fibres. *Cell Tiss. Res.*, **168**, 141–151

Dreifuss, J. J., Akert, K., Sandri, C. and Moor, H. (1976). Specific arrangements of membrane particles at sites of exo- and endocytosis in the freeze-etched neurohypophysis. *Cell Tiss. Res.*, **165**, 317–325

Dubé, D., Leclerc, R., Pelletier, G., Arimura, A. and Schally, A. V. (1975). Immunohistochemical detection of growth hormone-release inhibiting hormone (somatostatin) in the guinea-pig brain. *Cell Tiss. Res.*, **161**., 385–392

Dyer, R. G. (1973). An electrophysiological dissection of the hypothalamic regions which regulate the pre-ovulatory secretion of luteinising hormone in the rat. *J. Physiol. (Lond).*, **234**, 421–442

Dyer, R. G. and Cross, B. A. (1972). Antidromic identification of units in the preoptic and anterior hypothalamic areas projecting directly to the ventromedial and arcuate nuclei. *Brain Res.*, **43**, 254–258

Edwards, S. B. and de Olmos, J. S. (1976). Autoradiographic studies of the projections of the midbrain reticular formation: ascending projections of nucleus cuneiformis. *J. comp. Neurol.*, **165**, 417–432

Elde, R. P. and Parsons, J. A. (1975). Immunocytochemical localization of somatostatin in cell bodies of the rat hypothalamus. *Amer. J. Anat.*, **144**, 541–548

Engeland, W. C. and Dallman, M. F. (1975). Compensatory adrenal growth is neurally mediated. *Neuroendocrinology*, **19**, 352–362

Engeland, W. C. and Dallman, M. F. (1976). Neural mediation of compensatory adrenal growth. *Endocrinology,* **99**, 1659–1662

Eränkö, O. (1955). The histochemical demonstration of noradrenaline in the adrenal medulla of rats and mice. *J. Histochem. Cytochem.*, **4**, 11

Falck, B., Hillarp, N. A., Thieme, G. and Torp, A. (1962). Fluorescence of catecholamines and related compounds condensed with formaldehyde. *J. Histochem. Cytochem.*, **10**, 348–354

Field, P. M. (1972). A quantitative ultrastructural analysis of the distribution of amygdaloid fibres in the preoptic area and the ventromedial hypothalamic nucleus. *Exp. Brain Res.*, **14**, 527–538

Gerendai, I. and Halász, B. (1976). Hemigonadectomy-induced unilateral changes in the protein-synthesizing activity of the rat hypothalamic arcuate nucleus. *Neuroendocrinology* **21**, 331–337

Gerendai, I., Kiss, J., Molnár, J. and Halász, B. (1974). Further data on the existence of a neural pathway from the adrenal gland to the hypothalamus. *Cell Tiss. Res.*, **153**, 559–564

Gerendai, I., Marchetti, B. and Scapagnini, U. (1977). Chronic degeneration of norepinephrinergic (NE) ovarian endings and compensatory ovarian hypertrophy (OCH). *Proc. Int. Union Physiol. Sci.*, **Vol. XIII**, Actes du Congres, P.262

Goldsmith, P. C. and Ganong, W. F. (1975). Ultrastructural localization of luteinizing hormone-releasing hormone in the median eminence of the rat. *Brain Res.*, 97, 181–193

Halász, B. (1969). The endocrine effects of isolation of the hypothalamus from the rest of the brain. In W. F. Ganong and L. Martini (eds.), *Frontiers in Neuroendocrinology*, Oxford University Press, New York, 307–342

Halász, B. and Szentágothai, J. (1959). Histologischer Beweis einer nervösen Signalübermittlung von der Nebennierenrinde zum Hypothalamus. *Z. Zellforsch.*, **50**, 297–306

Harris, M. C., Makara, G. B. and Spyer, K. M. (1971). Electrophysiological identification of neurones of the tubero-infundibular system. *J. Physiol. (Lond.)*., **218**, 86–87

Harris, M. C. and Sanghera, M. (1974). Projection of medial basal hypothalamic neurones to the preoptic anterior hypothalamic areas and the paraventricular nucleus in the rat. *Brain Res.*, **81**, 401–411

Haymaker, W., Anderson, E. and Nauta, W. J. H. (1969). *The Hypothalamus*, Charles C. Thomas, Springfield, Illinois, USA

Heimer, L. (1975). Olfactory projections to the diencephalon. In W. E. Stumpf and L. D. Grant (eds.), *Anatomical Neuroendocrinology*, Karger, Basel, 30–39

Heimer, L. and Nauta, W. J. H. (1969). The hypothalamic distribution of the stria terminalis in the rat. *Brain Res.*, **13**, 284–297

Hendrickson, A. E., Wagoner, N. and Cowan, W. M. (1972). An autoradiographic and electron microscopic study of retino-hypothalamic connections. *Z. Zellforsch.*, **135**, 1–126

Hökfelt, T., Efendic, S., Hellerström, C., Johansson, O., Luft, R. and Arimura, A. (1975*b*). Cellular localization of somatostatin in endocrine-like cells and neurones of the rat with special references to the Al-cells of the pancreatic islets and to the hypothalamus. *Acta Endocrinol.*, Suppl. 200, **80**, 1–41

Hökfelt, T., Efendic, S., Johansson, O., Luft, R. and Arimura, A. (1974). Immunohistochemical localization of somatostatin (growth hormone release-inhibiting factor) in the guinea-pig brain. *Brain Res.*, **80**, 165–169

Hökfelt, T., Elde, R., Johansson, O., Luft, R., Nilsson, G. and Arimura, A. (1976). Immunohistochemical evidence for separate populations of somatostatin-containing and substance P-containing primary afferent neurones in the rat. *Neuroscience*, **1**, 131–136

Hökfelt, T., Fuxe, K., Johansson, O., Jeffcoate, S. and White, N. (1975*a*). Distribution of thyrotrophin-releasing hormone (TRH) in the central nervous system as revealed with immunohistochemistry. *Europ. J. Pharmacol.*, **34**, 389-392

Hyyppä, M., Motta, M. and Martini, L. (1971). "Ultrashort" feed-back control of follicle-stimulating hormone-releasing factor secretion. *Neuroendocrinology*, **7**, 227

Keefer, D. A. and Stumpf, W. E. (1975). Atlas of oestrogen-concentrating cells in the central nervous system of the squirrel monkey. *J. comp. Neurol.*, **160**, 419–442

King, J. C., Gerall, A. A., Fishback, J. B. and Elkind, K. E. (1975). Growth hormone-release inhibiting hormone (GH-RIH) pathway of the rat hypothalamus revealed by the unlabelled antibody peroxidase–antiperoxidase method. *Cell Tiss. Res.*, **160**, 423–430

Koritsánszky, S. and Köves, K. (1976). Data on the absence of axon terminals of medial preoptic area neurones in the surface zone of the median eminence. *J. Neural Transm.*, **38**, 159–167

Köves, K. and Réthelyi, M. (1976). Direct neural connection from the medial preoptic area to the hypothalamic arcuate nucleus of the rat. *Exp. Brain Res.*, **25**, 529–539

Krey, L. C., Butler, W. R. and Knobil, E. (1975*a*). Surgical disconnection of the medial basal hypothalamus and pituitary function in the rhesus monkey. I: Gonadotrophin secretion. *Endocrinology*, **96**, 1073–1087

Krey, L. C., Lu, K. H., Butler, W. R. and Hotchkiss, J. (1975*b*). Surgical disconnection of the medial basal hypothalamus and pituitary function in the rhesus monkey. II: GH and cortisol secretion. *Endocrinology*, **96**, 1088–1093

Lammers, H. J. (1972). The neural connections of the amygdaloid complex in mammals. In B. E. Eleftheriou (ed.). *Advances in Behavioral Biology*, vol. 2., Plenum Press, New York, 123–144

Leonardelli, J., Barry, J. and Dubois, M. P. (1973). Mise en evidence par immunofluorescence d'un constituant immunologiquement apparenté au LH-RF dans l'hypothalamus et l'éminence médiane chez les mammifères. *C. R. Acad. Sci. Paris*, **276**. 2043–2046

Léránth, Cs., Záborszky, L., Marton, J. and Palkovits, M. (1975). Quantitative studies on the supraoptic nucleus in the rat. I: Synaptic organization. *Exp. Brain Res.*, **22**, 509–523

Lisk, R. D. (1962). Diencephalic placement of oestradiol and sexual receptivity in the female rat. *Am. J. Physiol.*, **203**, 493–496

McBride, R. L. and Sutin, J. (1976). Projections of the locus coeruleus and adjacent pontine tegmentum in the cat. *J. comp. Neurol.*, **165**, 265–284

McGuire, J. L. and Lisk, R. D. (1969). Localization of oestrogen receptors in the rat hypothalamus. *Neuroendocrinology*, **4**, 289–295

Makara, G. B., Harris, M. C. and Spyer, K. M. (1972). Identification and distribution of tubero-infundibular neurones. *Brain Res.*, **40**, 283–290

Makara, G. B. and Hodács, L. (1975). Rostral projections from the hypothalamic arcuate nucleus. *Brain Res.*, **84**, 23–29

Mason, C. A. and Lincoln, D. W. (1976). Visualization of the retino-hypothalamic projection in the rat by cobalt precipitation. *Cell Tiss. Res.*, **168**, 117–131

Matsumoto, A. and Arai, Y. (1976). Effect of oestrogen on early postnatal development of synaptic formation in the hypothalamic arcuate nucleus of female rats. *Neurosci. Lett.*, **2**, 79–82

Meibach, R. C. and Siegel, A. (1977). Efferent connections of the septal area in the rat: an analysis utilizing retrograde and anterograde transport methods. *Brain Res.*, **119**, 1–20

Mess, B. (1969). Site and onset of production of releasing factors. In *Progress in Endocrinology*, Internat. Congr. Ser. 184, Excerpta Medica, Amsterdam, 564

Moore, R. Y. and Lenn, N. J. (1972). A retinohypothalamic projection in the rat. *J. comp. Neurol.*, **146**, 1–14

Moss, R. L. (1975). Unit responses in preoptic and arcuate neurones related to anterior pituitary function. In L. Martini and W. F. Ganong (eds.), *Frontiers in Neuroendocrinology*, Raven Press, New York, 95–128

Nauta, W. J. H. (1960). Limbic system and hypothalamus: anatomical aspects. *Physiol. Rev.*, **40**, 102–104

Nauta, W. J. H. and Haymaker, W. (1969). Hypothalamic nuclei and fiber connections. In W. Haymaker, E. Anderson and W. J. H. Nauta (eds.), *The Hypothalamus*, Charles C. Thomas, Springfield, Illinois, USA, 136–209

Pasquier, D. A. and Reinoso-Suarez, F. (1976). Direct projections from hypothalamus to hippocampus in the rat demonstrated by retrograde transport of horseradish peroxidase. *Brain Res.*, **108**, 165–169

Pelletier, G., Labrie, F., Puviani, R., Arimura, A. and Schally, A. V. (1974). Immunohistochemical localization of luteinizing hormone-releasing hormone in the rat median eminence. *Endocrinology*, **95**, 314–317

Pfaff, D. W., Gerlach, J. L., McEwen, B. S., Ferin, M., Carmel, P. and Zimmerman, E. A. (1976). Autoradiographic localization of hormone-concentrating cells in the brain of the female rhesus monkey. *J. comp. Neurol.*, **170**, 279–294

Pfaff, D. and Keiner, M. (1973). Atlas of oestradiol-concentrating cells in the central nervous system of the female rat. *J. comp. Neurol.*, **151**, 121–157

Poulain, P. and Partouche, C. (1967). Neural connections from the preoptic region and septum to the arcuate nucleus and median eminence. *C. R. Acad. Sci.* (Paris), **277**, 737–739

Raisman, G. (1966). The connexions of the septum. *Brain*, **89**, 317–348

Raisman, G., Cowan, W. M. and Powell, T. P. S. (1965). The extrinsic afferent, commissural and association fibres of the hippocampus. *Brain*, **88**, 963–995

Raisman, G., Cowan, W. M. and Powell, T. P. S. (1966). An experimental analysis of the efferent projection of the hippocampus. *Brain*, **89**, 83–108

Raisman, G. and Field, P. M. (1971). Anatomical considerations relevant to the interpretation of neuroendocrine experiments. In L. Martini and W. F. Ganong (eds.), *Frontiers in Neuroendocrinology*, Oxford University Press, New York, 3–44

Raisman, G. and Field, P. M. (1973). Sexual dimorphism in the neuropil of the preoptic area of the rat and its dependence on neonatal androgen. *Brain Res.*, **54**, 1–29

Renaud, L. P. (1976*a*). Tubero-infundibular neurones in the basomedial hypothalamus of the rat: electrophysiological evidence for axon collaterals to hypothalamic and extra-hypothalamic areas. *Brain Res.*, **105**, 59–72

Renaud, L. P. (1976*b*). An electrophysiological study of amygdalohypothalamic projections to the ventromedial nucleus of the rat. *Brain Res.*, **105**, 45–58

Renaud, L. P. and Hopkins, D. A. (1977). Amygdala afferents from the mediobasal hypothalamus: an electrophysiological and neuroanatomical study in the rat. *Brain Res.*, **121**, 201–213

Réthelyi, M. and Halász, B. (1970). Origin of the nerve endings in the surface zone of the median eminence of the rat hypothalamus, *Exp. Brain Res.*, **11**, 145–158

Saper, C. B., Loewy, A. D., Swanson, L. W. and Cowan, W. M. (1976*b*). Direct hypothalamo-autonomic connections. *Brain Res.*, **117**, 305–312

Saper, C. B., Swanson, L. W. and Cowan, W. M. (1976*a*). The efferent connections of the ventromedial nucleus of the hypothalamus of the rat. *J. comp Neurol.*, **169**, 409–442

Sar, M. and Stumpf, W. E. (1975). Distribution of androgen-concentrating neurones in rat brain. In W. E. Stumpf and L. D. Grant, *Anatomical Neuroendocrinology*, Karger, Basel, 120–133

Scharrer, E. and Scharrer, B. (1954). Die Neurosekretion. In *Hdb. d. Mikrosk. Anat. d. Menschen,* W. v. Möllendorff and W. Bargmann (eds.), Springer, Berlin-Göttingen-Heidelberg, VI, 5

Sétáló, G. (1977). Anatomy, using new immunohistological methods. In V. H. T. James (ed.), *Endocrinology*, Excerpta Medica, Amsterdam-Oxford, 100-104

Sétáló, G., Vigh, S., Schally, A. V., Arimura, A. and Flerkó, B. (1975). GH-RIH containing neural elements in the rat hypothalamus. *Brain Res.*, **90**, 352–356

Sétáló, G., Vigh, S., Schally, A. V., Arimura, A. and Flerkó, B. (1976). Immunohistological study of the origin of LH-RH containing nerve fibres of the rat hypothalamus. *Brain Res.*, **103**, 597–602

Siverman, A. J. and Desmoyers, P. (1976). Ultrastructural immunocytochemical localization of luteinizing hormone-releasing hormone (LH-RH) in the median eminence of the guinea-pig. *Cell Tiss. Res.*, **169**, 157–166

Sousa-Pinto, A. (1970). Electron microscopic observations on the possible retinohypothalamic projection in the rat. *Exp. Brain Res.*, **11**, 528–538

Szentágothai, J. (1964). The parvicellular neurosecretory system. In W. Bargmann and J. P. Schadé (eds.), *Progress in Brain Research*, Elsevier Publishing Company, Amsterdam, 135-146

Szentágothai, J. (1968). Anatomical considerations. In J. Szentágothai, B. Flerkó, B. Mess and B. Halász (eds.), *Hypothalamic Control of the Anterior Pituitary*, Akadémiai Kiadó, Budapest, 22–109

Szentágothai, J., Flerkó, B., Mess, B. and Halász, B. (1968). Hypothalamic control of the anterior pituitary. An experimental-morphological study. Akadémiai Kiadó, Budapest

Vandesande, F. and Dierickx, K. (1975). Identification of the vasopressin-producing and of the oxytocin-producing neurones in the hypothalamic magnocellular neurosecretory system of the rat. *Cell Tiss. Res.*, **164**, 153–162

Vandesande, F., Dierickx, K. and de Mey, J. (1975*a*). Identification of the vasopressin-neurophysin II and the oxytocin-neurophysin I-producing neurones in the bovine hypothalamus. *Cell Tiss. Res.*, **156**, 189–200

Vandesande, F., Dierickx, K. and de Mey, J. (1975*b*). Identification of the vasopressin-neurophysin-producing neurones of the rat suprachiasmatic nuclei. *Cell Tiss. Res.*, **156**, 377–380

Wakefield, C. and Hall, E. (1974). Hypothalamic projections to the amygdala in the cat. A light and electron-microscopic study. *Cell Tiss. Res.*, **151**, 499–508

Watkins, W. B. and Choy, V. J. (1977). Immunocytochemical study of the hypothalamo-neurohypophyseal system. III: Localization of oxytocin- and vasopressin-containing neurones in the pig hypothalamus. *Cell Tiss. Res.*, **180**, 491–503

Wenisch, J. J. C. (1976). Retinohypothalamic projection in the mouse: electron microscopic and iontophoretic investigations of hypothalamic and optic centres. *Cell Tiss. Res.*, **167**, 547–561

Yagi, K. and Sawaki, Y. (1970). On the localization of neurosecretory cells controlling adenohypophyseal function. *J. Physiol. Soc. Jap.*, **32**, 621–622

Yagi, K. and Sawaki, Y. (1975). Recurrent inhibition and facilitation: demonstration in the tubero-infundibular system and effects of strychnine and picrotoxin. *Brain Res.*, **84**, 155–159

Discussion

Schnieden (Manchester)
Could you tell us more about the olfactory projections to the hypothalamus? What is the pathway and where does it enter the hypothalamus?

Halász
The bulbus olfactorius has direct connections to the pyriform cortex which then connect with the hypothalamus. There are also connections from the bulbus olfactorius to the amygdala and from the cortico-medial part of this region there are projections which go into various regions of the hypothalamus.

Martini (Milan)
Could you comment further on the unilateral hypertrophy of the ovary which follows unilateral ovariectomy? What are the nervous components of this phenomenon?

Halász
The evidence I quoted was based on morphological studies but since that time my co-workers have performed the following experiment. Local administration of 6-hydroxydopamine to the ovary on one side causes hypertrophy of the ovary on the other side. If you remove one ovary and treat the other side locally with 6-hydroxydopamine you can prevent the compensatory ovarian hypertrophy. This suggests there are both afferent and efferent connections between the gonads and the hypothalamus.

Besser (London)
We are taught that the hypothalamus controls the pituitary and consequently pituitary hormones are released. However it is well established clinically that in pituitary or pituitary stalk lesions all the hormones do not disappear at once. Indeed, LH goes first and GH also quickly disappears, but ACTH and TSH disappear slowly. What is the evidence that anatomical or geographical distributions of the capillaries which deliver the hypothalamic hormones to the pituitary could explain these observations?

Halász
It is possible to observe this phenomenon experimentally. A small partial lesion of the stalk at the middle median eminence level results in gonadal and thyroid atrophy but pituitary ACTH will only decrease if the lesion is large. It has been suggested that specific portal vessels supply a certain region of the pituitary and an experimental partial lesion of the median eminence at pituitary level causes partial infarction of the pituitary. However, we do not yet know whether there is a differential distribution of terminals within the median eminence, for example LH-RH terminals in the lateral part and somatostatin terminals in the medial region and so on. This could be the explanation but at present I am not sure.
(See chapter 2 (editors))

Wilson (Dublin)
You mentioned that the retina which gives appreciation of light and dark has a connection with the hypothalamus. You also mentioned the ovaries and their connections with the hypothalamus. To what extent does the hypothalamus control biological rhythms?

Halász
Retinal fibres terminate on the suprachiasmatic nucleus and lesions of this nucleus interfere with the diurnal ACTH rhythm. In lesions which interrupt all connections to the hypothalamus there are no biological rhythms, which results in no diurnal fluctuation of pituitary ACTH secretion and the sexual cycle disappears. However, light itself is not an essential factor for biological rhythms as these do occur in the blind. Therefore sight probably only modulates the intrinsic rhythm.

Flack (Harlow)
We have become accustomed to think of the neuro-vascular system as controlling the anterior pituitary but there are reports of sympathetic nerves from the hypothalamus to the anterior pituitary, mainly in birds and in amphibia. Does a similar situation exist in mammals?

Halász
I did not cover the sympathetic innervation of the pituitary vessels but there is a large change in pituitary blood flow under various experimental conditions such as stress. Thus these aspects could be extremely important.

2
The transmitters of the hypothalamus

K. Fuxe*, T. Hökfelt*, K. Andersson*, L. Ferland*, O. Johansson*, D. Ganten†, P. Eneroth‡, J.-A. Gustafsson∮, P. Skett∮, S. I. Said¶ and V. Mutt″

INTRODUCTION

Monoamines, GABA and acetylcholine are the best-known transmitters of the hypothalamus (Fuxe and Hökfelt, 1969; Fuxe and Jonsson, 1974: Björklund *et al.*, 1973; Carlsson *et al.*, 1962; Dahlström and Fuxe, 1964; Hökfelt *et al.*, 1974*b*; Ungerstedt, 1971; Roberts, 1975; Roberts *et al.*, 1975; Tappaz *et al.*, 1976; Vogt, 1954; Snyder and Taylor, 1972; Brownstein *et al.*, 1975*b*; Fuxe *et al.*, 1976*b*; Hökfelt *et al.*, 1978). Recently evidence has accumulated that hypothalamic hormones such as luteinising hormone releasing hormone (LH-RH), somatostatin, thyrotrophin releasing hormone (TRH) and other types of peptides such as substance P and enkephalin can also subserve a transmitter or a modulator role (changing the action of a transmitter) at synaptic junctions within the hypothalamus (Barry and Dubois, 1974; Brownstein *et al.*, 1975*a*; Brownstein *et al.*, 1974; Elde *et al.*, 1976*a*; von Euler and Gaddum, 1931; Fuxe *et al.*, 1976*a*; Fuxe *et al.*, 1976*b*; Ganten *et al.*, 1975; Giachetti *et al.*, 1976; Guillemin, 1978; Guillemin *et al.*, 1976; Elde *et al.*, 1976*b*; Hökfelt *et al.*, 1974*a*; Hökfelt *et al.*, 1975*b*; Hökfelt *et al.*, 1975*c*; Hökfelt *et al.*, 1975*e*; Reichlin *et al.*, 1976; Renaud *et al.*, 1975; Said and Rosenberg, 1976 Vanderhaeghen *et al.*, 1975; Zimmerman, 1976; Larsson *et al.*, 1976; Fuxe *et al.*, 1977*d*; Hökfelt *et al.*, 1978). In the present article the distribution of the various monoamine-, GABA-, acetylcholine- and peptide-containing neurones within the hypothalamus will be described. The article

* Department of Histology, Karolinska Institutet, Stockholm, Sweden
†Department of Pharmacology, University of Heidelberg, Heidelberg, West Germany
‡Hormone Laboratory, Department of Obstetrics and Gynecology, Karolinska Hospital, Stockholm, Sweden
∮Department of Medical Chemistry, Karolinska Institutet, Stockholm, Sweden
¶Departments of Internal Medicine and Pharmacology, The University of Texas, Health Science Center, Dallas, Texas, USA
″Department of Biochemistry, Karolinska Institutet, Stockholm, Sweden

will then describe how monoamine systems can control activity in hypothalamic hormone-containing pathways and in other types of peptide-containing pathways. Finally, how the various types of transmitters and/or modulators can influence activity in the catecholamine (CA) systems of the hypothalamus, will be discussed with particular reference to the dopamine (DA) systems in the median eminence.

MAPPING OF HYPOTHALAMIC NEURONE PATHWAYS BASED ON THEIR CONTENTS OF TRANSMITTER AND/OR MODULATOR

Monoamine-Containing Pathways

Dopamine pathways

The best-known systems are the DA systems projecting to the external layer of the median eminence, to the posterior lobe and to the intermediate lobe of the pituitary gland. The highest density of DA terminals is found within the external layer of the median eminence and of the infundibular stalk (figure 2.1). Particularly dense networks are found within the lateral palisade zone of the median eminence.

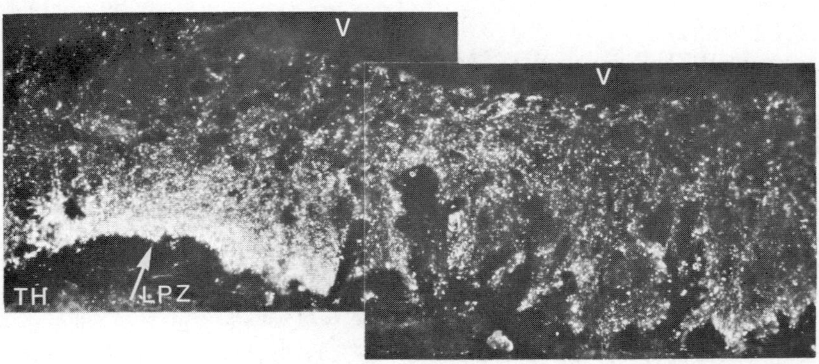

Figure 2.1 Tyrosine hydroxylase (TH) immunofluorescence within the median eminence of the normal male rat. A high density of TH-positive nerve terminals is found in the external layer particularly within the lateral palisade zone (LPZ; →). Magnification × 400; V = third ventricle

Figure 2.2 *A*, Catecholamine (CA) fluorescence within the nucleus anterior hypothalami. Two types of CA nerve terminals are observed; one type possesses thick intensely fluorescent varicosities (see arrows), the other type possesses fine weakly fluorescent varicosities; the former type probably contains noradrenaline and the latter type probably dopamine; magnification × 300. *B*, Tyrosine hydroxylase (TH) immunofluorescence within the medial basal hypothalamus of normal male rat; TH-positive dopamine nerve cell bodies are found within the ventral periventricular region (I→) and within the most ventrolateral part of the arcuate nucleus (II→); a high density of TH-positive nerve terminals is found within the external layer of the median eminence (→); V = third ventricle; magnification × 120. *C*, Tyrosine

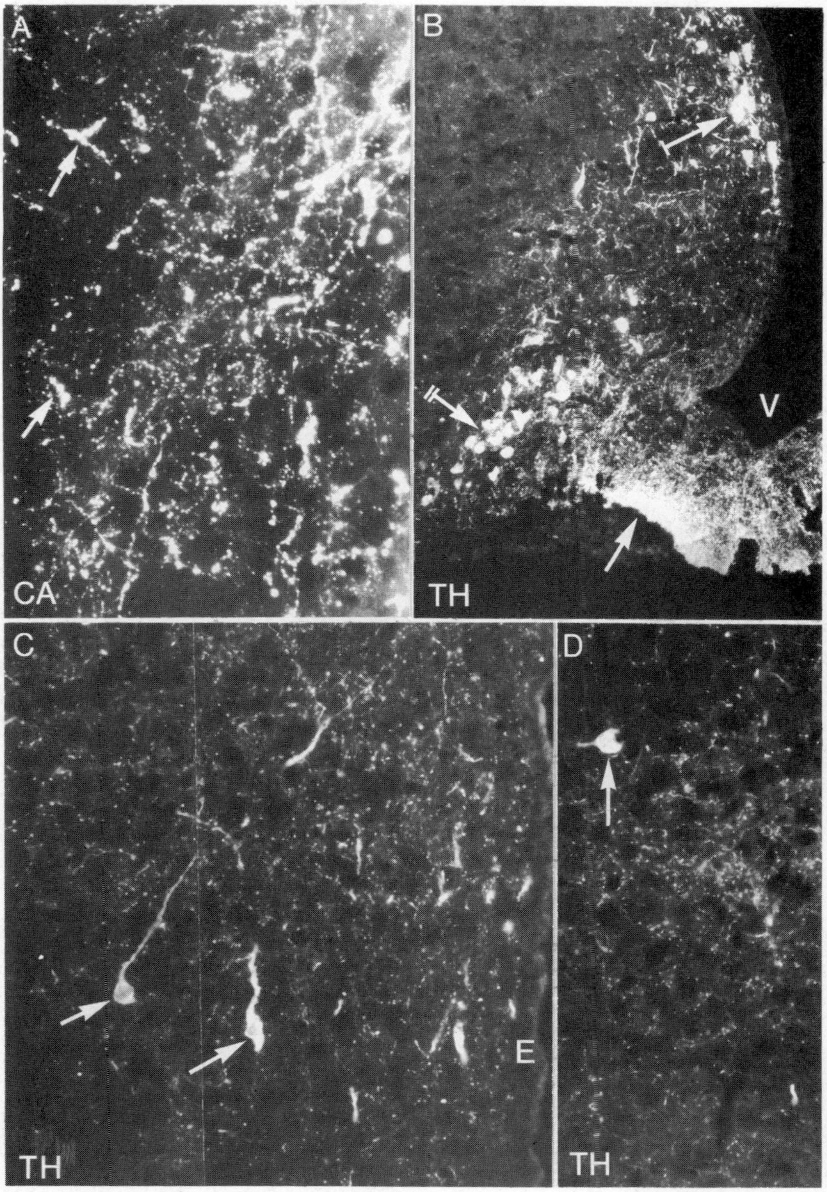

hydroxylase (TH) immunofluorescence within the border area between the nucleus anterior hypothalami (lower left) and the paraventricular hypothalamic nucleus (upper right); TH-positive nerve cell bodies are seen (→) and TH-positive fine nerve terminals, E = ependyma; magnification × 300. *D*, Tyrosine hydroxylase immunofluorescence within the nucleus anterior hypothalami; a fine plexus of TH nerve terminals is found probably representing dopamine (DA) nerve terminals; a DA nerve cell body showing TH immunoreactivity is seen (→); magnification × 300

The DA cell bodies responsible for the innervation described above are mainly located within the arcuate nucleus and within the ventral periventricular hypothalamic areas (figure 2.2). However, large numbers of DA cell bodies are found along the entire periventricular system of the hypothalamus and the preoptic area, and many are also found within the zona incerta. These DA cell bodies may project to various types of hypothalamic and preoptic nuclei such as the anterior hypothalamic nucleus (figure 2.2). Thus the DA terminals within the hypothalamus mainly originate from intrahypothalamic DA cell bodies (Hökfelt *et al.*, 1975*d*; Hökfelt *et al.*, 1976; Björklund *et al.*, 1973; Carlsson *et al.*, 1962; Dahlström and Fuxe, 1964; Fuxe, 1964, 1965; Fuxe and Hökflet, 1969; Hökfelt *et al.*, 1975*b*).

Noradrenaline pathways

Unlike DA nerve cell bodies, no NA nerve cell bodies have been identified in the hypothalamus. Rich networks of NA nerve terminals exist within most of the hypothalamic nuclei, and they appear to belong to the non-locus coeruleus NA system.

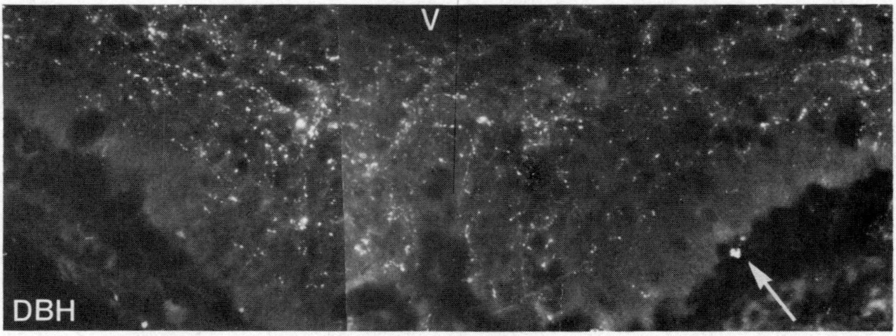

Figure 2.3 Dopamine-β-hydroxylase (DBH) immunofluorescence in the rat median eminence of a normal animal. A plexus of DBH-positive nerve terminals is found within the subependymal and the internal layer of the median eminence. A few are also seen reaching into the medial palisade zone. Arrow points to lateral palisade zone which hardly contains any DBH-positive terminals at all. V = third ventricle; magnification × 300

This system originates from NA cell bodies located in the reticular formation of the medulla oblongata and pons. Within the median eminence, NA is found mainly within the subependymal layer, but some is also present in the medial palisade zone of the median eminence (figure 2.3). However, in this latter area the majority of the terminals contain DA (Löfström *et al.*, 1976*a*; Andén *et al.*, 1966; Fuxe, 1965; Björklund *et al.*, 1973; Carlsson *et al.*, 1962; Dahlström and Fuxe, 1964; Fuxe and Hökfelt, 1969; Fuxe *et al.*, 1976*b*).

Adrenaline pathways

Using antibodies against phenylethanolamine-*N*-methyl-transferase (PNMT) it has been possible to identify in immunohistochemical studies PNMT-containing nerve

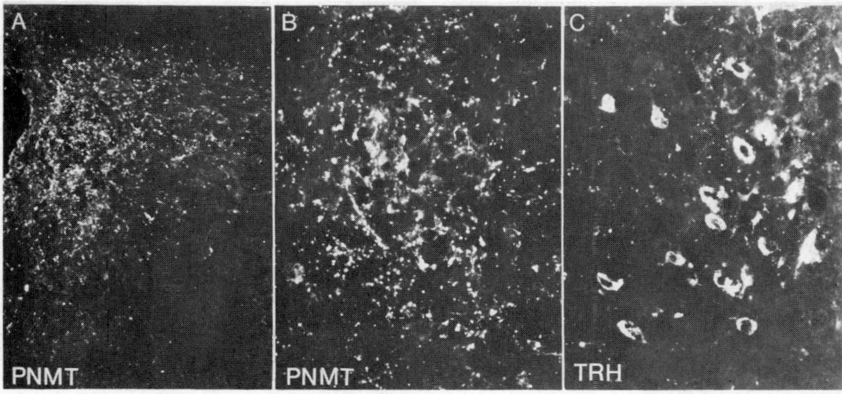

Figure 2.4 *A*, Phenylethanolamine-*N*-methyl transferase (PNMT) immunofluorescence of the parvicellular and the medial part of the magnocellular paraventricular nucleus; a high density of PNMT-positive nerve terminals is found within this area, whereas only a few are found within the lateral part of the magnocellular paraventricular nucleus; magnification × 120. *B*, Higher magnification of *A*; a dense plexus of PNMT-positive terminals is seen within the parvicellular paraventricular nucleus; magnification × 300. *C*, Thyrotrophin-releasing hormone (TRH) immunofluorescence in the parvicellular part of the paraventricular hypothalamic nucleus; an intraventricular injection of colchicine (50 μg/20 μl) had been performed 24 h before killing; a large number of TRH-positive nerve cell bodies is found within this area which is rich in PNMT-positive nerve terminals; magnification × 300

terminals within the hypothalamus and the preoptic area. The PNMT-positive terminals probably represent adrenaline nerve terminals. The highest concentrations of terminals are found within the dorsomedial hypothalamic nucleus, in the perifornical region, and in the parvicellular and medial part of the magnocellular paraventricular nucleus (figure 2.4). Large numbers also exist within periventricular areas. Within the arcuate nucleus the adrenaline nerve terminals have a lateral and ventral position and only few such terminals are present within the median eminence and then exclusively within the subependymal layer and in the internal layer (Hökfelt *et al.*, 1974*b*; Fuxe *et al.*, 1976*b*; Hökfelt *et al.*, 1978).

5-Hydroxytryptamine pathways

The insensitivity of the Falck–Hillarp procedure has made it difficult to demonstrate the 5-hydroxytryptamine (5-HT) nerve terminal networks in the hypothalamus and the preoptic area. Histochemically dense networks of 5-HT terminals can only be demonstrated within the suprachiasmatic nucleus. Biochemical studies, however, demonstrate a wide distribution of 5-HT nerve terminals within the hypothalamus. The 5-HT terminals mainly originate from 5-HT cell bodies located in the nucleus raphe dorsalis and medianus and within the raphe nuclei of the pons (Dahlström and Fuxe, 1964; Fuxe, 1965; Fuxe and Jonsson, 1974). Like the NA and adrenaline axons, the hypothalamic 5-HT axons run in the lateral hypothalamic area but have a more ventral position. Recently indoleamine nerve cell bodies have also been demonstrated within the hypothalamus itself, especially within periventricular areas and within the area overlying the nucleus suprachiasmaticus (Chan-Palay, 1977). These indoleamine nerve cell bodies seem in part to be identical to the dopadecarboxylase (DDC)-positive but tyrosine hydroxylase (TH)-negative nerve cell bodies which have been demonstrated in these areas (Hökfelt *et al.*, 1973; Hökfelt *et al.*, 1978; Lidbrink *et al.*, 1974). It should be pointed out that the antibodies against DDC also demonstrate the 5-hydroxytryptophan (5-HTP) decarboxylase in the 5-HT neurones, since these enxymes are antigenically identical (Christenson *et al.*, 1972). It is possible that the indoleamine present in the hypothalamic nerve cell bodies is different from 5-HT and could be 5-methoxytryptamine, which has been identified in the hypothalamus (Koslow, 1974).

Amino Acid Transmitter-Containing Pathways

GABA pathways

Roberts and collaborators (Roberts, 1975; Roberts *et al.*, 1975) have mapped out GABA neurones in several parts of the central nervous system using antibodies against glutamic acid decarboxylase (GAD) in immunohistochemical studies. In collaboration with Drs. Pérez de la Mora, Possani and Tapia we have been able to map out GAD-positive nerve terminals within the hypothalamus using an antiserum against GAD. Very dense networks of GAD-positive nerve terminals were observed within most hypothalamic nuclei (Hökfelt *et al.*, 1978; Fuxe *et al.*, 1976*b*), and some GAD-positive nerve terminals were present in the external layer of the median eminence. A few were also found in the internal layer. The present findings are in good agreement with the high levels of GABA and GAD reported in biochemical studies in the hypothalamus (Tappaz *et al.*, 1976). In our analysis, GAD-positive nerve cell bodies could not be demonstrated either in the hypothalamus or in other regions. Studies on the effect of hypothalamic deafferentation of GAD activity indicate that the GABA nerve terminals in several hypothalamic nuclei must have their nerve cell bodies located outside the hypothalamus. However, the GABA nerve terminals within the median eminence appear to have an origin within the hypothalamus (Brownstein *et al.*, 1976).

Peptide-Containing Pathways

The distribution of various peptide-containing nerve cell bodies and terminals in the hypothalamus has been summarised in previous articles (Hökfelt *et al.*, 1978; Fuxe *et al.*, 1976*b*). Only a brief description with some illustrations will therefore be given below.

Most of the luteinising hormone-releasing hormone (LH–RH)-positive nerve cell bodies have been identified within the pre- and suprachiasmatic areas. Recently, however, Hoffman and collaborators (Hoffman *et al.*, 1976) have also described LH–RH-positive cell bodies within the arcuate nucleus. In this study they used an antiserum which was directed towards a different part of the LH–RH molecule compared with previous antisera (Hoffman *et al.*, 1976). The highest densities of

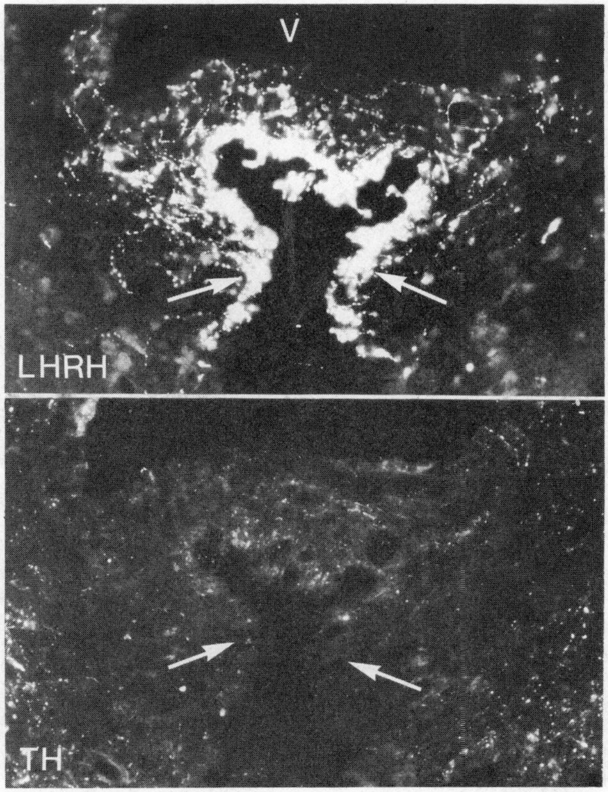

Figure 2.5 Luteinising hormone-releasing hormone (LH-RH; upper part) and tyrosine hydroxylase (TH; lower part) immunofluorescence is observed within the organum vasculosum of normal rat. A dense plexus of LH-RH-positive nerve terminals is found within the organum vasculosum (→), whereas no TH-positive nerve terminals are present within this organ. V = third ventricle; magnification ×300

LH–RH-positive nerve terminals are found within the lateral palisade zone of the median eminence and within the organum vasculosum (figure 2.5). Deafferentation experiments demonstrate that most of the LH–RH-positive nerve terminals present within the median eminence originate from descending pathways from the pre- and suprachiasmatic area.

Thyrotrophin- releasing hormone (TRH)-positive nerve cell bodies have been demonstrated by Johansson and collaborators (Johansson, Hökfelt and Jeffcoate, personal communication) with the dorsomedial hypothalamic nucleus, within the perifornical area and within the parvicellular part of the paraventricular nucleus (figure 2.4). Some TRH-positive nerve cell bodies also exist within periventricular areas of the hypothalamus. TRH-positive nerve terminals are found in high densities within the medial part of the external layer of the median eminence. Nerve termi-

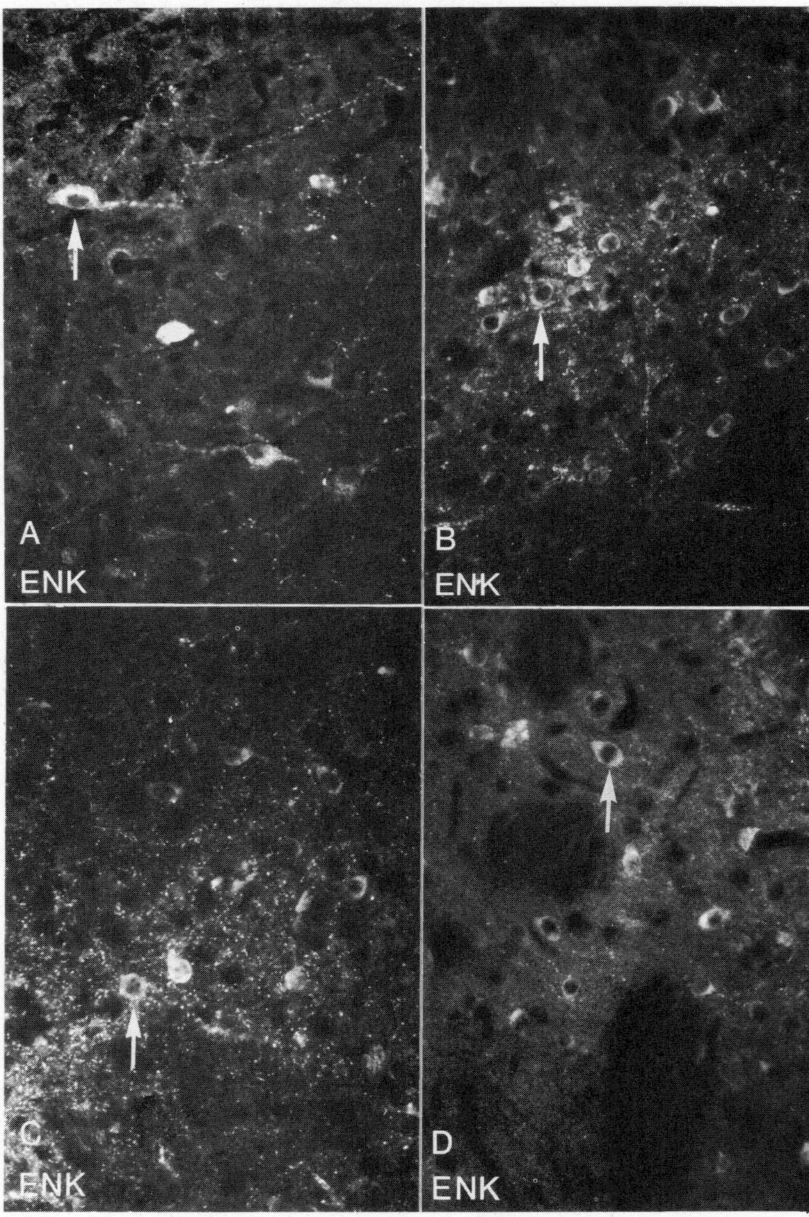

Figure 2.6 Enkephalin-positive (ENK) nerve cell bodies are seen *A* within the medial part of the magnocellular paraventricular hypothalamic nucleus, *B* within the central amygdaloid nucleus, *C* within the ventral part of the nucleus caudatus putamen and *D* within the nucleus caudatus. Arrows point to enkephalin-positive nerve cell bodies. Within the ventral part of the nucleus caudatus putamen a dis-

nals containing TRH-like immunoreactivity also exist within many other hypothalamic nuclei such as the dorsomedial hypothalamic nucleus and the perifornical area (Hökfelt *et al.*, 1975*c*). The median eminence innervation probably originates from the paraventricular region (Hökfelt *et al.*, 1978).

A large group of somatostatin-positive nerve cells exists within the anterior periventricular hypothalamic area. This group innervates the external layers of the median eminence, where a high density of somatostatin-positive terminals exists (Elde *et al.*, 1976*a*; Hökfelt *et al.*, 1978; Hökfelt *et al.*, 1974*a*). High densities are also found within many other hypothalamic areas such as the arcuate nucleus, the ventromedial hypothalamic nucleus and the suprachiasmatic nucleus.

Recent work (Hökfelt *et al.*, 1977*a*; Fuxe, Ganten and Hökfelt, unpublished data) has shown the existence of enkephalin-positive nerve cell bodies within several hypothalamic nuclei such as the arcuate nucleus, the ventromedial hypothalamic nucleus, the paraventricular hypothalamic nucleus (figure 2.6) and the pre-

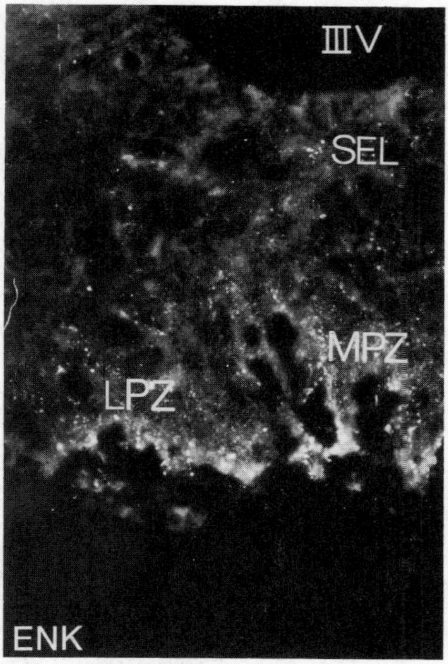

Figure 2.7 Enkephalin (ENK) immunofluorescence within the median eminence. Enkephalin-positive nerve terminals are observed within the medial palisade zone (MPZ) and lateral palisade zone (LPZ), whereas very few, if any, are found within the subependymal layer (SEL). IIIV = third ventricle; magnification × 300. (After Fuxe, Ganten and Hökfelt, unpublished data)

---

tinct plexus of enkephalin-positive nerve terminals is also observed. Within the nucleus caudatus and within the central amygdaloid nucleus the enkephalin-positive nerve terminals are so fine and sparse that they are difficult to observe in the picture. Magnification × 300. (After Fuxe, Ganten and Hökfelt, unpublished data)

mamillary nuclei. Some are also present within the perifornical region. Enkephalin-positive nerve terminals are found in fairly high densities in many hypothalamic nuclei, and some terminals are also found within the external layer of the median eminence (figure 2.7) (Elde *et al.*, 1976*b*; Hökfelt *et al.*, 1978; Fuxe, Ganten and Hökfelt, unpublished data). These findings suggest that intrahypothalamic enkephalin neurone systems may exist.

The existence of enkephalin-positive nerve cells within the medial part of the paraventricular hypothalamic nucleus, magnocellular part, indicates the existence of a paraventricular-infundibular pathway containing enkephalin. On the other hand, only scattered β-endorphin-positive nerve terminals have been found within the hypothalamus, while large numbers of β-endorphin-positive gland cells are found in the anterior pituitary (figure 2.8) in agreement with the work of Bloom and collaborators (Bloom *et al.*, 1977).

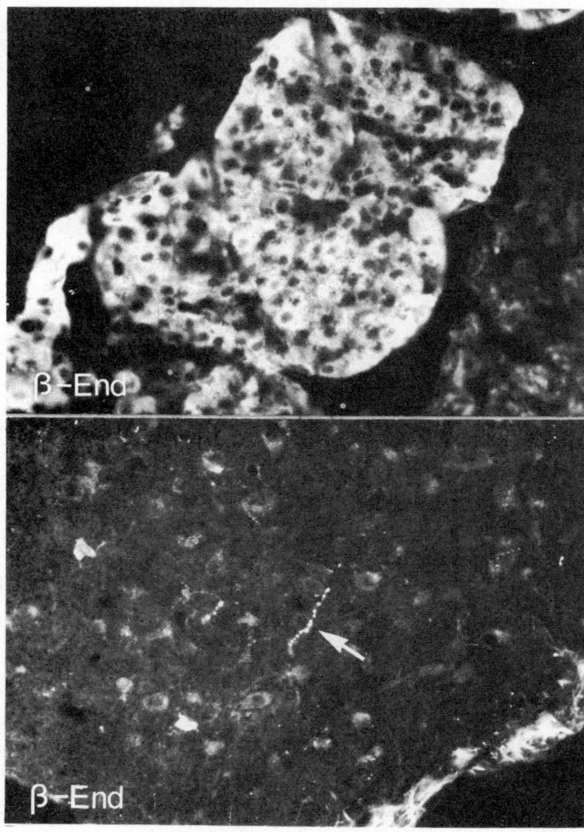

Figure 2.8 β-Endorphin-like (β-End) immunoreactivity within the pars intermedia of the adenohypophysis (upper part) and within the hypothalamus (lower part). A few β-endorphin-positive nerve terminals (→) are observed within the most ventral part of the hypothalamus close to the surface of the brain; magnification × 300. (After Fuxe, Ganten and Hökfelt, unpublished data)

Hökfelt and collaborators (Hökfelt *et al.*, 1975*e*; Hökfelt *et al.*, 1978) have described substance P-positive nerve cell bodies within the premamillary nucleus, dorsomedial hypothalamic nucleus and the ventromedial hypothalamic nucleus. High densities of substance P-positive nerve terminals are found all over the hypothalamus and the preoptic area with particularly high densities in the medial preoptic nucleus. Within the rat median eminence only a few substance P-positive nerve terminals exist. It is also of interest to note that the arcuate nucleus contains a dense innervation by substance P terminals.

Recent immunohistochemical results suggest that angiotensin II-like peptides may have a neuronal location, since angiotensin II-like immunoreactivity was found to be present in nerve cell bodies and nerve terminals of the hypothalamus (Fuxe *et al.*, 1976*a*; Fuxe *et al.*, 1976*b*; Ganten *et al.*, 1978). Angiotensin II-positive nerve cells have been found within the medial part of the magnocellular paraventricular nucleus (figure 2.9), and within the perifornical area. The highest

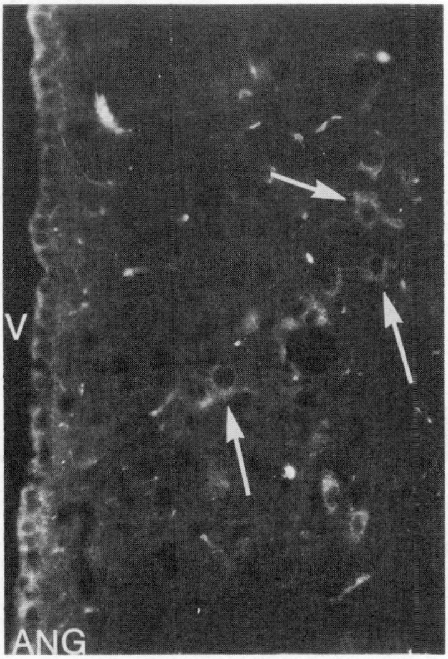

Figure 2.9 Angiotensin II-like (ANG) immunoreactivity within the medial part of the magnocellular paraventricular nucleus. Arrows point to angiotensin II-positive nerve cell bodies. V = third ventricle; magnification × 300

density of angiotensin II-positive nerve terminals exists within the external layer of the median eminence (figure 2.10). This latter system probably originates in the paraventricular region. Relatively high densities are also found within the dorsomedial hypothalamic nucleus, and scattered terminals are found in several hypothalamic nuclei. The present findings suggest the existence of an angiotensin II-positive paraventricular-infundibular pathway, which could have important

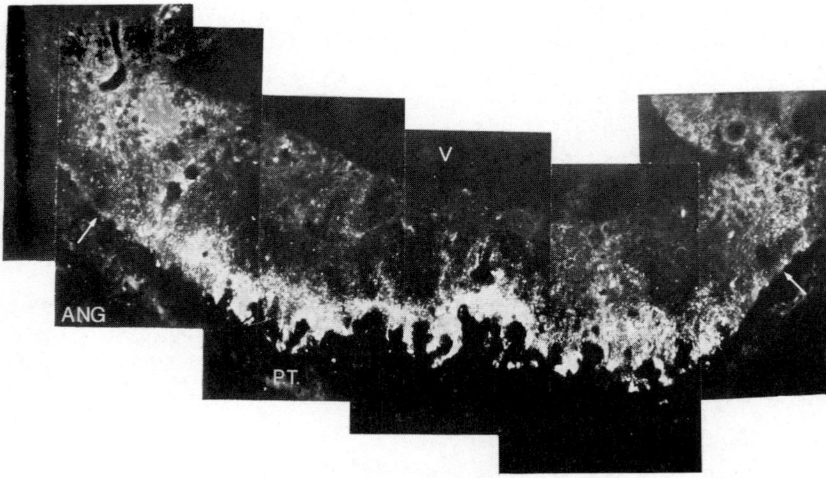

Figure 2.10 Angiotensin II- positive (ANG) immunofluorescence within the median eminence. A dense plexus of angiotensin II-positive nerve terminals is found within the external layer of the median eminence particularly within the medial palisade zone. Relatively few are found within the lateral palisade zone (→). V = third ventricle; P.T. = pars tuberalis of the anterior pituitary; magnification × 200

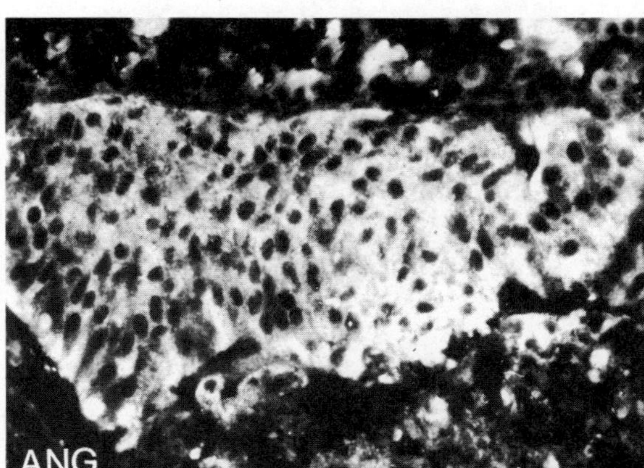

Figure 2.11 Angiotensin II-positive (ANG) immunofluorescence within the pars intermedia of the anterior pituitary. Strongly fluorescent angiotensin II-positive gland cells are found to build up the intermediate lobe; magnification × 120

neuroendocrine functions (Fuxe *et al.*, 1976*a*; Ganten *et al.*, 1978). Angiotensin II-positive gland cells are found in the anterior pituitary, especially within the pars intermedia (figure 2.11).

Immunoreactive vasoactive intestinal polypeptide (VIP) has been demonstrated within synaptosomes (Giachetti *et al.*, 1976; Said and Giachetti, 1977; Said and Rosenberg, 1976). VIP-positive nerve cell bodies within the brain have so far been

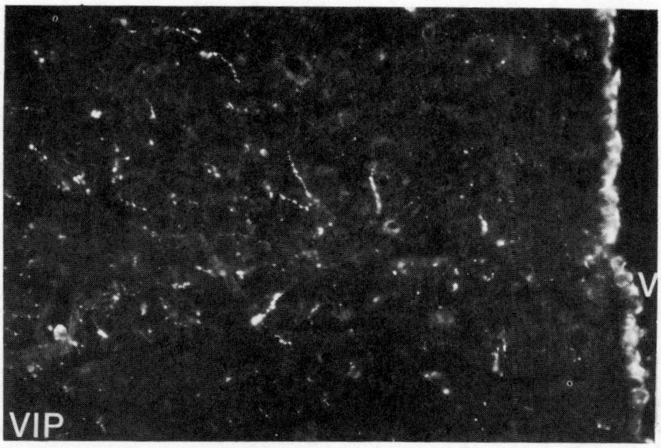

Figure 2.12 Vasoactive intestinal polypeptide (VIP)-like immunoreactivity within the dorsal part of the dorsomedial hypothalamic nucleus. Scattered VIP-positive nerve terminals are found within this area. V = third ventricle. Unspecific fluorescence is observed within the ependymal cells. Magnification × 120

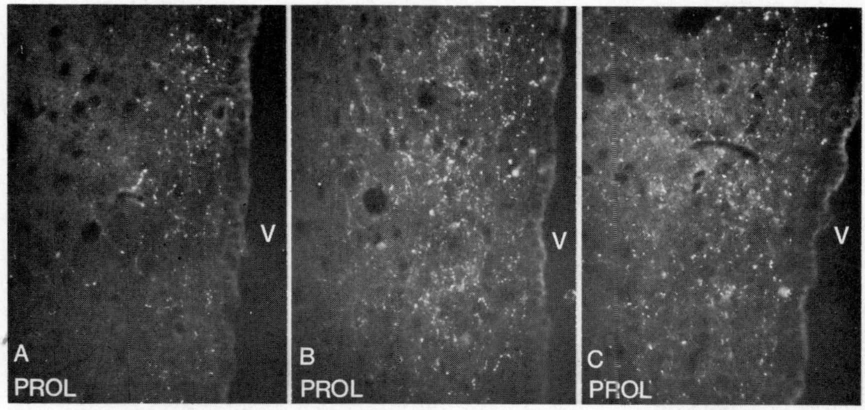

Figure 2.13 Prolactin-like (PROL) immunoreactivity within the periventricular area of the hypothalamus. *A* = ventral periventricular area, *B* = middle periventricular area and *C* = dorsal periventricular area. A plexus of prolactin-positive nerve terminals is found within the periventricular region. V = third ventricle; magnification × 300

observed mainly in cortical regions where VIP exists within association neurones (Fuxe *et al.*, 1977*d*). However, VIP-positive nerve terminals are present within the suprachiasmatic, medial preoptic and anterior hypothalamic nuclei as well as in the dorsal and posterior part of the hypothalamus (Larsson *et al.*, 1976; Fuxe *et al.*, 1977*d*) (figure 2.12). In the median eminence only scattered nerve terminals, if any, have been observed. Most of the VIP-positive nerve terminals are present within the old and new part of the cerebral cortex (Fuxe *et al.*, 1977*d*).

By means of antibodies against rat prolactin it has been possible to demonstrate nerve terminal networks in periventricular hypothalamic areas (figure 2.13). Within the arcuate nucleus, nerve terminal networks containing prolactin-like immunoreactivity have also been demonstrated (Fuxe *et al.*, 1977*c*). These fibres do not disappear after hypophysectomy and they are also present within the internal layers of the median eminence (Fuxe *et al.*, 1977*c*).

DOPAMINERGIC CONTROL OF PEPTIDE-CONTAINING SYSTEMS

Interaction with Luteinising Hormone-Releasing Hormone (LH–RH) Containing Pathways

A large number of experiments indicate that DA terminals located in the lateral palisade zone of the median eminence exert an inhibitory influence on the release of LH-RH from this region via an axo-axonic interaction (Fuxe *et al.*, 1977*e*; Fuxe *et al.*, 1978; Fuxe and Hökfelt, 1969). The evidence also indicates that at least part of the inhibitory central feedback actions of oestrogens and of androgens is mediated via an activation of the DA terminals in the lateral palisade zone (Fuxe *et al.*, 1977*f*; Löfström *et al.*, 1977; Fuxe *et al.*, 1978).

The Role of Dopamine (DA) Systems in the Control of Prolactin Secretion

The available evidence indicates that DA can be released from the median eminence into the primary capillary plexus to act as a prolactin inhibitory factor on DA receptors located on the anterior pituitary gland cells (Macleod and Lehmeyer, 1974). Experiments in this laboratory suggest that DA terminals in the medial palisade zone of the median eminence may largely represent the prolactin inhibitory factor-containing terminals (Fuxe *et al.*, 1977*e*; Fuxe *et al.*, 1977*f*). However, repeated systemic injections of prolactin not only increase DA turnover in the medial palisade zone but also DA turnover in the lateral palisade zone (Fuxe *et al.*, 1977*e*; Fuxe *et al.*, 1976*c*). This observation is of considerable interest, since it is known that hyperprolactinaemia is associated with a reduction of LH secretion and with a blockade of ovulation. It is thus possible that an increased release of DA in the lateral palisade zone by prolactin could in part be responsible for the low activity in the pituitary-gonadal axis in hyperprolactinaemic states in view of the inhibitory influence of DA on LH-RH release reported above. It seems possible that the actions of prolactin on DA turnover are mediated via centrally located prolactin receptors in the hypothalamus, although so far no such receptors have been demonstrated. In this connection it is of considerable interest to point out the presence of nerve terminal plexa containing prolactin-like immunoreactivity within periventricular hypothalamic areas and within the arcuate nucleus (Fuxe *et al.*, 1977*c*). It is conceivable that prolactin acts on receptors belonging to these nerve terminal networks. Since DA nerve cell bodies and dendrites exist in these areas, it is possible that the prolactin receptors could be located on the DA nerve cells themselves.

An acute intravenous injection of rat prolactin into the hypophysectomised rat produces a preferential increase of DA turnover within the dotted type of DA terminals located in the dorsal and posterior part of the nucleus accumbens (Fuxe *et al.*, 1977*b*). Thus it is also possible that these DA terminal networks are involved in the control of prolactin secretion. In support of this view it has been found that lesions of the nucleus accumbens produce impairment of maternal behaviour and of lactational performance (Smith and Holland, 1975). In a separate study it has also been shown that a negative intra-individual correlation exists between serum prolactin levels and activity in this DA terminal system of the nucleus accumbens (Löfström *et al.*, 1977; Fuxe *et al.*, 1977*f*).

On the Role of Dopamine (DA) Terminals in the Pituitary Gland

In view of the existence of DA terminals in the posterior pituitary it is conceivable that DA receptors could be located on oxytocin- and vasopressin-containing nerve terminals and thus that dopaminergic mechanisms could be involved in controlling oxytocin and vasopressin secretion.

Discoveries by Bloom and collaborators (Bloom *et al.*, 1977) that the gland cells of the intermediate lobe and the pars distalis may store endorphins suggest that DA mechanisms can also control the secretion of large opioid peptides from the anterior pituitary gland. It is also known that α-MSH is stored in the pars intermedia gland cells and can be controlled by inhibitory DA nerve terminals (Lichtensteiger *et al.*, 1977).

Possible Interactions with Somatostatin (GH–RIH) and Growth Hormone Releasing Hormone (GH–RH) Pathways

In many circumstances DA receptor agonists have been shown to increase the secretion of growth hormone (GH) (Mueller *et al.*, 1976; Brown *et al.*, 1974) and these findings suggest that inhibitory DA receptors could be located on the somatostatin nerve terminals in the median eminence. In agreement with this view, large numbers of somatostatin-positive nerve terminals have been discovered in the medial palisade zone and also in the lateral palisade zone of the median eminence, where large numbers of DA terminals exist (Hökfelt *et al.*, 1975*a*; Hökfelt *et al.*, 1974*a*). Furthermore it has recently been discovered that rat GH given to hypophysectomised rats produces a marked reduction of DA turnover in the medial and lateral palisade zones of the median eminence (figure 2.16 later) (Andersson *et al.*, 1977). It is thus possible that the inhibitory feedback of GH could involve an action on the brain and specifically a reduction of DA turnover, allowing an increased secretion of somatostatin to occur.

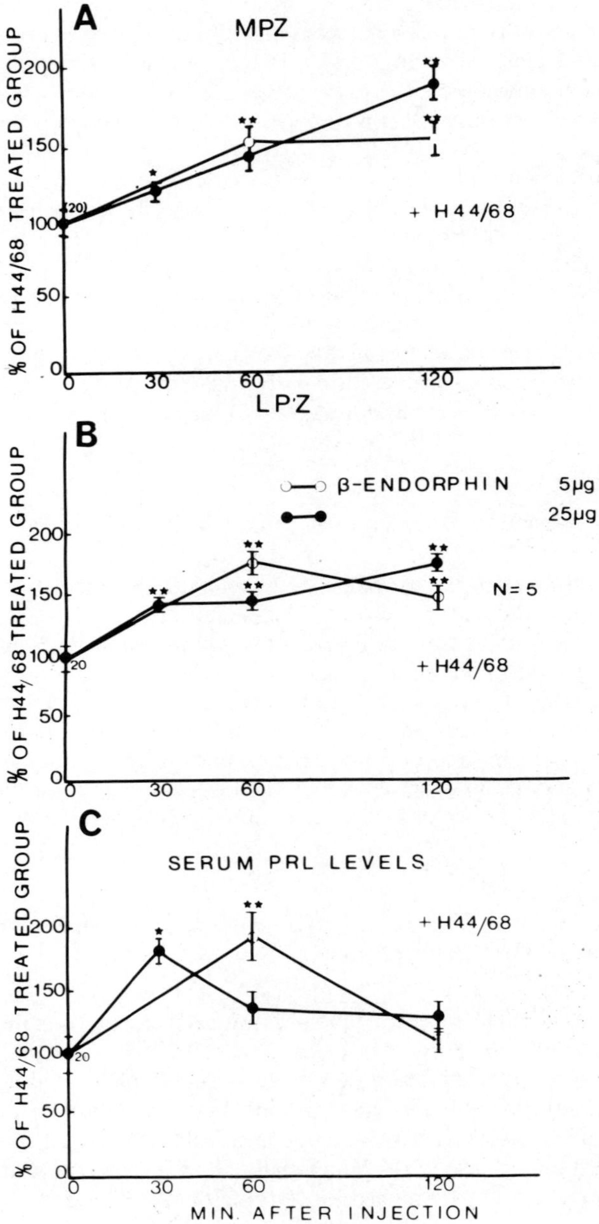

Figure 2.14 The effect of β-endorphin on the H44/68-induced disappearance of dopamine (DA) fluorescence in *A* the medial (MPZ), *B* the lateral palisade zone (LPZ) of the median eminence and *C* on serum prolactin levels (PRL) of the normal male rat. β-endorphin was given intraventricularly to unanaesthetised male rats in a dose of 5 or 25 μg/rat. The time after injection of β-endorphin is shown on

Under certain circumstances DA receptor agonists have been shown to reduce GH secretion (Muller *et al.*, 1968). Also it has recently been shown in our laboratory that enkephalin- and β-endorphin-induced increases of GH secretion and prolactin (Dupont *et al.*, 1977*a*; Dupont *et al.*, 1977*b*; Lien *et al.*, 1976; Rivier *et al.*, 1977) are correlated with a reduction of DA turnover in the median eminence (Ferland *et al.*, 1977; Fuxe *et al.*, 1978) (figure 2.14). In order to explain these findings it seems necessary to postulate in addition the existence of DA receptors located on putative GH-RH-containing nerve terminals in the median eminence. The final outcome of a reduction of DA release or of DA receptor activation by means of DA receptor agonists would then be dependent upon the sensitivity of the DA receptor sites located on the somatostatin and GH releasing hormone-containing nerve terminals, respectively. Obviously further work is essential in order to evaluate these speculations.

ON THE NORADRENERGIC CONTROL OF PEPTIDE-CONTAINING PATHWAYS

Interactions with Luteinising Hormone-Releasing Hormone (LH–RH) Containing Pathways

The work of Sawyer and collaborators (Sawyer, 1975) has demonstrated that the NA pathways in the hypothalamus exert a facilitatory influence on the LH-RH-containing pathways. Work in this laboratory (see Fuxe *et al.*, 1977*e*; Agnati *et al.*, 1977; Löfström, 1977) has shown that the NA nerve terminals within the subependymal layer of the median eminence and within the medial preoptic area are also activated in the critical period of the ovarian cycle during which the LH-RH pathways are activated. These results support the view of Sawyer and also indicate that the NA terminals can control activity in LH-RH cell bodies present both in the preoptic area and in the basal hypothalamus. In the castrated animal, in which a hypersecretion of LH and FSH exists, these NA pathways are not activated (Löfström *et al.*, 1977; Löfström *et al.*, 1976*b*). Thus studies on turnover changes in discrete NA nerve terminal systems of the hypothalamus indicate that the NA systems are mainly involved in the phasic control of LH-RH secretion in relation to the critical period leading to the peak secretion of LH and finally ovulation.

Possible Interactions with Putative Growth Hormone-Releasing Hormone (GH–RH) Containing Systems

The studies of Martin and co-workers indicate the existence of a facilitatory noradrenergic mechanism in the control of GH secretion (Martin *et al.*, 1977). In recent studies on the effect of rat GH on NA turnover in the hypothalamus of hypophysectomised animals, it was possible to demonstrate a selective reduction of

---

the *x*-axis. On the *y*-axis the fluorescence values and the serum prolactin levels are expressed as a percentage of the values found in the respective H44/68 alone group. Means ± s.e.m. are shown. All statistical comparisons have been made with the H44/68 alone group. Student's t-test. *: $p < 0.05$ and **: $p < 0.01$

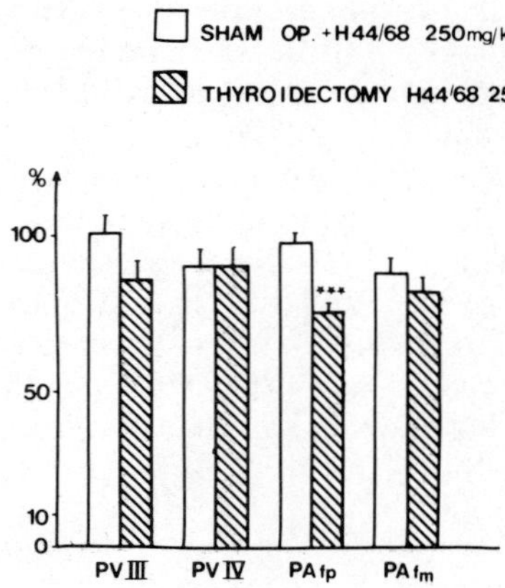

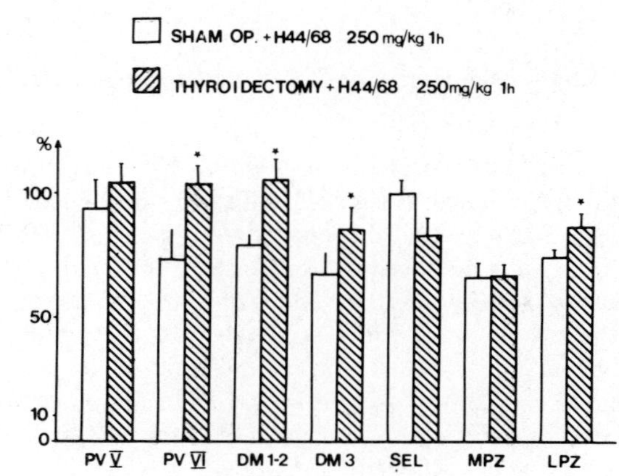

Figure 2.15 The effects of thyroidectomy on the H44/68-induced disappearance of (CA) catecholamine fluorescence in various hypothalamic nuclei. Thyroidectomy was performed 4 weeks before the day of the experiment. H44/68 was administered 1 h before killing in a dose of 250 mg/kg. CA fluorescence was measured in the following areas. PV III = anterior periventricular hypothalamic region dorsal part; PV IV = anterior periventricular hypothalamic nucleus ventral part; PA f_p = the parvicellular part of the paraventricular hypothalamic nucleus; PA f_m = the magnocellular part of the paraventricular hypothalamic nucleus; PV V = posterior periventricular hypothalamic nucleus dorsal part, PV VI = posterior periventricular hypothalamic nucleus ventral part; DM 1–2 = nucleus dorsomedialis hypothalami; DM 3 = the area located ventrally of the fornix; SEL = subependymal layer; MPZ = medial palisade zone; LPZ = lateral palisade zone. The fluorescence values are expressed as a percentage of the sham-operated group treated with H44/68. Means ± s.e.m. are shown out of 5 rats. Student's t-test. *:$p < 0.05$, **:$p < 0.01$ and ***:$p < 0.001$

NA turnover within the posterior periventricular region of the hypothalamus. These results support the existence of a facilitatory NA mechanism in the control of GH secretion. In view of the fact that the somatostatin-positive nerve cell bodies are located within the anterior periventricular hypothalamus (Hökfelt *et al.*, 1978; Hökfelt *et al.*, 1975*a*; Hökfelt *et al.*, 1975*d*), it seems possible that the reduction of NA turnover observed in the posterior periventricular hypothalamus under the influence of the inhibitory feedback action of GH could reflect the existence of a facilitatory noradrenergic mechanism controlling activity within GH-RH-containing neurones in this area.

Interactions with Thyrotrophin Releasing Hormone (TRH)-Containing Pathways

The work of several groups indicates the existence of facilitatory noradrenergic mechanisms in the control of the TRH-containing pathways (Krulich *et al.*, 1977; Reichlin *et al.*, 1976). In support of this view, morphological analysis reveals that there is a dense innervation, by NA nerve terminals, of the TRH nerve cell body-rich areas of the hypothalamus such as the parvicellular part of the paraventricular hypothalamic nucleus and the dorsomedial hypothalamic area. In order to understand further the role of NA systems in the control of the TRH neurones, the NA turnover was evaluated in various hypothalamic nuclei after thyroidectomy. In this study a selective increase of NA turnover was observed in the parvicellular part of the paraventricular nucleus, while in the other TRH cell body-rich areas a reduction of NA turnover was observed (figure 2.15) (Andersson, personal communication). It seems likely from lesion experiments (Hökfelt and Johansson, unpublished data) that the TRH cell bodies in the paraventricular nucleus innervate the external layer of the median eminence. Therefore the present analysis indicates the existence of a facilitatory noradrenergic mechanism controlling the activity in the TRH nerve cells of the parvicellular part of the paraventricular nucleus which discharge TRH into the hypophyseal portal system. The reduction of NA turnover observed within the dorsomedial hypothalamic nucleus and within the perifornical region may be related to the changes in temperature regulation and food intake which occur following thyroidectomy. NA hypothalamic systems are known to be involved in the control of such hypothalamic functions (Leibowitz, 1975; Myers and Waller, 1975). In the present experiments after four weeks of thyroidectomy no changes in DA turnover were observed in the medial palisade zone of the median eminence rich in TRH nerve terminals. Thus at the present time it is unclear to what extent the DA terminals in the median eminence are involved in the control of the activity of TRH release at the nerve terminal level.

Possible Interactions with Corticotrophin-Releasing Factor-Containing Systems

The work of several groups (see Ganong, 1973; Smelik, 1977) indicates the existence of an inhibitory noradrenergic mechanism in the control of the secretion of adrenocorticotrophic hormone (ACTH). In order to study further the role of the NA hypothalamic systems in the control of ACTH secretion, the effects of

ACTH (1-24) were studied on NA turnover in various hypothalamic nuclei of hypophysectomised animals. In this analysis it was found that ACTH (1-24) preferentially increased NA turnover within the medial palisade zone and the subependymal layer of the median eminence (figure 2.16), while leaving the hypothalamic areas unaffected. These results point to a role of the median eminence NA systems in the control of the secretion of corticotrophin- releasing hormone.

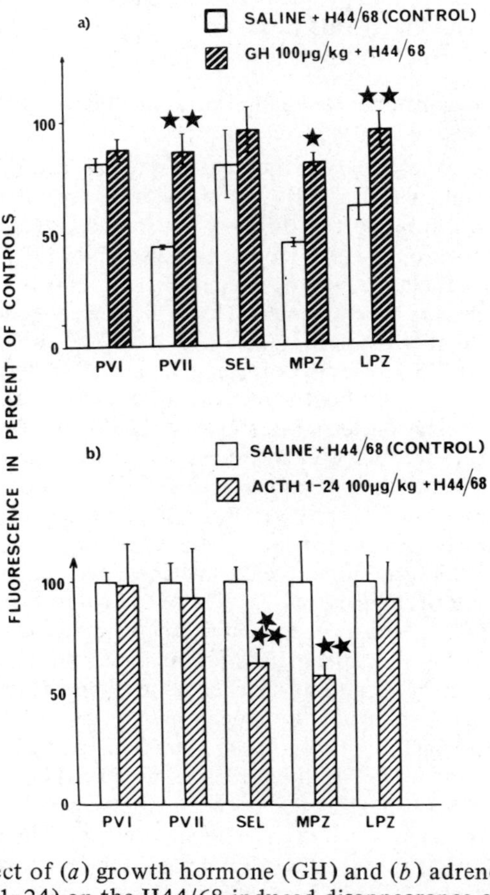

Figure 2.16 Effect of (*a*) growth hormone (GH) and (*b*) adrenocorticotrophic hormone (ACTH 1-24) on the H44/68-induced disappearance of catecholamine (CA) fluorescence within the hypothalamus and the median eminence. ACTH 1-24 was given intravenously in a dose of 100 µg/kg 2 h before the H44/68 injection (250 mg/kg, i.p., 2 h before killing). The various areas measured are shown on the *x*-axis, and on the *y*-axis the fluorescence values are expressed as a percentage of the group treated with solvent and H44/68 alone (ACTH) or as a percentage of the CA fluorescence values at the time of the H44/68 injection (GH). Mean ± s.e.m. are shown from 4 to 6 rats. A selective increase in the H44/68-induced CA disappearance is observed within the subependymal layer (SEL) and the medial palisade zone (MPZ). Student's t-test. **: $p < 0.01$ and ***: $p < 0.001$. PV I = anterior periventricular region; PV II = posterior periventricular region; LPZ = lateral palisade zone

Since under the experimental conditions ACTH would be expected to exert an inhibitory feedback action, the results indicate that the inhibitory noradrenergic mechanisms postulated by Ganong and others (Ganong, 1973) may exist within the median eminence itself.

From the above it is clear that the DA and NA systems have an important role in the control of the peptide-containing systems of the hypothalamus. In most instances there is no doubt that the peptides and the monoamines are each located in different neurones and that they interact with each other by neuronal links as discussed above. However, Hökfelt *et al.*, (1977*b*) have recently discovered that somatostatin is present within peripheral adrenergic nerve cells and that substance P is present within 5-HT containing nerve cells of the lower brainstem (unpublished data). It cannot therefore be excluded that in certain instances both peptides and monoamines can be stored in the same hypothalamic nerve cell. However, the evidence so far indicates that such a mutual storage of a peptide and a monoamine within the hypothalamus is the exception rather than the rule. Like the monoamines, the peptides are probably stored in granules. However, their synthesis may be limited to the cell bodies, where synthesis can take place via cleavage from prohormones. In contrast, the monoamines are synthesised in the whole neurone.

It should also be pointed out when discussing the existence of peptide neurones containing angiotensin II-like peptides, somatostatin-like peptides or enkephalin-related peptides, etc. that unknown peptides may also exist which can react with the antisera and thus give false positive results in the immunohistochemical and immunochemical procedures. Further research will show whether or not such peptides exist within the hypothalamus.

ON THE NERVOUS CONTROL OF THE MEDIAN EMINENCE DOPAMINE SYSTEMS

Previous work (see Fuxe and Hökfelt, 1975; Hökfelt and Fuxe, 1972; Krieg and Sawyer, 1976) indicated that NA and/or adrenaline neurones could control activity in the tubero-infundibular DA systems. The α-adrenoceptor agonist clonidine reduced DA turnover within the median eminence, while an α-adrenergic blocking agent such as phenoxybenzamine increased the activity in the tubero-infundibular DA systems. These findings suggest the existence of an inhibitory α-adrenergic control of the tubero-infundibular DA systems. Such an interaction could be of considerable importance in the control of the peptide-containing systems, since in many cases NA and possibly adrenaline on the one hand and DA on the other hand seem to have an antagonistic role in the control of the peptide-containing systems. Thus, for example, in the case of the control of the LH-RH pathway, maximum facilitation of release of LH-RH is brought about by an activation of NA systems accompanied by reduced activity in the inhibitory DA systems in the lateral palisade zone of the median eminence. At the present time there is little evidence that 5-HT receptor activity can control the DA pathways of the median eminence. Thus, 5-HT-releasing agents such as the ergolene derivative PTR 17402 ((5R, 8R)-8-(4-*p*-methoxyphenyl-1-piperazinyl methyl)-6-methylergolene) cannot influence the DA turnover within the median eminence when administered in doses of 0.5–3 mg/kg (unpublished data).

Cholinergic mechanisms may also control the tubero-infundibular DA systems. Thus nicotine in high doses has consistently been shown to increase DA turnover within the medial and lateral palisade zone of the median eminence (Fuxe *et al.*, 1977*a*; Eneroth *et al.*, 1977*a, b*). Thus it is possible that the inhibitory effects of nicotine on LH and prolactin secretion could in part involve an activation of the DA nerve terminals in the median eminence. It is not known whether the nicotinic cholinergic receptors are located in the median eminence itself, on the DA terminals or in other areas of the hypothalamus such as the arcuate nucleus, where the DA cell bodies and dendrites are located. It has also been postulated that muscarinic, cholinergic receptors can influence the tubero-infundibular DA systems (Libertun and McCann, 1974; Lichtensteiger and Keller, 1974). Thus the inhibition of prolactin secretion caused by pilocarpine, which is a cholinergic muscarinic agonist, is blocked by pretreatment with a DA receptor blocking agent (Grandison and Meites, 1976). So far, however, in the normal male rat anticholinergic drugs such as scopolamine and atropine have not changed the DA turnover in the median eminence.

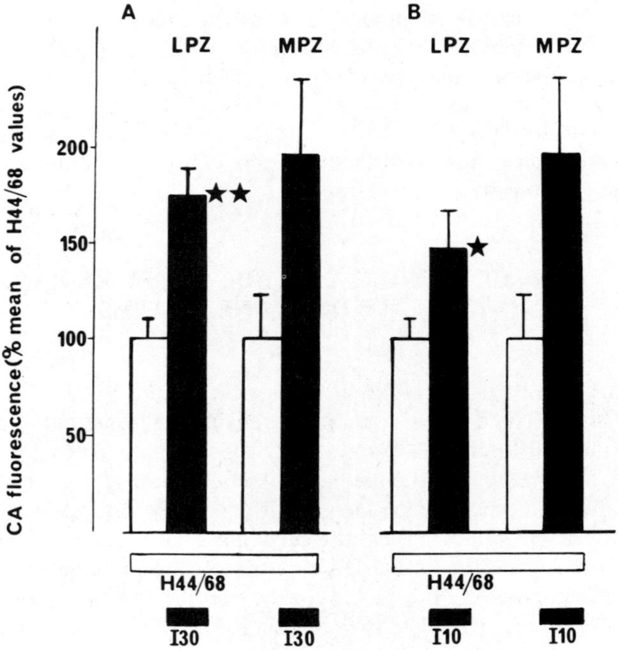

Figure 2.17 Effect of ibotenic acid (I) on the H44/68-induced disappearance of catecholamine (CA) fluorescence in the median eminence of rats. Ibotenic acid (10 or 30 mg/kg) was either given at the same time (*A*) or 15 min before (*B*) the H44/68 injection (250 mg/kg, i.p., 2 h before killing). The fluorescence values are given as a percentage of the respective H44/68 alone group mean value. Means ± s.e.m. are shown from 4 to 5 rats. A reduction of the H44/68-induced disappearance of CA fluorescence of observed. Significant changes were only observed in the lateral palisade zone (LPZ) in view of the large variability of values within the medial palisade zone (MPZ). *: $p < 0.1$ and **: $p < 0.01$

Dense networks of GABA terminals exist within the entire hypothalamus, and some terminals are also found in the external layer of the median eminence. Studies with a putative GABA receptor agonist such as ibotenic acid, which is converted to muscimol *in vivo*, show that such a compound can markedly reduce the DA turnover in the tubero-infundibular DA systems (figure 2.17). Thus under certain conditions GABA receptor activity seems to be capable of an inhibitory control of the DA systems in the median eminence. However, after pretreatment with neuroleptics, GABA-ergic drugs can increase DA turnover in the median eminence, indicating in addition the existence of a facilitatory control of the tubero-infundibular DA neurones.

Peptide-containing systems may also control the DA pathways. Thus enkephalin terminals have been demonstrated in the external layer of the median eminence and within various hypothalamic nuclei (see above). β-endorphin-positive nerve terminals also exist within the basal hypothalamus but in fewer numbers than the enkephalin-positive terminals (see Bloom *et al.*, 1977, and above). Intraventricular injections of met-enkephalin, β-endorphin and morphine have produced a marked reduction of DA turnover within the medial and lateral palisade zone of the median eminence (figure 2.14), suggesting an important control by opiate receptors of the tubero-infundibular DA systems (Ferland *et al.*, 1977; Fuxe *et al.*, 1978). In view of the distribution of the enkephalin-positive terminals, the opiate receptors may be located within the external layer of the median eminence (although in relatively low numbers), within the arcuate nucleus or within other regions of the hypothalamus. It should also be noted that in studies on DA release in the caudatus, opiate peptides were shown to inhibit markedly the release of DA (Loh *et al.*, 1976). Thus an axo-axonic interaction within the external layer of the median eminence between enkephalin-positive and DA terminals could in part explain the marked reductions of DA turnover observed.

The mapping out of somatostatin-positive and substance P-positive terminals within the hypothalamus, particularly within the arcuate nucleus, also makes it possible that these peptides could control the activity in the DA systems of the hypothalamus. In fact Hökfelt and collaborators (Hökfelt *et al.*, 1978) have shown the existence of dense substance P-positive nerve plexa around the DA cell bodies of the arcuate nucleus.

The results reported above indicate that a large number of putative transmitters and/or modulators can influence the activity within the tubero-infundibular DA neurones. Thus the tubero-infundibular DA system appears to be an important mediator of information to the anterior pituitary gland and to the various types of releasing and inhibitory hormone-containing nerve terminals within the median eminence.

CONCLUSIONS

The biochemical identification of various putative transmitters and/or modulators and of several hypothalamic hormones has been made possible using histochemical and immunohistochemical methods to map out the localisation and distribution of these compounds. The studies show that these putative transmitters and/or modulators have a neuronal location, each of them being located in a specific neurone system with a characteristic and unique distribution within the hypothalamus. The studies further indicate that the dopamine (DA) and noradrenaline (NA) pathways play an important role in the control of activity in the various peptide-containing pathways, and thus in the control of anterior pituitary functions. Similary, peptide-containing pathways can probably also control activity in the hypothalamic catecholamine (CA) pathways, since they themselves have a transmitter and/or modulator role in the hypothalamus.

Studies on the turnover of the various DA and NA terminal systems under the influence of various endocrine states and hormonal conditions suggest that the CA terminal systems can be differentially controlled, for example, having different types of receptor populations on their nerve cell membrane or by being controlled by axo-axonic synapses influenced directly or indirectly by hormone receptors. These studies emphasise the importance of local neuronal mechanisms for the control of the activity in nerve terminals of a specific area.

ACKNOWLEDGEMENTS

This work was supported by a grant (04X-715) from the Swedish Medical Research Council and by grants from Karolinska Institutets Fonder and Magn. Bergvalls Stiftelse.

REFERENCES

Agnati, L., Fuxe, K., Löfström, A. and Hökfelt, T. (1977). Dopaminergic drugs and ovulation: studies on PMS-induced ovulation and changes in median eminence DA and NE turnover in immature female rats. In E. Costa and G. L. Gessa (eds.), *Advances in Biochemical Psychopharmacology*, Vol. 16, Raven Press, New York, 159–168

Andén, N. E., Dahlström, A., Fuxe, K., Larsson, K., Olson, L. and Ungerstedt, U. (1966). Ascending monoamine neurones to the telencephalon and diencephalon. *Acta Physiol. Scand.*, **67**, 313–326

Andersson, K., Fuxe, K., Eneroth, P., Gustafsson, J-Å. and Skett, P. (1977). On the catecholamine control of growth hormone regulation. Evidence for discrete changes in dopamine and noradrenaline turnover following growth hormone administration. *Neuroscience Lett.*, **5**, 83–89

Barry, J. and Dubois, M. P. (1974). Immunofluorescence study of the preoptico-infundibular LH-RH neurosecretory pathway of the guinea-pig during the estrous cycle. *Neuroendocrinology*, **15**, 200–208

Björklund, A., Moore, R. Y., Nobin, A. and Stenevi, U. (1973). The organisation of the tubero-hypophyseal and reticulo-infundibular catecholamine neurone systems in the rat brain. *Brain Res.*, **51**, 171–191

Bloom, F., Battenberg, E., Rossier, J., Ling, N., Leppaluoto, J., Vargo, T. M. and Guillemin, R. (1977). Endorphines are located in the intermediate and anterior lobes of the pituitary gland, not in the neurohypophysis. *Life Sci.*, **20**, 40–48

Brown, W. A., Krieger, D. T., Van Woert, M. H. and Ambani, L. M. (1974). Dissociation of growth hormone and cortisol release following apomorphine. *J. clin. Endocr. Metab.*, **38**, 1127–1130

Brownstein, M., Arimura, A., Sato, H., Schally, A. V. and Kizer, J. S. (1975*a*). The regional distribution of somatostatin in the rat brain. *Endocrinology,* **96**, 1456–1461

Brownstein, M., Kobayashi, R., Palkovits, M. and Saavedra, J. M. (1975*b*). Choline acetyltransferase levels in diencephalic nuclei of the rat. *J. Neurochem.*, **24**, 35–38

Brownstein, M., Palkovits, M., Saavedra, J. M., Bassiri, R. and Utiger, R. D. (1974). Thyrotropin-releasing hormone in specific nuclei of rat brain. *Science,* **185**, 267–269

Brownstein, M., Palkovits, M., Tappaz, M., Saavedra, J. and Kizer, S. (1976). Effect of surgical isolation of the hypothalamus on its neurotransmitter content. *Brain Res.*, **117**, 287–295

Carlsson, A., Falck, B. and Hillarp, N. Å. (1962). Cellular localisation of brain monoamines. *Acta Physiol. Scand.*, **56** (Suppl. 196), 1–28

Chan-Palay, V. (1977). Indoleamine neurons and their processes in the normal rat brain and in chronic diet-induced thiamine deficiency demonstrated by uptake of ^{3}H-serotonin. *J. comp. Neurol.*, **176**, 467–494

Christenson, J. G., Dairman, W. and Udenfriend, S. (1972). On the identity of DOPA decarboxylase and 5-hydroxytryptophan decarboxylase. *Proc. Nat. Acad. Sci. (Wash.)*., **69**, 343–347

Dahlström, A. and Fuxe, K. (1964). Evidence for the existence of monoamine-containing neurons in the central nervous system. Part I: Demonstration of monoamines in the cell bodies of brainstem neurons. *Acta Physiol. Scand.*, **62** (Suppl. 232), 1–55

Dupont, A., Cusan, L., Garon, M., Labrie, F. and Li, C. H. (1977*a*). β-Endorphin: stimulation of growth hormone release *in vivo. Proc. Nat. Acad. Sci., USA*, **74**, 358–359

Dupont, A., Cusan, L., Labrie, F., Coy, D. H. and Li, C. H. (1977*b*). Stimulation of prolactin release in the rat by intraventricular injection of β-endorphin and methionine-enkephalin. *Biochem. Biophys. Res. Comm.*, **75**, 76–82

Elde, R. P., Hökfelt, T., Johansson, O., Efedić, O. and Luft, R. (1976*a*). Somatostatin containing pathways in the nervous system. *Neurosci. Abstr.*, **2**, 759

Elde, R. P., Hökflet, T., Johansson, O. and Terenius, L. (1976*b*). Immunohistochemical studies using antibodies to leucine-enkephalin. Initial observations on the nervous system of the rat. *Neuroscience*, **1**, 349–351

Eneroth, P., Fuxe, K., Gustafsson, J. -Å., Hökfelt, T., Löfström, A., Skett, P. and Agnati, L. (1977*a*). The effect of nicotine on central catecholamine neurons and gondatropin secretion. Part II: Inhibitory influence of nicotine on LH, FSH and prolactin secretion in the ovariectomised female rat and its relation to regional changes in dopamine and noradrenaline levels and turnover. *Med. Biol.*, **55**, 158–166

Eneroth, P., Fuxe, K., Gustafsson, J. -Å., Hökfelt, T., Löfström, A., Skett, P. and Agnati, L. (1977*b*). The effect of nicotine on central catecholamine neurons and gonadotropin secretion. Part III: Studies on prepubertal female rats treated with pregnant mare serum gonadotropin. *Med. Biol.*, **55**, 167–176.

von Euler, U. S. and Gaddum, J. H. (1931). An unidentified depressor substance in certain tissue extracts. *J. Physiol. (Lond.)*, **72**, 74–87

Ferland, L., Fuxe, K., Eneroth, P., Gustafsson, J. -Å. and Skett, P. (1977). Effects of methionine-enkephalin on prolactin release and catecholamine levels and turnover in the median eminence. *Europ. J. Pharmacol.*, **43**, 89–90

Fuxe, K. (1964). Cellular localisation of monoamines in the median eminence and the infundibular stem of some mammals. *Z. Zellforsch.*, **61**, 710–724

Fuxe, K. (1965). Evidence for the existence of monoamine neurons in the central nervous system. Part IV: The distribution of monoamine nerve terminals in the central nervous system. *Acta Physiol. Scand.*, **64** (Suppl. 247), 39–85

Fuxe, K., Agnati, L., Eneroth, P., Gustafsson, J. -Å., Hökfelt, T., Löfström, A., Skett, B. and Skett, P. (1977*a*). The effect of nicotine on central catecholamine neurons and gonadotropin secretion. Part I: Studies in the male rat. *Med. Biol.*, **55**, 148–157

Fuxe, K., Eneroth, P., Gustafsson, J. -Å., Löfström, A. and Skett, P. (1977*b*). Dopamine in the nucleus accumbens: preferential increase of DA turnover by rat prolactin. *Brain Res.*, **122**, 177–182

Fuxe, K., Ferland, L., Andersson, K., Eneroth, P., Gustafsson, J. -Å. and Skett, P. (1978). On the functional role of hypothalamic catecholamine neurons in control of the secretion of hormones from the anterior pituitary, particularly in the control of LH and prolactin secretion. In D. Scott (ed.), *Brain-Endocrine Interaction*, Vol. III, Neural Hormones and Reproduction, 3rd Int. Symp. Würzburg, 1977, 172–182

Fuxe, K., Ganten, D., Hökfelt, T. and Bolme, P. (1976*a*). Immunohistochemical evidence for the existence of angiotensin II-containing nerve terminals in the brain and spinal cord in the rat. *Neurosci. Lett.*, **2**, 229–234

Fuxe, K. and Hökfelt, T. (1969). Catecholamines in the hypothalamus and the pituitary gland. In W. F. Ganong and L. Martini (eds.), *Frontiers in Neuroendocrinology*, Vol 1, Oxford University Press, New York, 47–96

Fuxe, K. and Hökfelt, T. (1975). The effects of hormones and psychoactive drugs on the tubero-infundibular neurons. In *Some Aspects of Hypothalamic Regulation of Endocrine Functions*, F. K. Schattauer Verlag, Stuttgart/New York, 51–61

Fuxe, K., Hökfelt, T., Eneroth, P., Gustafsson, J. -Å. and Skett, P. (1977*c*). Prolactin-like immunoreactivity: localisation in nerve terminals of rat hypothalamus. *Science*, **196**, 899–900

Fuxe, K., Hökfelt, T., Johansson, O., Ganten, D., Goldstein, M., Perez de la Mora, M., Possani, L., Tapia, R., Teran, L., Palacios, R., Said, S. and Mutt, V. (1976*b*). Monoamine neuron systems in the hypothalamus and their relation to the GABA and peptide containing neurons. In R. Mornex and J. Barry (eds.), *Colloque de Synthèse des Actions Thèmatiques 22 et 35. Neuromediateurs et Polypeptides Hypothalamiques à Action Relàchante ou Inhibitrice*, Institut National de la Santé et de la Recherche Médicale, Paris, 17–40

Fuxe, K., Hökfelt, T., Löfström, A., Johansson, O., Agnati, L., Everitt, B., Goldstein, M., Jeffcoate, S., White, N., Eneroth, P., Gustafsson, J. -Å. and Skett, P. (1976*c*). On the role of neurotransmitters and hypothalamic hormones and their interactions in hypothalamic and extrahypothalamic control of pituitary function and sexual behavior. In F. Naftolin, K. J. Ryan and J. Davies (eds.), *Subcellular Mechanisms in Reproductive Neuroendocrinology*, Elsevier Scientific Pub. Co., Amsterdam, 193–246

Fuxe, K., Hökfelt, T., Said, S. I. and Mutt, V. (1977*d*). Vasoactive intestinal polypeptide and the nervous system: immunohistochemical evidence for localisation in central and peripheral neurons, particularly intracortical neurons of the cerebral cortex. *Neurosci. Lett.*, **5**, 241–246

Fuxe, K. and Jonsson, G. (1974). Further mapping of central 5-hydroxytryptamine neurons: studies with the neurotoxic dihydroxytryptamines. In E. Costa, G. L. Gessa and M. Sandler (eds.), *Advances in Biochemical Psychopharmacology*, Vol. 10, Serotonin: New Vistas, Histochemistry and Pharmacology, Raven Press, New York, 1–12

Fuxe, K., Löfström, A., Agnati, L., Hökfelt, T., Johansson, O., Eneroth, P., Gustafsson, J. -Å., Skett, P., Jeffcoate, S. and Fraser, H. (1977*e*). Functional morphology of the median eminence. On the involvement of catecholamines in the control of FSH, LH and prolactin secretion. In P.O. Hubinont, M. L'Hermite and C. Robyn (eds.), *Progress in Reproductive Biology*, Vol. 2. Clinical Reproductive Neuroendocrinology, Karger, Basel, 41–53

Fuxe, K., Löfström, A., Eneroth, P., Gustafsson, J. -Å., Skett, P., Hökfelt, T., Wiesel, F. -Å. and Agnati, L. (1977*f*). Involvement of central catecholamines in the feedback actions of 17 β-estradiolbenzoate on luteinizing hormone secretion in the ovariectomized female rat. *Psychoneuroendocrinology*, **2**, 203–225

Ganong, W. F. (1973). Catecholamines and the secretion of renin. ACTH and growth hormone. In E. Usdin and S. Snyder (eds.), *Frontiers in Catecholamine Research*, Pergamon Press, New York, 819–824

Ganten, D., Fuxe, K., Phillips, M. I., Mann, J. F. E. and Ganten, U. (1978). The brain iso-renin-angiotensin system: biochemistry, localization and possible role in drinking and blood-pressure regulation. In W. F. Ganong and L. Martini (eds.), *Frontiers in Neuroendocrinology*, Raven Press, New York, 61–99

Ganten, D., Hutchinson, J. S., Schelling, P., Ganten, U. and Fischer, H. (1975). The iso-renin angiotensin systems in extrarenal tissue. *Clin. exp. Pharmacol. Physiol.*, **2**, 127–151

Giachetti, A., Rosenberg, R. N. and Said, S. I. (1976). Vasoactive intestinal polypeptide in brain synaptosomes. *Lancet*, **ii**, 741–742

Grandison, L. and Meites, J. (1976). Evidence for adrenergic mediation of cholinergic inhibition of prolactin release. *Endocrinology*, **99**, 775–779

Guillemin, R. (1978). Biochemical and physiological correlates of hypothalamic peptides. The new endocrinology of the neuron. In S. Reichlin, R. J. Baldessarini and J. B. Martin (eds.). *The Hypothalamus* (ARNMD, Vol. 56), Raven Press, New York, 155–194

Guillemin, R., Ling, N. and Burgus, R. (1976). Endorphines, peptides d'origine hypothalamique et neurohypophysaire à activité morphinomimétique. Isolement et structure moléculaire de 1L-endorphine. *C. R. Acad. Sci., Paris*, **282**, 783–785

Hoffman, G. E., Moynihan, J. A. and Knigge, K. M. (1976). Immunocytochemical localization of luteinizing hormone-releasing hormone (LH-RH). Differences with different antisera. *Neurosci. Abstr.*, **2**, 673

Hökfelt, T., Efendić, S., Hellerström, C., Johansson, O., Luft, R. and Arimura, A. (1975*a*). Cellular localisation of somatostatin in endocrine-like cells and neurons of the rat with special references to the A_1 cells of the pancreatic islets and to the hypothalamus. *Acta Endocrinol.* (Kbh.), Suppl. 200, 5–41

Hökfelt, T., Efendić, S., Johansson, O., Luft, R. and Arimura, A. (1974*a*). Immunohistochemical localization of somatostatin (growth hormone release-inhibiting factor) in the guinea-pig brain. *Brain Res.*, **80**, 165–169

Hökfelt, T., Elde, R., Fuxe, K., Johansson, O., Ljungdahl, Å., Goldstein, M., Luft, R., Efendic, S., Nilsson, G., Terenius, L., Ganten, D., Jeffcoate, S. L., Rehfeld, J., Said, S., Perez de la Mora, M., Possani, L., Tapia, R., Teran, L. and Palacios, R. (1978*a*). Aminergic and peptidergic pathways in the nervous system with special reference to the hypothalamus. In S. Reichlin, R. J. Baldessarini and J. B. Martin (eds.), *The Hypothalamus*, Raven Press, New York, 69–135

Hökfelt, T., Elde, B., Johansson, O., Terenius, L. and Stein, L. (1977*a*). Distribution of enkephalin immunoreactive cell bodies in the rat central nervous system. *Neurosci. Lett.*, **5**, 25–33

Hökfelt, T., Elfvin, L. G., Elde, R., Schultzberg, M., Goldstein, M. and Luft, R. (1977*b*). Occurrence of somatostatin-like immunoreactivity in some peripheral sympathetic nor-adrenergic neurons. *Proc. Natl. Acad. Sci.*, **74**, 3587–3591

Hökfelt, T. and Fuxe, K. (1972). On the morphology and the neuroendocrine role of the hypothalamic catecholamine neurons. In K. M. Knigge, D. E. Scott and W. Weindl (eds.) *Brain-Endocrine Interaction. Median Eminence: Structure and Function*, Int. Symp., Munich, Karger, Basel, 181–223

Hökfelt, T., Fuxe, K. and Goldstein, M. (1973). Immunohistochemical studies on monoamine-containing cell systems. *Brain Res.*, **62**, 461–469

Hökfelt T., Fuxe, K., Goldstein, M. and Johansson, O. (1974*b*). Immunohistochemical evidence for the existence of adrenaline neurones in the rat brain. *Brain Res.*, **66**, 235–251

Hökfelt, T., Fuxe, K., Goldstein, M., Johansson, O., Fraser, H. and Jeffcoate, S. (1975*b*). Immunofluorescence mapping of central monoamine and releasing hormone (LRH) systems. In W. E. Stumpf and L. D. Grant (eds.), *Anatomical Neuroendocrinology*, Karger, Basel, 381–392

Hökfelt, T., Fuxe, K., Johansson, O., Jeffcoate, S. L. and White, N. (1975*c*). Distribution of thyrotropin-releasing hormone (TRH) in the central nervous system as revealed with immunohistochemistry. *Europ. J. Pharmacol.*, **34**, 389–392

Hökfelt, T., Johansson, O., Fuxe, K., Goldstein, M. and Park, D. (1976). Immunohistochemical studies on the localisation and distribution of monoamine neurone systems in the rat brain. Part I: Tyrosine hydroxylase in the mes- and diencephalon. *Med. Biol.*, **54**, 427–453

Hökfelt, T., Johansson, O., Fuxe, K., Löfström, S., Goldstein, M., Park, D., Ebstein, R., Fraser, H., Jeffcoate, S., Efendic, S., Luft, R. and Arimura, A. (1975*d*). Mapping and relationship of hypothalamic neurotransmitters and hypothalamic hormones. In J. Tuomisto and M. K. Paasonen (eds.), *CNS and Behavioural Pharamacology*, Proc. Sixth Int. Congr. Pharmacol., Vol. 3, Forssan Kirjapaino, Oy, 93–110

Hökfelt, T., Kellerth, J-O., Nilsson, G. and Pernow, B. (1975*e*). Substance P: localization in the central nervous system and in some primary sensory neurons. *Science*, **190**, 889–890

Koslow, S. H. (1974). 5-Methoxytryptamine: a possible central nervous system transmitter. In *Advances in Biochemical Psychopharmacology*, Vol. II, Eds. E. Costa, G. L. Gessa and M. Sandler, New York, Raven Press, 95–100

Krieg, R. J. and Sawyer, C. H. (1976). Effects of intraventricular catecholamines on luteinizing hormone release in ovariectomized-steroid-primed rats. *Endocrinology*, **99**, 411–419

Krulich, L., Giachetti, A., Marchlenska-Koj, A., Hefco, E. and Jameson, H. E. (1977). On the role of the central noradrenergic and dopaminergic systems in the regulation to TSH secretion in the rat. *Endocrinology*, **100**, 496–505

Larsson, L. -I., Fahrenkrug, J., Schaffalitzky de Muckadell, O., Sundler, F., Hakanson, R. and Rehfeld, J. F. (1976). Localisation of vasoactive intestinal polypeptide (VIP) to central and peripheral neurones. *Proc. Nat. Acad. Sci. (Wash.)*, **73**, 3197–3200.

Leibowitz, S. F. (1975). Pattern of drinking and feeding produced by hypothalamic norepinephrine injection in the satiated rat. *Physiology and Behavior*, **14**, 731–742

Libertun, C. and McCann, S. M. (1974). Further evidence for cholinergic control of gonadotropin and prolactin secretion (38374). *Proc. Soc. Exp. Biol. Med.*, **147**, 498–504

Lichtensteiger, W. and Keller, P. J. (1974). Tubero-infundibular dopamine neurons and the secretion of luteinizing hormone and prolactin: extrahypothalamic influences, interaction with cholinergic systems and the effect of urethane anesthesia. *Brain Res.*, **74**, 279–303

Lichtensteiger, W., Lienhart, R. and Kopp, H. G. (1977). Peptide hormones and central dopamine neuron systems. *Psychoneuroendocrinology*, **2**, 237–248

Lidbrink, P., Jonsson, G. and Fuxe, K. (1974). Selective reserpine resistant accumulation of catecholamines in central dopamine neurons after dopa administration. *Brain Res.*, **67**, 439–456

Lien, E. L., Fenichel, R. L., Garsky, V., Sarantakis, D. and Grant, N. H. (1976). Enkephalin-stimulated prolactin release. *Life Sci.*, **19**, 837–840

Löfström, A. (1977). Catecholamine turnover alterations in discrete areas of the median eminence of the 4– and 5–day cyclic rat. *Brain Res.*, **120**, 113–131

Löfström, A., Eneroth, P., Gustafsson, J. -Å. and Skett, P. (1977). Effects of estradiol benzoate on catecholamine levels and turnover in discrete areas of the median eminence and the limbic forebrain, and on serum luteinizing hormone, follicle stimulating hormone and prolactin concentrations in the ovariectomized female rat. *Endocrinology*, **101**, 1559–1569

Löfström, A., Jonsson, G. and Fuxe, K. (1976*a*). Microfluorimetric quantitation of catecholamine fluorescence in rat median eminence. I: Aspects on the distribution of dopamine and noradrenaline nerve terminals. *J. Histochem. Cytochem.*, **24**, 415–429

Löfström, A., Jonsson, G., Wiesel, F. -A. and Fuxe, K. (1976*b*). Microfluorimetric quantitation of catecholamine fluorescence in rat median eminence. II: Turnover in hormonal states. *J Histochem. Cytochem.*, **24**, 430–442

Loh, H. H., Brase, D. A., Sampath-Khanna, S., Mar, J. B. and Way, E. L. (1976). β-Endorphine *in vitro* inhibition of striatal dopamine release. *Nature*, **264**, 567–568

Macleod, R. M. and Lehmeyer, J. E. (1974). Studies on the mechanism of the dopamine-mediated inhibition of prolactin secretion. *Endocrinology*, **94**, 1077–1085

Martin, J. B., Durand, D. and Saunders, A. (1977). Evidence for a role of catecholamines and serotonin in regulation of episodic growth hormone secretion in the rat. In V. H. T. James (ed.), *Endocrinology*, Vol. I., Excerpta Medica, Amsterdam/Oxford, 148–151

Mueller, G. P., Simpkins, J., Meites, J. and Moore, K. E. (1976). Differential effects of dopamine agonists and haloperidol on release of prolactin, thyroid-stimulating hormone, growth hormone and luteinizing hormone in rats. *Neuroendocrinology,* **20**, 121–135

Muller, E. E., Da Prada, P. and Pecile, A. (1968). Influence of brain neurohumors injected into the lateral ventricle of the rat on the growth hormone release. *Endocrinology,* **83**, 893–896

Myers, R. D. and Waller, M. B. (1975). Species continuity in the thermoregulatory responses of the pigtailed macaque to monoamines injected into the hypothalamus. *Comp. Biochem. Physiol.,* **51A**, 639–645

Reichlin, S., Saperstein, R., Jackson, I. M. D., Boyd, A. E. (III) and Patel, Y. (1976). Hypothalamic hormones. *Ann. Rev. Physiol.,* **38**, 389–424

Renaud, L. P., Martin, J. B. and Brazeau, P. (1975). Depressant action of TRH, LH-RH and somatostatin on activity of central neurones. *Nature,* **255**, 233–235

Rivier, C., Vale, W., Ling, N., Brown, M. and Guillemin, R. (1977). Stimulation *in vivo* of the secretion of prolactin and growth hormone by β-endorphin. *Endocrinology,* **100**, 238–241

Roberts, E. (1975). Immunocytochemistry of the GABA system—a novel approach to an old transmitter. In J. A. Ferrendelli, B. S. McEwen and S. H. Snyder (eds.), *Neuroscience Symposia,* Vol. I. Neurotransmitters, Hormones and Receptors: Novel Approaches, Society for Neuroscience, Bethesda, Maryland, 123–138

Roberts, E., Chase, T. N. and Tower, D. B. (eds.) (1975). GABA *in Nervous System Function*, Raven Press, New York

Said, S. I. and Giachetti, A. (1977). Vasoactive intestinal polypeptides: distribution in normal tissues and preliminary report on its subcellular localization in brain. In S. Bonfils, P. Fromageot and G. Rosselin (eds.), *First International Symposium on Hormonal Receptors in Digestive Tract Physiology,* INSERM Symposium No. 3, North-Holland Publ., Amsterdam, 417–423

Said, S. I. and Rosenberg, R. N. (1976). Vasoactive intestinal polypeptide: abundant immunoreactivity in neural cell lines and normal nervous tissue, *Science,* **192**, 907–908

Sawyer, C. H. (1975). First Geoffrey Harris Memorial Lecture. Some recent developments in brain–pituitary–ovarian physiology. *Neuroendocrinology,* **17**, 97–124

Smelik, P. G. (1977). Neurotransmitter control of ACTH release. In V. H. T. James (ed.). *Endocrinology,* Vol. I, Excerpta Medica, Amsterdam/Oxford, 158–162

Smith, M. O. and Holland, R. (1975). Effects of lesions of the nucleus accumbens on lactation and postpartum behavior. *Physiol. Psychol.,* **3**, 331–350

Snyder, S. H. and Taylor, K. M. (1972). Histamine in the brain: a neurotransmitter? In S. H. Snyder (ed.), *Perspectives in Neuropharmacology—a Tribute to Julius Axelrod*, Oxford University Press, New York, 43–73

Tappaz, M. L., Brownstein, M. J. and Palkovits. M. (1976). Distribution of glutamate decarboxylase in discrete brain nuclei. *Brain Res.,* **108**, 371–379

Ungerstedt, U. (1971). Stereotaxic mapping of the monoamine pathways in the rat brain, *Acta Physiol. Scand.,* Suppl. 367, 1–48

Vanderhaeghen, J. J., Signeau, J. C. and Gepts, W. (1975). New peptide in the vertebrate CNS reacting with antigastrin antibodies. *Nature,* **257**, 604–605

Vogt, M. (1954). The concentration of sympathin in different parts of the central nervous system under normal conditions and after the administration of drugs. *J. Physiol. (Lond.),* **123**, 451–481

Zimmerman, E. A. (1976). Localization of hypothalamic hormones by immunocytochemical techniques. In L. Martini and W. F. Ganong (eds.), *Frontiers in Neuroendocrinology,* Raven Press, New York, 25–62

Discussion

Szabadi (Manchester)
This is a question to both Professor Fuxe and Professor Halász. What is known about species differences in the fine anatomy of the hypothalamus? How far can the observations made in the rat be extrapolated to the human brain?

Fuxe
I can speak only of the monoamines. There seems no doubt that the principal features are very similar, although there are some interesting small differences.

Halász
That is also true for the peptides. For example, LH-RH-producing neurones are more restricted in the monkey than in the human. In the rat they are distributed over a larger area, but in principle the situation is very similar. Also, if you consider the anatomical control of the gonadotrophic hormones there are small differences. In the rat the medial part of the hypothalamus is not capable of maintaining cyclic gonadotrophin functions, yet if you isolate the medial part of the hypothalamus of a female monkey then that monkey has an irregular cycle.

Martini (Milan)
You said that there is a total separation between the neurones which release hormones and those which manufacture catecholamines. Are there results to show where the sex steroids or glucocorticoids are captured by the hypothalamus and is there any connection between the releasing neurones, the catecholaminergic neurones and the neurones which capture the steroids?

Fuxe
There is some data to show that some dopamine cell bodies have the oestrogen cytosol receptor and this appears also to be true for the noradrenaline cell bodies in ganglia of the reticular formation of the medulla. We also know that dopamine and noradrenaline are involved in inhibitory and facilitatory actions of the oestrogens. Regarding the possibility of cytosol oestrogen receptors within the LH-RH cell body, I do not know of any data yet but I am sure it will come very shortly.

Biggs (London)
You did not mention histamine. Do you have any comments on its role in the hypothalamus?

Fuxe
I did not mention histamine because I know so little about it. Of course there is evidence to support its role as a neurotransmitter but unfortunately we have not been able to contribute morphologically to this problem. Data is accumulating that histamine blockers can have effects on the hypophyseal hormones which could be related to H_1 and H_2 receptors. (See chapter 3 *editors*)

Crook (London)
What is the experimental evidence for dopamine release into the portal vessels?

Fuxe
The evidence now looks pretty good. Firstly, there are dopamine receptors in the pituitary. This is circumstantial evidence but if there were no receptors there would be no point in having dopamine released. Secondly, it has now been demonstrated that dopamine is present within the hypothalamic portal system. Thus you have the receptors and the terminals to release it into the plexus. It is this kind of evidence which suggests that dopamine may be a prolactin release inhibiting factor.

McCracken (Bath)
Professor Halász mentioned male and female differences in the number of axonal contacts in the hypothalamus. I wonder if you have seen any differences in the monoamine patterns between the sexes?

Fuxe
Yes, we are at present involved in such a project but we have not been able to show any exciting differences so far.

3
Pharmacology of hypothalamic neurones

J. S. Kelly* and L. P. Renaud†

INTRODUCTION

In this chapter our aim is not so much to review the current literature on the neurophysiology and neuropharmacology of the hypothalamus, but rather to draw attention to the way in which the action of oxytocin on the contractile tissues of the lactating breast has allowed the neurophysiologist to explore the manner in which electrical activity in the magnocellular neurohypophyseal system is transformed into hormone release. Although there may be a similar causal relationship between electrical activity in the parvicellular or tubero-infundibular system and the release of gonadotrophins (Dufy *et al.*, 1974; Wuttke, 1974), an effect of these neurones on the release of other trophic or inhibitory substances involved in the regulation of adenohypophyseal secretion has yet to be established.

Unfortunately several factors hinder this work. Firstly there are few neuroendocrine reflexes as simple to quantitate as the release of oxytocin during milk ejection. Secondly, where a particular behavioural event is associated with pulsatile hormone release, the need to determine plasma hormone levels at frequent intervals, to show a correlation between hormone release and neuronal activity, may alter the original behaviour of the animal. Thirdly, the protracted time sequence needed for determination of plasma hormone levels using radioimmunoassay techniques necessitates delayed analysis of these correlations. Finally, one should be aware of the possibility that the intermittent release of trophic or inhibitory factors from median eminence terminals of tubero-infundibular neurones may not be directly associated with their invasion by action potentials but may rather result from changes in their membrane potential initiated by the release of putative transmitters from adjacent nerve terminals.

We also review neurophysiological evidence indicating that almost all neurones which project either to the neurohypophysis or the median eminence

*MRC Neurochemical Pharmacology Unit, Department of Pharmacology, Medical School, Hills Road, Cambridge, CB2 2QD, UK
†Montreal General Hospital, Division of Neurology, 1650 Cedar Avenue, Montreal Quebec, H3G 1A4, Canada

are inhibited by a recurrent pathway. In all probability this involves inhibitory interneurones which reside locally within each of the individual hypothalamic nuclei. Thus there is a close association between inhibitory interneurones and projecting neurones in each of the nuclei. This, together with the microiontophoretic data which show that individual neurones are consistently either excited or inhibited by the same putative neurotransmitter or neurohormones, increases the difficulty with the interpretation of experiments involving the microinjection of substances into small regions of the hypothalamus.

We also draw attention to the presence of trophic and inhibitory neuroendocrine factors throughout the brain and their possible association with the branching axons of the tubero-infundibular neurones even though the presence of peptides within this far-reaching neuronal network has yet to be established. The idea that there is a branching network of peptide neurones originating in the hypothalamus is particularly attractive in view of the potent behavioural effects of parenteral or intra-cerebral injections of the peptides.

NEUROHYPOPHYSEAL SYSTEM

The extensive anatomical and functional investigations which clearly established the dual neural and endocrine nature of the supraoptic and paraventricular nuclei of the hypothalamus and their outflow, the hypothalamo-neurohypophyseal

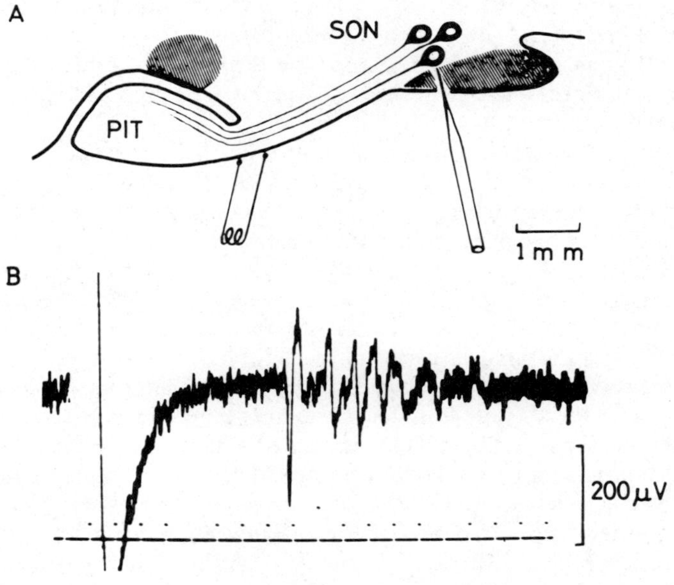

Figure 3.1 *A* schematic representation of a bipolar stimulating electrode positioned on the pituitary (PIT) stalk and a micropipette penetrating the ventral hypothalamus near the supraoptic nucleus (SON); *B* typical extracellular recording of a compound field potential from the region of the supraoptic nucleus of the rat evoked by a 9 V shock to the pituitary stalk; time calibration 1 ms. (Reproduced with permission from Dreifuss and Kelly, 1972*a*)

tract, have been described in some detail by Haymaker *et al*. (1969).

The initial intracellular studies from the equivalent neuroendocrine cells of the goldfish preoptic nucleus by Kandel (1964) demonstrated that these cells could develop action potentials in response to both orthodromic and antidromic stimulation and generate a short-latency inhibitory response to electrical stimulation of the pituitary stalk. Although a clear demonstration of the ability of mammalian neurosecretory cells to propagate action potentials was made earlier by Yagi *et al*. (1966), it was not until the early 1970s that antidromic stimulation became an established technique for the identification of extracellular records from the cells of these nuclei (figure 3.1) (Dyball and Koizumi, 1969; Kelly and Dreifuss, 1970; Novin *et al*., 1970; Yamashita *et al*., 1970; Barker *et al*., 1971*c*; Dyball, 1971). The release of vasopressin and oxytocin has been shown to be quantitatively related to the number and frequency of propagated action potentials elicited by electrical stimulation of the hypothalamo-neurohypophyseal tract both *in vivo* (Harris *et al*., 1969) and *in vitro* (Ishida, 1970; Dreifuss *et al*., 1971).

The action potentials responsible for the release of neurohormones from the nerve endings of the neurohypophysis appear to be no different from those of other neurones of the same shape and size and their frequency is probably modulated by synaptic events on the soma and dendrites of the supraoptic and paraventricular neurones.

Electrophysiological Identification

The technique of antidromic activation is widely used for the identification of certain hypothalamic neurones and neural mechanisms; therefore the methods currently in use will be described in some detail (*cf*. Kelly *et al*., 1975). Impulses initiated by the electrical stimulation of the neurohypophyseal fibres are conducted in retrograde fashion to the cell soma where they initiate fully developed action potentials. Action potentials of antidromic origin can be distinguished from those of spontaneous origin using the following properties as criteria: (1) irrespective of the suprathreshold stimulus strength, single, all-or-none action potentials follow each single stimulus at constant latency; (2) trains of high-frequency stimuli are followed by action potentials at constant latencies; and (3) an antidromic action potential is cancelled by a spontaneous or orthodromic action potential (see figure 3.2).

While the first two criteria can be tested in neurones which fail to display any spontaneous activity and are commonly used to facilitate localisation of neurosecretory neurones, the third and probably most definitive of these criteria requires the presence of some form of ongoing activity. In practice, it is best to use action potentials from supraoptic and paraventricular neurones which are either spontaneously active or excited by microiontophoretic application of an excitatory substance, such as L-glutamate, DL-homocysteic acid or acetylcholine, or reflexly excited by hyperosmolar stimuli, vaginal distension or haemorrhage, to trigger the stimulator used to stimulate the neurohypophyseal tract. As the interval between the orthodromically-evoked spike and the antidromic stimulus is progressively shortened, a critical interval is reached where the antidromic spike fails and only the orthodromic spike is recorded from the cell body. Since the collision of the two spikes can occur anywhere along the neurohypophyseal axon, the critical

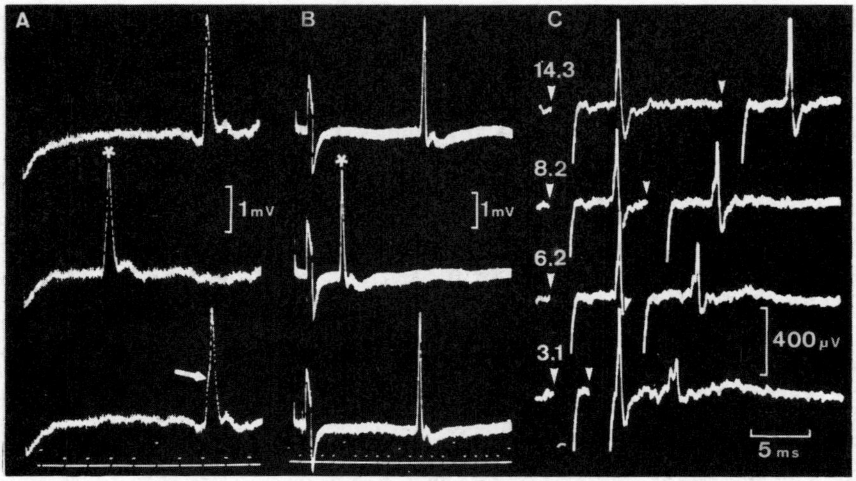

Figure 3.2 Antidromically evoked action potentials. *A* and *B* show two control records of antidromic spikes at constant latency and the middle records show the spontaneous action potential responsible for the cancellation of the antidromic action potential. In *A* the characteristic notch on the positive-rising phase of the action potential is marked by an arrow. Spontaneous as opposed to antidromic action potentials are marked by *. *C*, two-shock study in which the shock artefacts are marked by arrows and the numbers near the origin of each trace show the intershock interval in milliseconds; time calibrations in *A* and *B*, 1 ms. (Reproduced with permission from Dreifuss and Kelly, 1972*a*)

period will exceed the time taken for the orthodromic spike to travel from the cell body to the exciting electrodes placed on the pituitary stalk by the duration of the absolute refractory period of the axon under study. When the spontaneous firing rate is high, however, it is much simpler to stimulate the pituitary stalk at regular intervals and to film each oscilloscope sweep until a collision occurs (figure 3.2). The majority of the oscilloscope sweeps will show an antidromic spike at fixed latency from the stimulus artefact and only where the spontaneous spike occurs during the critical period will the fixed latency antidromic spike be absent. However, Kelly *et al.* (1975) have also pointed out that although clear notching of the rising phase (the initial segment/somatodendritic (IS/SD) break) is often visible on both orthodromic and antidromic spikes, only antidromic spikes can be fractionated into IS and SD components by high-frequency stimulation. In the absence of spontaneous or orthodromic activity this is a valuable and practical test (figures 3.2, 3.13, 3.14 and 3.16).

As we shall see later, recordings obtained from supraoptic and paraventricular neurones identified by antidromic invasion following pituitary stalk stimulation have served to characterise the electrical properties of neurosecretory cells under a wide variety of conditions (Arnauld *et al.*, 1974, 1975; Dreifuss *et al.*, 1976*a*, *b*; Dyball, 1971; Hayward and Jennings, 1973; Lincoln and Wakerley, 1974, 1975; Wakerley and Lincoln, 1973; Wakerley *et al.*, 1975) and show their discharge frequency to be modulated by synaptic excitatory and inhibitory events on their cell bodies and dendrites, similar to those on other neurones within the central nervous sytem. Clearly these events are important in the neural reflexes which modulate the functions of the neuroendocrine axis and indicate in electrophysiological terms that the peptidergic neurosecretory supraoptic and paraventricular neurones behave in an identical manner to other central neurones.

Recurrent inhibition and axon collaterals in the neurohypophyseal pathway

In 1964, by recording intracellularly from the equivalent neurosecretory neurones in the preoptic nucleus of the goldfish, Kandel not only demonstrated that these neurones generated action potentials which could be evoked by both orthodromic and antidromic stimulation, but also the presence of short-latency hyperpolarising inhibitory postsynaptic potentials evoked by electrical stimulation of the pituitary stalk. This raised the possibility that axons of the neurosecretory cells had terminals not only in the posterior pituitary, but also axon collaterals which formed part of a recurrent inhibitory feedback pathway to the parent neurone. This pathway could either be direct with collaterals synapsing with their own neurones of origin, or indirect through trans-synaptic excitation of local inhibitory interneurones whose axons synapsed with the neurosecretory cells. Later Kelly and Dreifuss (1970) (cf. Dreifuss and Kelly, 1972*a*) were able to show that in all antidromically-identified cells, antidromic stimulation was followed by a period of reduced probability of firing which proved indistinguishable from recurrent inhibition seen elsewhere in the central nervous system when the appropriate tests were applied (Gordon and Jukes, 1964). More specifically, inhibition resulted from antidromic shock subthreshold for the test axon and could not, therefore, be the result of excitability changes of the cell induced by invasion of an antidromic spike. Such inhibition must therefore have arisen from the activity of adjacent axons excited by the shock to the pituitary stalk.

In addition, Kelly and Dreifuss (1970) showed that high-frequency stimulation of the pituitary stalk caused a long lasting depression of the spontaneous spike discharge. In figure 3.3, continuous records *D* and *E* show the spontaneous discharge of an identified cell of the supraoptic nucleus to be abolished for 6s by a short-stimulus train applied to the pituitary stalk at a frequency of 100Hz. In contrast, an antidromic action potential evoked by a near-threshold single shock was followed by a period of silence and reduced excitability lasting less than a second. Presumably the ease with which these inhibitory phenomena can be elicited accounts for some of the difficulties experienced during earlier attempts to identify hypothalamic nuclei by stimulation of the pituitary stalk (Cross and Green, 1959; Cross and Silver, 1966). The inhibitory period could result from a direct collateral pathway with temporal dispersion of the presynaptic volleys converging on a particular supraoptic neurone. The extremely large difference in conduction velocity between the fastest and slowest fibres of the neurohypophyseal tract makes this a real but unprecedented possibility. However, an inhibitory period of approximately 80 msec in duration evoked by stimulus intensities near threshold for the test axon yet well below threshold for invasion of the majority of cells in the nucleus by an antidromic action potential can best be attributed to interneurones interpolated between the terminals of

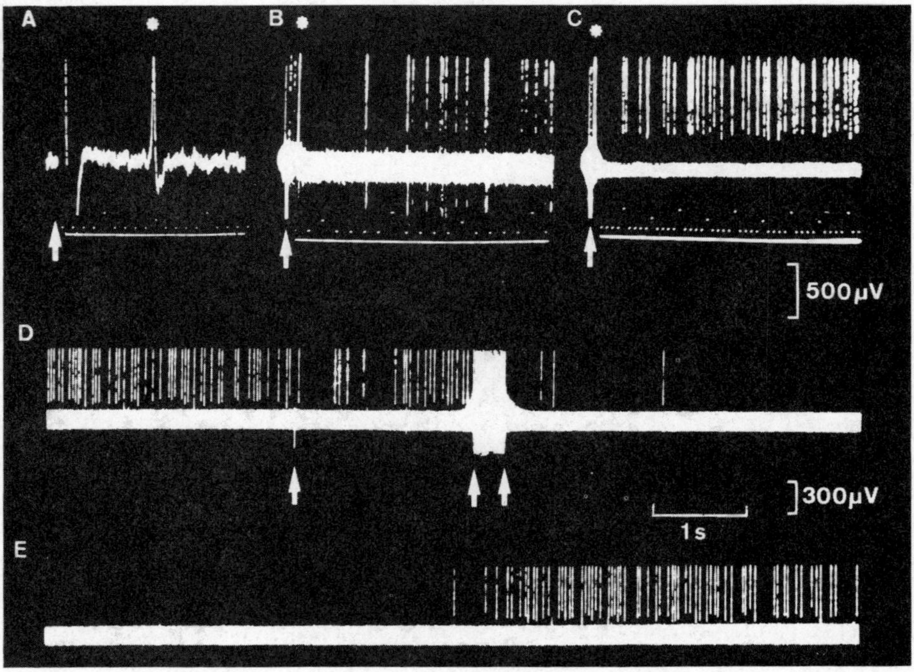

Figure 3.3 Depression of the spontaneous discharge by antidromic stimulation. *A*, control antidromic action potential at a fixed latency of 9.5 ms. *B–C*, 10 oscilloscope traces superimposed photographically to show suppression of the spontaneous firing by stimulation of the pituitary stalk. *D* and *E* are continuous records from moving film to show the recovery of the spontaneous discharge following a single shock and a train of shocks at 100 Hz, marked by single and double arrows respectively. Stimulus artefacts are marked by arrows; in *A*, *B* and *C* antidromic action-potentials by *, and time calibrations are 1 ms for *A* and 10 ms for *B–C*. (Reproduced with permission from Dreifuss and Kelly, 1972*a*)

the collaterals and supraoptic nucleus. By analogy with the Renshaw inhibition of spinal motoneurones (Eccles *et al*., 1954) the fibres of the neurohypophyseal tract may be presumed to have collateral branches which act in a recurrent fashion on the nucleus through one or more interneurones. Although later neurophysiological studies have yielded evidence for recurrent inhibition in other vertebrate species (Barker *et al*., 1971*c*; Koizumi and Yamashita, 1972; Negoro and Holland, 1972), problems have, however, arisen during attempts to locate the cells considered to be local inhibitory interneurones in this recurrent inhibitory pathway. Thus Barker *et al*. (1971*c*) were unable to demonstrate local trans-synaptic events within these nuclei following pituitary stalk stimulation. They therefore suggested that the axon terminals of neurohypophyseal cells synapsed directly on the neurosecretory neurones, even though they could not exclude the possibility of local inhibitory interneurones. On the other hand, Koizumi and Yamashita (1972) described small Renshaw-type cells in the vicinity of the supraoptic nucleus which responded to stalk stimulation with high-frequency repetitive discharges, and suggested that these could be the local inhibitory interneurones. Other investigators have similarly described neurones which show trans-synaptic activation following stimulation of the neural lobe (Negoro and Holland, 1972 ; Negoro *et al.*, 1973).

Morphologically, both Cajal (1911) and Christ (1966) have demonstrated the presence of axon collaterals in the neurohypophyseal pathway. There is also evidence from degeneration studies in the rat in favour of the existence of recurrent collaterals from the axons of supraoptic and paraventricular neurones (Olivecrona, 1957). On the other hand, Leontovich (1970) was unable to observe such collaterals in his Golgi preparation of the dog brain. Recent electron microscopic studies in the rat supraoptic nucleus (Léránth *et al.*, 1975; Zaborszky

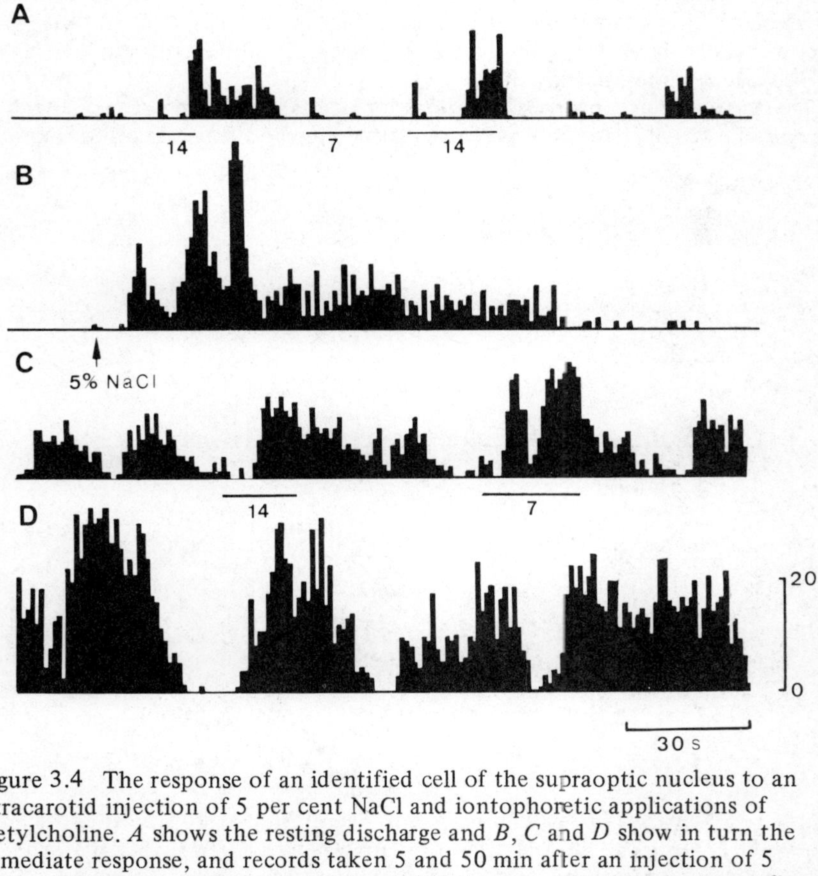

Figure 3.4 The response of an identified cell of the supraoptic nucleus to an intracarotid injection of 5 per cent NaCl and iontophoretic applications of acetylcholine. *A* shows the resting discharge and *B*, *C* and *D* show in turn the immediate response, and records taken 5 and 50 min after an injection of 5 per cent NaCl into the contralateral carotid artery. In this and subsequent figures the duration of the iontophoretic ejecting currents are represented by horizontal bars and the current intensity expressed in nanoamperes negative or positive with respect to the inside of the electrode barrel. Vertical calibration indicates the number of spikes per bin. (Reproduced with permission from Dreifuss and Kelly, 1972*b*)

et al., 1975) indicate that nearly two-thirds of the synapses on these neurones appear to be intranuclear, that is, of local origin. Presumably some of these terminals arise from interneurones located in and around the supraoptic nucleus and/or from intranuclear axon collaterals. However, it is important to realise that there is still no definitive electron microscopic evidence for axon collaterals of neurosecretory neurones since none of the synaptic endings described within the supraoptic nucleus exhibit classical neurosecretory granules.

From the functional point of view recurrent inhibition could presumably serve as a rate-limiting or negative feedback mechanism which might explain the observed phasic or episodic discharge characteristics of some supraoptic or paraventricular neurones (figure 3.4). For example, some form of inhibition must account for the long-lasting reduction of spontaneous discharges which follows osmotically-induced augmentation of supraoptic nucleus cellular discharges and the similar inhibitions observed in paraventricular neurones after the milk ejection reflex (figure 3.5) (Hayward and Jennings, 1973; Lincoln and Wakerley, 1974; Wakerley and Lincoln, 1973).

There are of course other explanations for the silence of these neurones and the episodic activity of others, perhaps involving cyclic changes in cell membrane

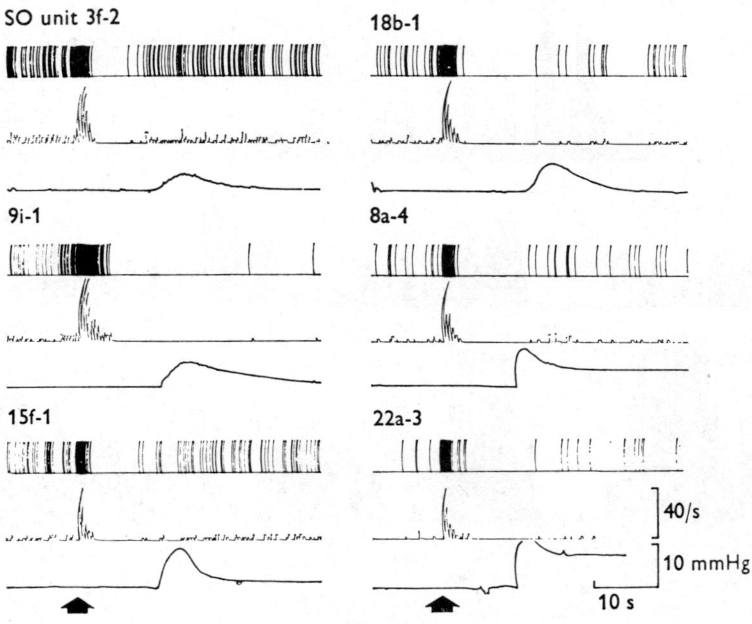

Figure 3.5 Polygraph records derived from six antidromically identified supraoptic neurones to illustrate the correlation between unitary activity and milk ejection. Each vertical deflection in the top traces represents a single action potential, the middle trace represents an integration of discharge frequency and the lower traces represent intramammary pressure. Note the acceleration of activity 10–13 s before the rise in intramammary pressure, and the silent period following the burst of unit activity. (Reproduced with permission from Lincoln and Wakerley, 1974)

excitability or synaptic activity from other sources. Presumably if axon collaterals do exist in this peptidergic neurosecretory pathway, Dale's hypothesis (1935) would suggest that either vasopressin or oxytocin would be liberated and function at the peri- or intra-nuclear synapses involved (Barker *et al.*, 1971*a,c*). However, the transmitter substance involved in recurrent inhibition still remains to be identified. Although vasopressin was shown by Nicoll and Barker (1971*b*) to depress the firing of supraoptic neurones when applied by microiontophoresis, recurrent inhibition (Dreifuss *et al.*, 1973) is still present in the Brattleboro rat (Dyball, 1974) which cannot synthesise vasopressin. Similarly it is unlikely that oxytocin could serve as a candidate for an inhibitory transmitter since it tends to have an excitant effect on cells in the supraoptic and paraventricular nuclei (Moss *et al.*, 1972*a*).

Of course it is possible that peptidergic neurones can synthesise and release more than one substance and the ability of a neurone to secrete more than one agent has been discussed (Brownstein *et al.*, 1974*a*; Burnstock, 1976; Mroz *et al.*, 1976).

Recurrent facilitation

Neurophysiological studies in the isolated hypothalamo-hypophyseal system of the bullfrog suggest that both recurrent inhibition *and* facilitation can co-exist in this species (Koizumi *et al.*, 1973). The threshold for facilitation was observed to be lower than for inhibition, indicating that cellular activity in the neurohypophyseal pathway might activate a facilitatory pathway prior to the activation of an inhibitory one. The neural network involved remains obscure. Presumably some form of direct interaction between adjacent neurosecretory neurones could occur either through axon collaterals or direct contact between the cell somata, dendrites or axons. Such an interaction would, for example, provide a powerful means by which numerous neurosecretory cells could be activated synchronously or simultaneously to ensure a sufficiently plentiful output of oxytocin during the milk-ejection reflex (Lincoln and Wakerley, 1974; Wakerley and Lincoln, 1973) or of vasopressin during haemorrhage (Wakerley *et al.*, 1975).

Afferent pathways

Single cell recording has shown that septal, amygdaloid and reticular formation stimulation influences the firing frequency of supraoptic and paraventricular neurones (Negoro and Holland, 1972; Negoro *et al.*, 1973). In the case of the septum, the predominant response from 75 per cent of neurosecretory paraventricular neurones, antidromically identified after pituitary stalk stimulation, was a silent interval at a mean latency of 18 ms and a mean duration of 60 ms. Stimulation in the amgydala inhibited 64 per cent of paraventricular neurosecretory cells at a mean latency of 17 ms and a mean duration of 60 ms. Initial excitatory responses from either stimulus site were unusual. This contrasted sharply with the responses of paraventricular non-neurosecretory neurones which often displayed orthodromic trans-synaptic excitatory responses (mean latency 15 ms) to stimulation of the pituitary stalk. These cells, which could function as interneurones in the recurrent inhibitory pathway, were excited by septal stimulation in the majority of instances (94 per cent). The majority of these cells (67 per cent) were also excited orthodromically by amygdaloid stimulation. From these observations one might suggest that recurrent and afferent inhibition of paraventricular neurosecretory neurones is mediated by similar, possibly the same inhibitory interneurones.

OXYTOCIN AND VASOPRESSIN RELEASE

The neurohypophyseal tracts from supraoptic and paraventricular nuclei comprise the final neuroendocrine pathway for oxytocin and vasopressin release in response to a variety of stimuli. As mentioned earlier, a number of studies have shown the neural activity of supraoptic and paraventricular neurones to be correlated with the release of oxytocin and vasopressin in the neural lobe terminals (Dreifuss *et al*., 1971; Harris *et al*., 1969). Several studies have reported accelerated firing of supraoptic and paraventricular neurones in response to chemical or neural stimuli known to influence the release of oxytocin and/or vasopressin (Cross and Green, 1959; Dreifuss and Kelly, 1972*b*; Dyball, 1971; Dyball and Koizumi, 1969; Hayward, 1975).

For instance, figure 3.4 shows the response of an identified supraoptic neurone to an injection of hypertonic sodium chloride (5 per cent NaCl; 0.2 ml into the contralateral carotid artery. After a short delay there was a clear-cut acceleration of the background discharge which lasted only a few minutes. However, five minutes after the initial excitation the discharge, which had been slow and rather irregular except when excited by acetylcholine, became distinctly periodic in nature. One hour later the background discharge rate was still increased and the periodic pattern was well established.

However, much of the information derived in this way was indirect (figure 3.4) and not precisely related either to the stimulus or to the amount or nature of the hormone released. In part this arose because the studies were conducted in anaesthetised preparations, and in part because the techniques for measuring the hormones were inadequate (Ginsburg, 1968; Ginsburg and Brown, 1956). We have therefore reviewed a number of recent attempts to circumvent these problems and to demonstrate directly an association between cellular activity and the release of oxytocin and vasopressin.

Neural Events Related to Oxytocin and Vasopressin Release

The milk ejection reflex, the response of the myoepithelial cells of the mammary glands producing a rise in intramammary ductal pressure and expulsion of milk, is considered to be regulated by episodic pulses of circulating oxytocin released from the neurohypophysis. In the lactating rat there is a close temporal relationship between suckling and milk ejection. Single unit recordings in the lactating rat have shown a periodic activation of supraoptic and paraventricular neurosecretory neurones to be related temporally to an increase in intramammary pressure (figure 3.5). A similar increase in intramammary pressure occurs during systemic injections of oxytocin (Cross *et al*., 1975; Dyball, 1971; Lincoln and Wakerley, 1974, 1975; Wakerley and Lincoln, 1973). The acceleration of firing which accompanies the reflex liberation of oxytocin always occurs in antidromically identified units which fire randomly. This never occurs in units which have a phasic firing pattern (see below) and presumably during lactation the phasic cells are inhibited. The pulses of random firing associated with milk release recur uniformly every 4–8 min and the afferent stimulus for this periodic activation of 'oxytocinergic' neurones during nursing is the suckling activity of the pups. In the lactating rat this response is resistant to surgical levels of anaesthesia, and is dependent in part on the number of suckling young, the degree of mammary distension, and time.

As mentioned earlier several workers have found that electrical stimulation of the pituitary stalk *in vivo* could evoke milk ejection in lactating rabbits or rats only if stimulation frequencies were greater than 20–30 Hz with 50 Hz being optimal (Dreifuss *et al*., 1971; Harris *et al*., 1969). Similarly, *in vitro* studies using incubated rat neurohypophysis have indicated that there is a greater release of oxytocin into the medium at higher frequencies of stimulation and that this neurosecretion is a calcium-dependent phenomenon (Dreifuss *et al*., 1971). Thus it would seem that only a rapid succession of action potentials arriving at the nerve terminal achieves a sufficient degree of depolarisation to release oxytocin. It has also been demonstrated that osmotic (dehydration) (Brimble and Dyball, 1977), volumetric (haemorrhage) (Poulain *et al*., 1977) and behavioural (drinking) (Arnauld *et al.*, 1975) factors interact to release vasopressin from the neural lobe of the pituitary.

However, there has been some difficulty in correlating blood levels of vasopressin with magnocellular neurosecretory neuronal activity, in part owing to conditions of anaesthesia which can alter firing patterns of central neural neurones and induce aberrant changes in blood vasopressin levels. Nonetheless several studies support a correlation between the activity of supraoptic and paraventricular neurosecretory cells and circulating levels of vasopressin (Arnauld *et al.*, 1974, 1975; Hayward, 1975; Hayward and Jennings, 1973; Wakerley *et al.*, 1975). Approximately three-quarters of identified supraoptic and paraventricular neurones fire regularly in a fairly random manner, whilst the other quarter display phasic bursts of activity, a pattern seldom observed from other hypothalamic neurones.

Although some phasic neurones respond to suckling and display bursts of activity associated with the milk ejection reflex (figure 3.3), most of the phasic units do not, and instead are responsive to other forms of stimuli such as osmotic injections, dehydration (Arnauld *et al*., 1974, 1975), carotid occlusion (Dreifuss *et al*., 1976*a*) and haemorrhage (Wakerley *et al*., 1975) which are thought to release vasopressin rather than oxytocin. Based on these observations it has been suggested that the phasic neurosecretory supraoptic and paraventricular neurones are 'vasopressinergic'.

Additional studies suggest that phasic discharges are but one functional state of activity of these neurones. For example, Arnauld *et al*. (1975) demonstrated that the predominant cell firing pattern of supraoptic neurones in the unanaesthetised monkey under conditions of dehydration changed from one of irregular low-frequency discharges to phasic discharges and then to a pattern of high-frequency continuous firing in concert with increasing plasma osmolarity. The reverse trend occurred upon rehydration.

The afferents responsible for the reflex activation of vasopressin release are not precisely known. The activity of neurosecretory neurones in the supraoptic nucleus changes in response to vagal or carotid sinus nerve stimulation (Barker *et al*., 1971*b*). Osmosensitive neurones in the supraoptic nucleus have been described but it is still not certain whether the neurosecretory cells themselves or adjacent cells behave as the specific osmoreceptors (Hayward and Jennings, 1973).

Recently attention has been drawn to changes in the synthesis, storage and release of neurohypophyseal hormones during the female sex cycle or after injections of ovarian steroids or gonadotrophins (Roberts, 1971, 1973, 1975; Roberts and Share, 1969, 1970) and to the question of how sex steroids might be involved in the increase in oxytocin release which occurs in response to vaginal dilatation. Previous biochemical work had shown that oestrogens could facilitate oxytocin release whilst the effect of progesterone was inhibitory (Roberts 1971, 1973). By recording from several different classes of paraventricular neurones Freund-Mercier and Richard (1977) showed that only the spontaneous activity of the neurones identified antidromically from the pituitary stalk is influenced by the endocrine state during the oestrous cycle and lactation (figure 3.6). Indeed the highest levels of spontaneous activity occurred during lactation and the first day of the oestrous cycle, whilst during the rest of the cycle activity decreased from the first day to the last.

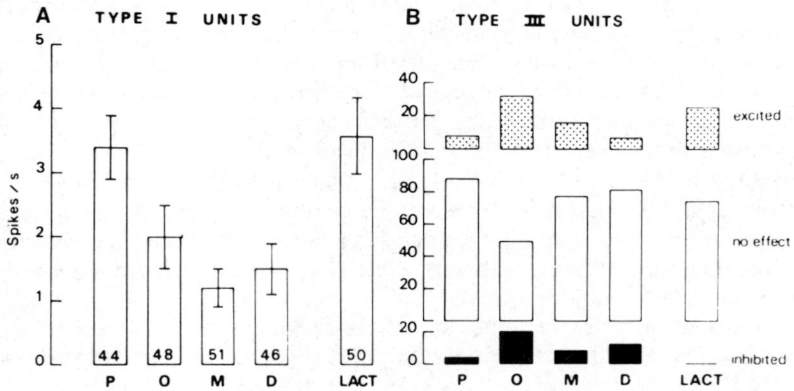

Figure 3.6 *A*, mean spontaneous firing rates of antidromically identified cells of paraventricular nucleus or type I cells during the oestrous cycle: P, pro-oestrus; O, oestrus; M, metoestrus; D, dioestrus; LACT, lactation. The vertical lines represent the standard errors and the figures at the bottom of the columns represent the number of cells monitored in each experimental situation. *B*, effects of vaginal dilatation on spontaneous firing of cells of the paraventricular nucleus which were unaffected by antidromic stimulation or type III cells in cycling and lactating rats. All type III cells monitored were spontaneously active. (Reproduced with permission from Freund-Mercier and Richard, 1977)

In lactating rats the firing rate of the antidromically identified cells could be increased by vaginal dilatation and this was thought to correlate with the release of oxytocin which is known to occur. However, a much stronger correlation was seen between vaginal stimulation and the oestrous cycle in cells which were unaffected by pituitary stalk stimulation (figure 3.6). Unfortunately, the exact reason for this is not known since a nociceptive stimulus which had earlier been shown not to release oxytocin exhibited a similar correlation.

NEUROPHARMACOLOGY OF SUPRAOPTIC AND PARAVENTRICULAR NEURONES

Microiontophoresis is a technique whereby minute quantities of putative neurotransmitters and other agents can be introduced directly into the extracellular environment while monitoring extracellular or intracellular excitability characteristics from neurones (Kelly, 1975; Kelly *et al.*, 1975). This technique has been usefully applied to determine the sensitivity of (and possible presence of receptor sites on) cells in various CNS regions for a variety of putative neurotransmitters.

Acetylcholine

In 1939 Pickford reported on the antidiuretic effect of acetylcholine (ACh). Subsequent studies have shown that the supraoptic nucleus contains high levels of choline acetyltransferase and acetylcholinesterase (Abrahams *et al.*, 1957; Feldberg and Vogt, 1948). Nerve fibres containing acetylcholinesterase have in fact been demonstrated in all regions of the hypothalamus (Shute, 1970). ACh injected into the carotid artery of unanaesthetised dogs has been shown to evoke release of oxytocin and vasopressin. The milk ejection reflex described earlier is blocked by anti-cholinergic drugs and by emotional stress, which involves a central inhibition of oxytocin release (Cross *et al.*, 1975)

The literature therefore long predicted the presence of ACh-excitable neurones within the supraoptic nucleus and the possible transmitter role which ACh might play. Most of this earlier work which showed a correlation between vasopressin release and the injection of ACh, cholinomimetics and anticholinesterases either directly into the nucleus or into its arterial blood supply has been summarised by Bisset (1968, 1976).

Eventually Dreifuss and Kelly (1970, 1972*b*) showed that both antidromically identified and adjacent non-projecting neurones from the region of the supraoptic nucleus were excited by iontophoretic applications of ACh. The proportion of ACh-excited cells was very much greater than was predicted in the earlier studies of the hypothalamus by Bloom *et al.* (1963), who found that less than 20 per cent of cells were excited by ACh.

Several features of the discharge of supraoptic neurones by iontophoretic ACh can perhaps be attributed to its interaction with postsynaptic membrane receptors located beneath the terminals of cholinergic fibres. Like Renshaw cells, where the role of ACh as a transmitter (Eccles *et al*., 1954; Eccles *et al*., 1956; Curtis and Eccles, 1958; Curtis and Ryall, 1966*a*, *b*) has been most clearly established, the majority of supraoptic neurones are excited by extracellular ACh concentrations attained with fairly modest currents.

The similarity between the ACh receptors of supraoptic neurones and those of Renshaw cells is also supported by the demonstration that ACh and nicotine are equipotent when tested on the same cells. In addition, an intracarotid injection of dihydro-β-erythroidine, which has been shown specifically to block the nicotinic receptors of Renshaw cells (Eccles *et al*., 1954; Eccles *et al*., 1956; Curtis and Eccles, 1958; Curtis and Ryall, 1966*a,b*) reversibly blocks the action of both ACh and nicotine on supraoptic neurones for several minutes.

Although it is difficult to deny the existence of nicotinic receptors for ACh on the cells of the supraoptic nucleus, some features of ACh-evoked discharge closely resemble the characteristics of ACh effects on cortical neurones which are usually attributed to muscarinic receptors (Krnjevic and Phillis, 1963*a,b*; Spehlmann, 1963; Crawford and Curtis, 1966). The onset of the excitation of supraoptic neurones is characteristically slow and begins 5-15s after the onset of the ACh release and continues for some seconds after its termination. However, the cells with the highest levels of spontaneous activity respond more readily to ACh than the more quiescent ones. Furthermore, the reduction in the response

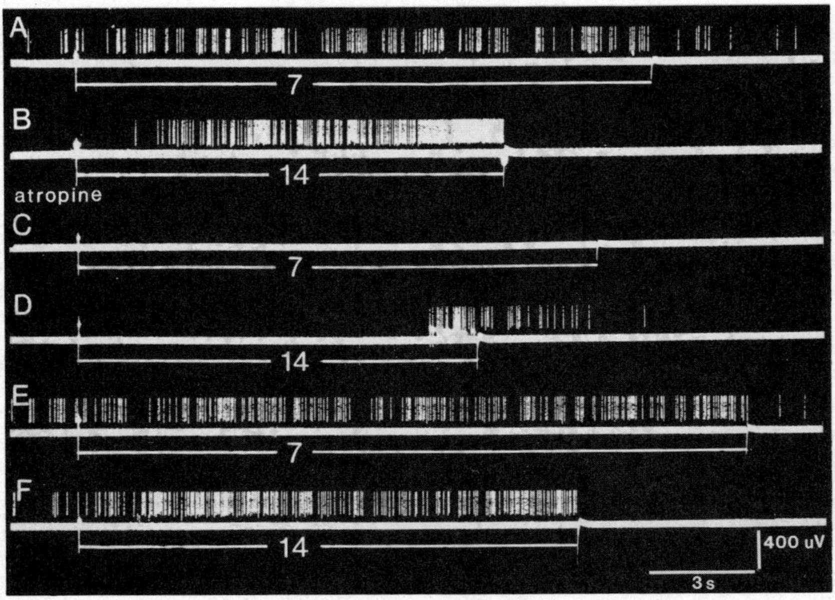

Figure 3.7 Effect of iontophoretic acetylcholine on the excitability of the identified neurone shown in figure 3.4 to be excited by 5 per cent NaCl. *A* and *B* are control records showing discharges evoked by acetylcholine released by currents of 7×10^{-9} A and 14×10^{-9} A. Records *C–D* and *E–F*, respectively, show the discharge evoked by similar amounts of acetylcholine at approximately 2 and 60 min after an injection of 40 mg of atropine into the contralateral carotid artery. (Reproduced with permission from Dreifuss and Kelly, 1972*b*)

to ACh after intravenous atropine (figure 3.7) and its enhancement during the accelerated discharge evoked by injections of hypertonic 5 per cent NaCl into the carotid artery can be attributed to a dependence of the ACh excitation on the existing excitability of the cell (figure 3.4). Since the reduction in ACh sensitivity after intracarotid atropine is accompanied by a marked reduction of the background discharge, it seems that there is no evidence of specific muscarinic receptors on supraoptic cells.

Recently depolarisation of the membranes of cortical cells by ACh has been shown by Krnjevic *et al*. (1971) to be associated with a decrease in potassium permeability. As a result, the ACh sensitivity of a particular neurone would be dependent on both the permeability of the membrane to sodium and the external sodium concentration. It is possible therefore that the differential

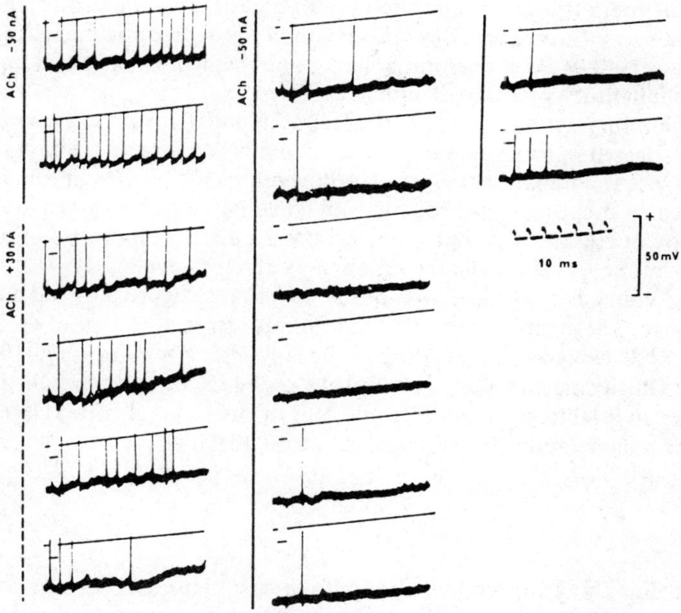

Figure 3.8 Effects of acetylcholine (ACh) on the membrane potential of a lateral hypothalamic neurone. The upper record in each pair of traces shows the zero potential level and the current pulse (0.2 nA, 7 ms) applied through the intracellular recording electrode. The lower trace shows the membrane potential. Records in each column were taken at 1.3 s intervals from top to bottom, being continuous from the left to right columns. Between the middle and right column 3.8 s elapsed. The two upper-left records show the intracellular potential before ACh (−50 nA braking current applied to ACh pipette). Lower-left records show the intracellular potential during ACh application (+30 nA). Progressive hyperpolarisation of the membrane potential and diminution of the firing rate can be seen in records of the middle column. These effects persisted even after termination of the ACh ejecting current. Recovery from the effects of ACh are shown in the right column. (Reproduced with permission from Oomura *et al.*, 1976)

sensitivity of the supraoptic neurones to ACh and the delayed onset of excitation may reflect some property of the membrane of these cells which changes during the regulation of the extracellular ionic environment, rather than the nature or distribution of ACh receptors. Earlier we suggested that the different spontaneous firing patterns of supraoptic neurones may also be correlated with the functional status of their membrane excitability.

Later reports from other laboratories confirmed the excitation of neurosecretory cells by iontophoretic ACh. The spontaneous discharges of almost all the paraventricular neurones of anaesthetised rabbits were found by Cross *et al.* (1971) to be accelerated by ACh; the response was delayed as described above. In the cat (Barker *et al.*,1971*a*, *b*), excitation of supraoptic neurones required much higher currents than described here and could be prevented by nicotinic antagonists. Still higher amounts led to a depression of the spontaneous discharge.

Since neither intravenous nor microiontophoretically applied atropine has been shown to influence recurrent inhibition in the supraoptic nucleus (Nicoll and Barker, 1971*b*), ACh does not appear to be the neurotransmitter in the recurrent inhibitory pathway described earlier.

The inhibitory action of ACh in the lateral hypothalamus has, however, been investigated in some detail by Oomura and his colleagues (Oomura *et al.*, 1970, 1976). The inhibitory effect of ACh was blocked by atropine and unaffected by dihydro-β-erythroidine, and could be mimicked by an iontophoretic application of eserine. Strychnine had no effect on the ACh-evoked inhibition. Intracellular records show the reduction in excitability evoked by ACh to be associated with a hyperpolarisation of the membrane (figure 3.8) and a decrease in resistance. The reversal potential for ACh was identical to that of an atropine-sensitive IPSP evoked by stimulation in the amydgala (Oomura *et al.*, 1970). Recently Dingledine and Kelly (1978) have reviewed the evidence which suggests ACh plays an inhibitory role on specific cells of the cerebral cortex, lateral geniculate nucleus and nucleus reticularis of the thalamus.

Amines

Noradrenaline (NA) applied by microiontophoresis depresses the activity of most neurosecretory neurones, but enhances the discharge frequency of a comparatively greater number of adjacent non-neurosecretory supraoptic and paraventricular fibres (Barker *et al.*, 1971*b*). Although NA-evoked depression is blocked by α-adrenoceptor blocking agents, recurrent inhibition in the supraoptic and paraventricular nuclei has proved resistant to treatment with 6-hydroxydopamine which destroys noradrenergic nerve endings and virtually abolishes all of the NA fluorescence in the supraoptic nucleus (Nicoll and Barker, 1971*b*). Dopamine and serotonin either enhance or depress the activity of both neurosecretory and non-neurosecretory supraoptic and paraventricular neurones (Moss *et al.*, 1972*b*). Since both nuclei contain relatively low levels of these substances and few dopaminergic or serotoninergic nerve endings (Fuxe and Hökfelt, 1967), these responses may be of pharmacological rather than of physiological importance.

Amino Acids

The amino acid L-glutamate enhances the activity of virtually all supraoptic and paraventricular neurones tested (Moss *et al*., 1972*b*), while glycine and gamma-aminobutyric acid (GABA) consistently depress their activity. Glycine-evoked depression is antagonised by strychnine, while picrotoxin and bicuculline appear to antagonise selectively GABA-evoked depression (Nicoll and Barker, 1971*b*). However, none of these convulsants administered by microiontophoresis or systemic injection appears to antagonise recurrent inhibition, except for bicuculline, which produces partial antagonism when administered systemically in convulsive doses (Nicoll and Barker, 1971*b*).

Histamine

Injections of histamine into the ventricles or the hypothalamus have been shown to cause an antidiuretic effect (Bennett and Pert, 1974), a fall in body temperature (Green *et al*., 1975) and an increase in the delta activity of the EEG (Wolf and Monnier, 1973). Lesions of the medial forebrain bundle reduce the levels of histamine and its associated enzymes in the hippocampus and cortex as well as in the hypothalamus (Garbarg *et al*., 1974; Schwartz, 1975). In most regions of the brain the application of histamine to single cells by microiontophoresis has a weak depressant effect (Phillis *et al*., 1968, 1975; Renaud, 1976*e*), whereas in the hypothalamus, where the highest levels of histamine are found (Brownstein *et al*., 1974*b*), supraoptic and other neurones are usually excited (figure 3.9) by histamine (Haas, 1974; Haas *et al*., 1975; Haas and Wolf, 1977).

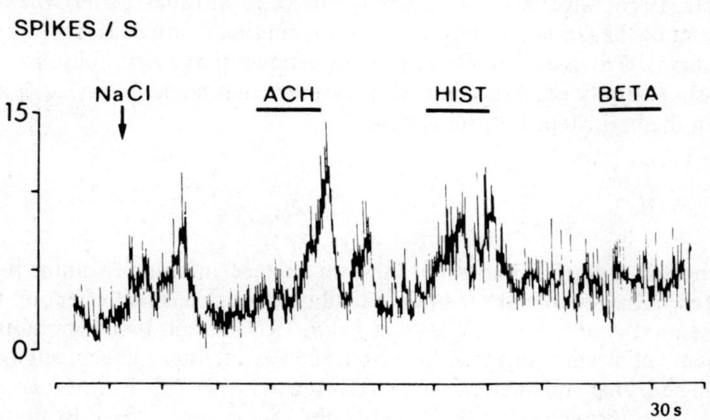

Figure 3.9 Osmosensitive neurone from the anterolateral hypothalamus of the cat. Ratemeter record shows excitation of the cell by NaCl, injection of 1 ml of 1M NaCl solution into the contralateral carotid artery and by micro-electrophoretically applied acetylcholine (ACH) and histamine (HIST). Betazole (BETA) an H_2 agonist, was ineffective. Ejecting currents for all substances, 90 nA. (Reproduced with permission from Haas and Wolf, 1977)

Although the depressant actions of histamine on some hypothalamic and most non-hypothalamic neurones appear to involve H_2 receptors and can be blocked with metiamide, the excitant actions of histamine on supraoptic neurones are unaffected and none of the H_1 antagonists in current use appears to be selective for histamine—as opposed to ACh-evoked excitations. The exact nature of the action of histamine on the neurones has yet to be elucidated and its release from supraoptic neurones has yet to be demonstrated.

Peptides

Much of the relevant pharmacology has been summarised in a recent review by Bisset (1976). Vasopressin applied by microiontophoresis depresses the activity of most responsive neurosecretory neurones in the supraoptic nucleus but only at relatively high microiontophoretic currents. On the other hand, oxytocin enhances the firing frequency of paraventricular neurosecretory neurones (Moss *et al*., 1972*a*). The physiological significance of these observations remains to be determined but most of the evidence suggests that these peptides are not involved in the recurrent inhibitory pathways. Vasopressin, however, appears to influence animal behaviour and to be important in memory trace consolidation (Wimersma Greidanus *et al*., 1975). It is therefore plausible that there are central oxytocin- and vasopressin-containing nerve endings and that these peptides may subserve some other role in brain tissue separate from their systemic actions. In fact several investigators have now used immunohistochemical techniques to demonstrate the presence of neurophysin-vasopressin containing fibres not only in the neurohypophysis but also in the median eminence (Parry and Livett, 1976; Silverman 1976; Zimmerman and Antunes, 1976), the organum vasculosum of the lamina terminalis and suprachiasmatic nucleus (see Zimmerman 1976). Swanson (1977) has also shown that a neurophysin-vasopressin pathway extends from the paraventricular nuclei to several autonomic centres in the brainstem and spinal cord.

Angiotensin II

Angiotensin II, a peptide important in water-balance and blood-volume regulation, is reported to have a role in the brain, possibly through a direct effect on neural tissue (Bennett *et al*. 1976; Phillips and Felix, 1976). Brain tissue contains all the components of a renin–angiotensin system such as angiotensinogen, angiotensin-converting enzyme and angiotensinase (Ganten *et al*., 1976). Furthermore angiotensin receptors have been demonstrated (Sakai *et al*., 1974; Bennett *et al*., 1976; Sirett *et al*., 1977) and there is immunohistochemical evidence for angiotensin-containing nerve terminals in the brain and spinal cord (Fuxe *et al*., 1976) (see chapter 2).

Direct application of angiotensin II to supraoptic neurosecretory cells increases their discharge rate (Nicoll and Barker, 1971*a*) and intracranial injection of the peptide into anterior hypothalamic, septal or preoptic regions causes

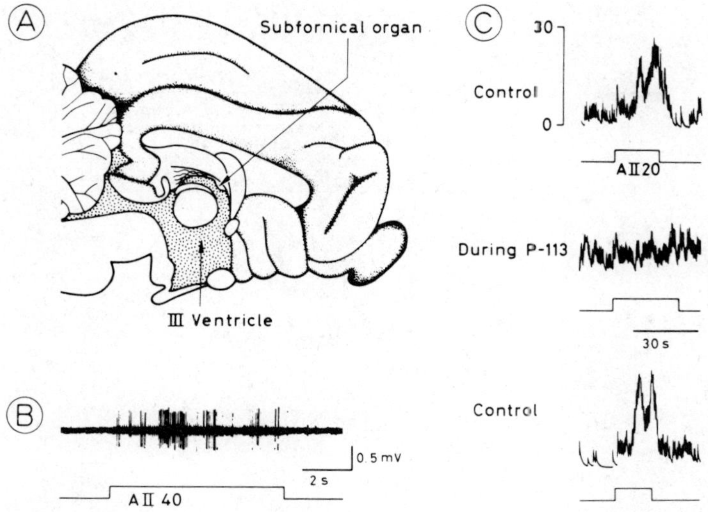

Figure 3.10 *A* the effect of angiotensin II on the cat subfornical organ. *B* effect of angiotensin II on the firing rate. *C* angiotensin II antagonism by Sar-1-ala-8 angiotensin II (P-113, saralasin). Ejecting currents given in nanoamperes. (Reproduced with permission from Felix and Phillips, 1978)

animals in a normal state of water-balance to drink (Epstein *et al.*, 1970). The suggestion that the subfornical organ (figure 3.10) might be a likely receptor site for the dipsogenic action of angiotensin II is supported by experiments in which angiotensin II, applied intraventricularly close to the organ, produced a short-latency drinking response (Simpson and Routtenberg, 1973). Lesions of the organ temporarily abolished drinking induced by injections of angiotensin II (Buggy *et al.*, 1975) and microiontophoretic studies (Felix and Akert, 1974; Felix, 1976) on the subfornical organ showed the existence of neurones specifically sensitive to angiotensin II (figure 3.10). In the iontophoresis experiments the effect was dose-dependent and could also be evoked by intravenous injections of angiotensin II. The response appeared to be specific, since the neurones were not responsive to the nonapeptides bradykinin, eledoisin or physalaemin, whilst the same neurones were excited by ACh. Only the response to angiotensin II was blocked (figure 3.10) by the competitive angiotensin II antagonist, Sar-1-ala-8 angiotensin II (P-113, saralasin) (Phillips and Felix, 1976). Structure–activity studies showed that the heptapeptide, angiotensin II (2–8) was more active than angiotensin II, whilst the tetrapeptide (5–8) angiotensin II was somewhat less potent (figure 3.11); the effect of both these substances was blocked by saralasin. In contrast, the tripeptide angiotensin (6–8) was without effect. However, these actions of angiotensin II might not be quite so specific for the subfornical organ since microiontophoretically applied angiotensin II has been shown to have effects on the thalamus (Wayner *et al.*, 1973), medial preoptic area (Gronan and York, 1976) and cerebral cortex (Phillis and Limacher, 1974).

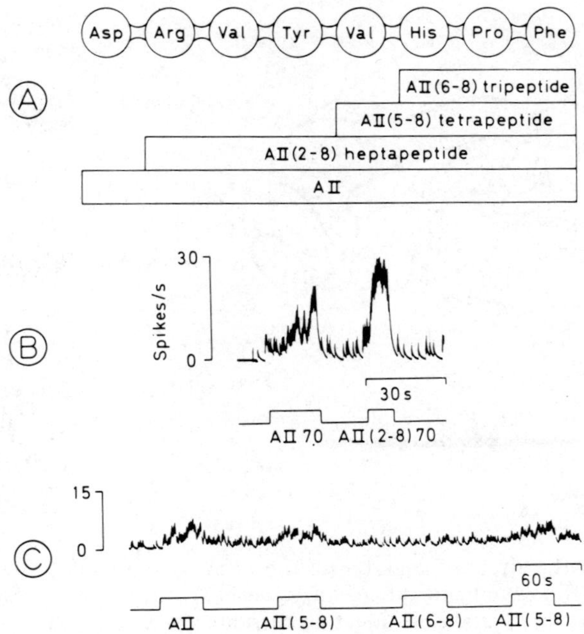

Figure 3.11 *A* amino acid sequence of angiotensin II and the peptide fragments described in the text. *B* differential effect of angiotensin II (2-8). *C* effect of a series of angiotensin II fragments. Ejecting currents in nanoamperes. (Reproduced with permission from Felix and Phillips, 1978)

THE 'PARVICELLULAR TUBERO-INFUNDIBULAR' SYSTEM

Tubero-infundibular neurones are defined anatomically as parvicellular neurosecretory neurones located within the hypophysiotrophic area of the mediobasal hypothalamus (Halász, 1969; Szentágothai *et al.*, 1968). Their fine axons course independently from the larger neurohypophyseal fibres and terminate in the surface zona of the median eminence in close relation to the primary capillaries of the portal system (figure 3.12). These terminals are distinct from those in the neural lobe and contain two vesicular populations: a plethora of synaptic vesicles of the usual type and somewhat larger vesicles with dense osmiophilic cores. The electrophysiological definition of a tubero-infundibular neurone refers to cells located in the hypothalamus and basal forebrain area which can be activated antidromically following stimulation of the surface of the median eminence (Makara *et al.*, 1972; Sawaki and Yagi, 1973).

It is widely believed that a system of tubero-infundibular neurosecretory neurones located within the medial and basal hypothalamus is the final common pathway for the neural regulation of adenohypophyseal secretion (Harris, 1955; Szentágothai *et al.*, 1968). The final messengers in this system are peptides which

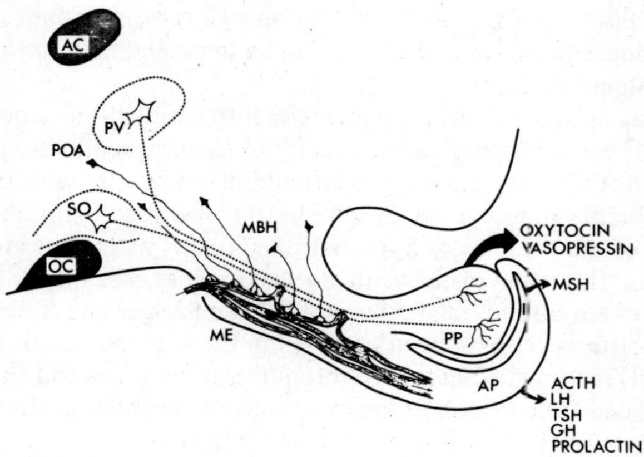

Figure 3.12 A simplified sketch of the hypothalamus to illustrate the neurohypophyseal tract (interrupted lines) and the magnocellular paraventricular (PV) and supraoptic (SO) neurones to the posterior pituitary (PP) responsible for liberation of oxytocin and vasopressin. The parvicellular tubero-infundibular pathway is also shown. It originates from cells in the medial basal hypothalamus (MBH) and preoptic area (POA) and terminates on the portal capillaries in the median eminence (ME) which are responsible for elaborating the hypothalamic releasing factors which regulate anterior pituitary (AP) secretion. Abbreviations: AC, anterior commissure; I, intermediate lobe; OC, optic chiasma. (Reproduced with permission from Renaud, 1978*a*)

either evoke or inhibit the secretions of the adenohypophyseal cells, and it is assumed that these peptides are released into the median eminence portal capillary plexus from nerve endings of the tubero-infundibular neurones (Blackwell and Guillemin, 1973; Schally *et al.*, 1973). Presumably, following their release in response to appropriate neural stimuli, these peptides are carried in portal blood to the adenohypophysis (Porter *et al.*, 1971). However, it is not at all clear whether this release requires an increase in neuronal discharges in the tubero-infundibular system analogous to events which accompany secretion of posterior pituitary hormones.

Electrical Characteristics of Tubero-infundibular Neurones

In the first electrophysiological study of hypothalamic tubero-infundibular neurones in the rat the majority of neurones were located in the arcuate and ventromedial nuclei, while a smaller number were noted in the suprachiasmatic, anterior periventricular and dorsal premamillary nuclei (Makara *et al.*, 1972). Subsequent investigators (Harris and Sanghera, 1974; Moss *et al.*, 1972*a*; Renaud, 1976*a,b*; Renaud and Martin, 1975) have confirmed and extended these intial observations to include the anterior hypothalamic area and medial preoptic area. In each instance electrophysiological recordings have generally been brief, implying the presence of small somata. While it is noteworthy that many of these tubero-infundibular neurones are recognised within the 'hypophysiotrophic area' described by Halász (1969), a substantial number of these neurones are located over a more elongated region of the basal forebrain extending into the rostral preoptic area (figure 3.13).

By converting antidromic invasion latencies into conduction velocities, Renaud (1978*b*) has confirmed earlier reports of impulse conduction velocities of under 2.0 m/s. Curiously, the tubero-infundibular neurones with the slowest conduction velocities (under 0.2 m/s) were located less than 0.5mm from the ventral surface of the mediobasal hypothalamus, that is within the arcuate nucleus and ventral portion of the ventromedial nucleus, and may therefore be dopaminergic tubero-infundibular neurones (Lichtensteiger and Keller, 1974).

No notable differences have been observed in the spontaneous or antidromic action potentials recorded from tubero-infundibular neurones and the action potentials recorded from neighbouring hypothalamic or other central neurones. Frequently they display an inflexion on their rising phase.

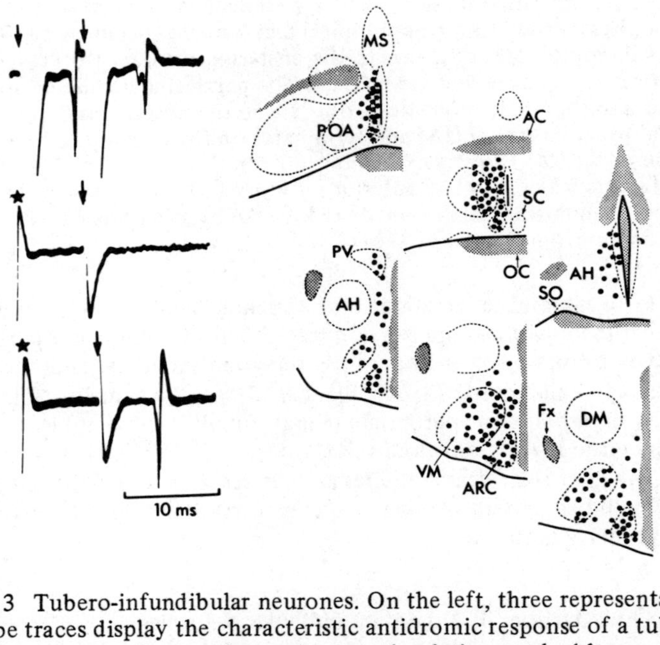

Figure 3.13 Tubero-infundibular neurones. On the left, three representative oscilloscope traces display the characteristic antidromic response of a tubero-infundibular neurone after median eminence stimulation marked by arrowheads. The upper trace displays constant latency responses at high frequency; the lower traces illustrate collision at the appropriate intervals between a spontaneous action potential (represented by a star) and the antidromic spike. On the right, the approximate locations of medial basal hypothalamic and preoptic tubero-infundibular neurones are indicated by black dots. All cells displayed antidromic invasions after median eminence stimulation. Abbreviations: AC, anterior commissure; AH, anterior hypothalamic area; ARC, arcuate nucleus; DM, dorsomedial hypothalamic nucleus; Fx, fornix; MS, medial septum; OC, optic chiasma; POA, preoptic area; SC, suprachiasmatic nucleus; SO, supraoptic nucleus; VM, ventromedial nucleus. (Reproduced with permission from Renaud, 1976*a*)

One of the interesting observations derived from the electrophysiological studies on hypothalamic neurones in general, and on tubero-infundibular neurones in particular, is the ability of these cells to conduct antidromic impulses (and presumably orthodromic impulses) at frequencies up to 400–500Hz. This appears somewhat surprising for axons whose conduction velocities are under 2.0 m/s and whose fibre diameters are often less than 1 μm (Brawer, 1972; Brawer and Sonnenschein, 1975). Although spontaneously active tubero-infundibular cells rarely discharge for long periods of time at high frequencies, some of these cells do indeed produce bursts of impulses with intraburst frequencies approaching 400Hz. It remains to be determined whether such activity is a prerequisite for the liberation of releasing factors similar to the events associated with the liberation of oxytocin and vasopressin described earlier.

Recurrent inhibition

In an anaesthetised preparation many tubero-infundibular cells display little or no spontaneous activity and those which are active either discharge randomly or in bursts. Stimulation of the median eminence at stimulus intensities subthreshold for antidromic invasion is usually followed by a decrease in neuronal excitability lasting 40–50 ms (figure 3.14). An increase in the stimulus intensity to twice threshold usually prolongs the inhibitory period to 100–130 ms. Since

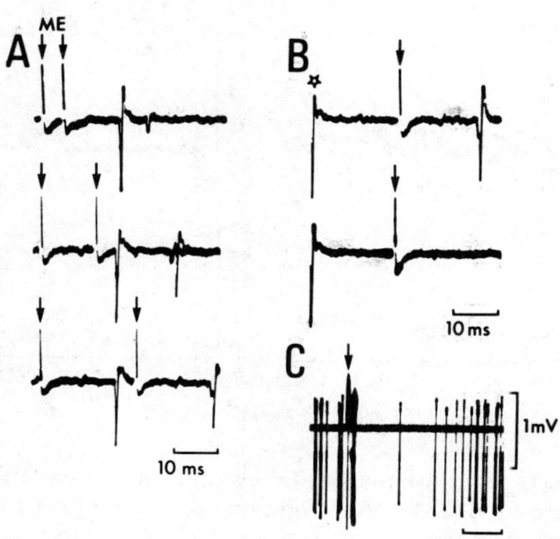

Figure 3.14 Identification of tubero-infundibular neurones by antidromic stimulation of the median eminence (ME). *A* paired shocks to show the occlusion of the second antidromic spike by the first. *B* two traces to show the ablation of the antidromic spike evoked by a stimulus to the median eminence by the spontaneous action potential at the origin of the sweep (marked by a star). *C* superimposed sweeps to show the inhibitory pause which follows the antidromic stimulation of the median eminence marked by an arrow. (Reproduced from the unpublished work of Renaud)

this inhibition has been observed after both spontaneous and glutamate-evoked activity, it must be considered a postsynaptic process analogous to that in supraoptic and paraventricular neurones (Renaud, 1976*a,b*; Sawaki and Yagi, 1973, 1976; Yagi and Sawaki, 1974). Presumably recurrent inhibition could operate either directly through recurrent axon collaterals terminating on the cells of origin (tubero-infundibular neurones) or indirectly through a local inhibitory interneurone. The latter explanation appears more tenable since Renaud (1976*a*) has described the trans-synaptic excitation of local non-tubero-infundibular neurones following stimulation of the median eminence. A disynaptic pathway is also suggested by observations which show that where recurrent inhibition can be demonstrated at stimulus intensities subthreshold for antidromic invasion, the latency of the onset of the inhibitory period is usually equal to or somewhat greater than the latency for the antidromic spikes.

In accordance with Dale's hypothesis (1935), the presence of recurrent inhibition in a putative peptidergic pathway raises the suspicion that peptides can be

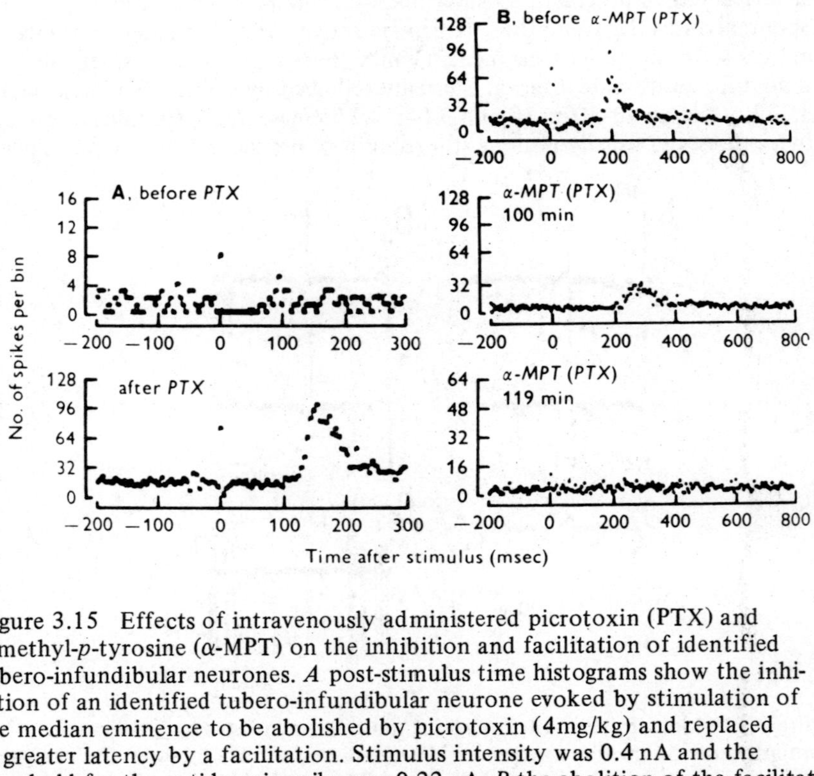

Figure 3.15 Effects of intravenously administered picrotoxin (PTX) and α-methyl-*p*-tyrosine (α-MPT) on the inhibition and facilitation of identified tubero-infundibular neurones. *A* post-stimulus time histograms show the inhibition of an identified tubero-infundibular neurone evoked by stimulation of the median eminence to be abolished by picrotoxin (4mg/kg) and replaced at greater latency by a facilitation. Stimulus intensity was 0.4 nA and the threshold for the antidromic spike was 0.22 nA. *B* the abolition of the facilitation evoked by stimulation of the median eminence following pretreatment with (PTX) by an injection of α-MPT (250 mg/kg). The time after the injection is shown on each histogram. Stimulus intensity was 0.5 mA and the threshold for the antidromic spike 0.14 mA. (Reproduced with permission from Sawaki and Yagi, 1976)

released from the same neurone both in the median eminence portal capillary plexus and at central synapses where they could act as conventional neurotransmitters. Since it is apparent that the action of the inhibitory transmitter in the final common pathway of the recurrent loop is picrotoxin-sensitive (figure 3.15), it may therefore be GABA (Yagi and Sawaki, 1974; Sawaki and Yagi, 1976). However, should recurrent inhibition utilise a disynaptic pathway, the transmitter at the initial synapse could be excitatory. Microiontophoretic studies with a number of different hypothalamic peptides have shown that they can evoke both excitatory and inhibitory effects on central neurones, although the most frequently observed pattern is a decrease in excitability (Dyer and Dyball, 1974; Kawakami and Sakuma, 1974, 1976; Moss, 1977; Renaud *et al.*, 1975, 1976).

Recurrent facilitation

In recent studies of the rat tubero-infundibular system, Sawaki and Yagi have observed recurrent facilitation following median eminence stimulation (Sawaki and Yagi, 1976; Yagi and Sawaki, 1974) and many more cells were seen to be facilitated following the blockade of recent inhibition. In these experiments facilitation was abolished by treatment of the animals with α-methyl-*p*-tyrosine, or was not observed following pretreatment of the animals with reserpine, suggesting that catecholaminergic neurones may be involved in the pathways which mediate facilitation of tubero-infundibular cells during stimulation of the median eminence (figure 3.15).

Axon collaterals in the tubero-infundibular system

In his attempts to define the electrophysiology of the tubero-infundibular system, Renaud (1978*b*) has shown that some tubero-infundibular neurones exhibited unequivocal antidromic invasion (figure 3.16) not only from the median eminence but also from other sites within the brain; the anterior hypothalamic area, the medial preoptic area, thalamic midline structures in and around the nucleus medialis dorsalis and the amygdala (figure 3.17). In addition, evidence from cancellation experiments suggests that the axons of the tubero-infundibular neurones branch very close to their site of origin (see also Harris and Sanghera, 1974; Dyer *et al.*, 1976).

The existence of such widely distributed axon collaterals in a putative peptidergic system heightens speculation as to the role of these axon terminals in intrahypothalamic and extrahypothalamic brain regions. It is plausible that some of these connections are feedback circuits which govern activity in the tubero-infundibular system. Conversely these axon collaterals may simply inform other brain regions of the status of activity in the tubero-infundibular system and hence of the likely hormonal status of the adenohypophysis. In any event, the axonal arborisation patterns described above suggest that the cells of the tubero-infundibular system are involved in complex integrative tasks in neural function which involve reciprocal relations with many brain regions.

Of equal interest is the possibility that the neurotransmitter agent(s) or peptides contained in the central terminals of these tubero-infundibular cells are released at central synaptic sites outside the hypothalamus. Radioimmunoassay

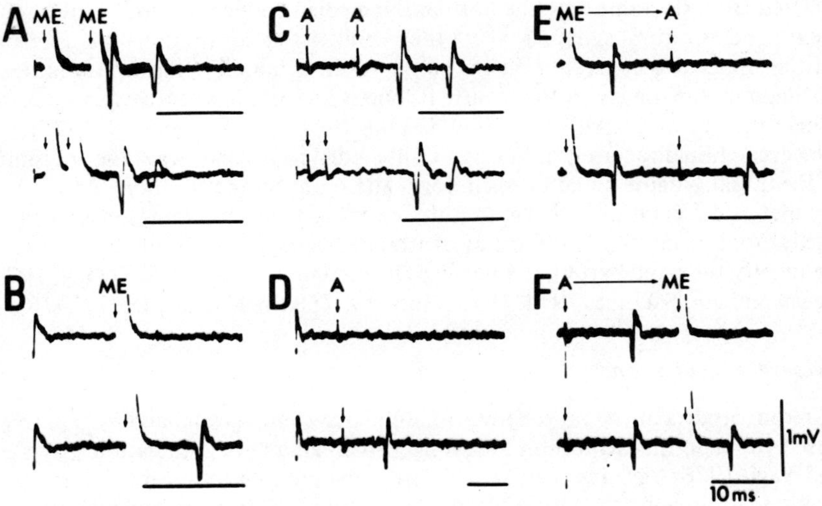

Figure 3.16 Axon collateral pathways in the tubero-infundibular system, illustrating axon branching to the median eminence (ME) and the amygdala (A). *A* and *C* high-frequency stimulation of the median eminence (ME) and amygdala (A), respectively, to show fractionation of the antidromic spike. *B* and *D* the cancellations between spontaneous and antidromic spikes from both median eminence and amygdala. *E* and *F* cancellation by antidromic invasion evoked by appropriately timed suprathreshold stimuli to the other branch of this neurone. (Reproduced with permission from Renaud, 1976*c*)

studies describe hypothalamic peptides in widespread areas of the brain and immunohistochemical studies show the presence of peptidergic neural pathways (Barry and Dubois, 1975; Barry *et al.*, 1974; Brownstein, 1977; Brownstein *et al.*, 1976; Dubois and Kolodziejczyk, 1975; Silverman, 1976). Numerous investigators have shown that the hypothalamic peptides exert profound behavioural effects through a direct action on brain tissues (Prange *et al.*, 1978; Severs and Daniels-Severs, 1973; De Wied *et al.*, 1975; Plotnikoff and Kastin, 1976; Wimersma Greidanus *et al.*, 1975). Subcellular fractionation studies have described hypothalamic peptides localised to synaptosomal fractions of brain (Epelbaum *et al.*, 1977; Ramirez *et al.*, 1975) and in certain instances high-affinity receptor binding for these peptides has been observed in brain tissue (Burt and Snyder, 1975). As we shall see later, the hypothalamic peptides TRH (thyrotrophin releasing hormone), LH-RH (luteinising hormone releasing hormone) and somatostatin (growth hormone release inhibiting hormone) appear to exert potent depressant effects on central neuronal activity when applied by microiontophoretic techniques (Dyer and Dyball, 1974; Kawakami and Sakuma, 1974, 1976; Moss, 1976, 1977; Renaud *et al.*, 1975, 1976). Thus there is indirect evidence to suggest that some peptides may in fact influence neuronal behaviour.

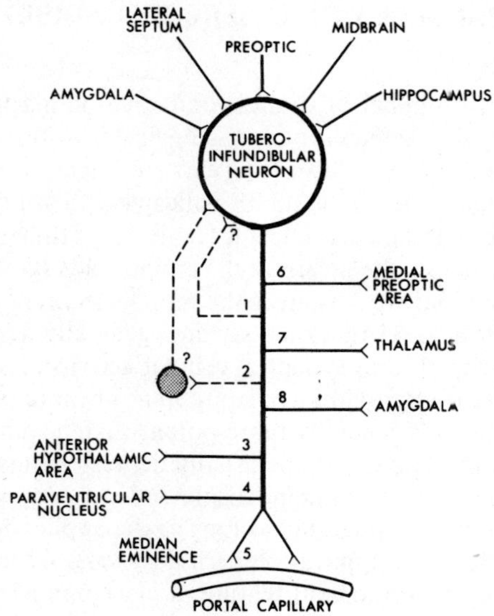

Figure 3.17 Summary of the collateral connections of mediobasal hypothalamic tubero-infundibular neurones based on neurophysiological data. The principal axon of a tubero-infundibular neurone is illustrated terminating on a median eminence portal capillary (No. 5). Other intrahypothalamic axon collaterals mediate recurrent inhibition either directly (1) or indirectly through an inhibitory interneurone (2), and may also reach the anterior hypothalamic area (3) or paraventricular nucleus (4). Extrahypothalamic axon collaterals reach the medial preoptic area (6), the midline thalamic nuclei (7) or amygdala (8). Afferents from the amygdala, preoptic area and hippocampus are also illustrated. (Reproduced with permission from Renaud, 1978*b*)

Afferent connections of tubero-infundibular neurones

Numerous stimulation and lesion experiments have demonstrated that extrahypothalamic regions, notably the amygdala, preoptic area and hippocampus, can influence patterns of adenohypophyseal secretion (Lichtensteiger and Keller, 1974; Zolovick, 1972). Renaud (1978*b*) has reviewed the electrophysiological evidence for synaptic connections between tubero-infundibular neurones and those brain areas known to modify different parameters of adenohypophyseal secretion. Indeed there appears to be some evidence that the amygdala can selectively modify the activity of tubero-infundibular neurones within the ventromedial nucleus. The hippocampus seems to modify cellular activity in the arcuate nucleus and periventricular region, and the anterior hypothalamic and medial preoptic areas modify the activity of certain tubero-infundibular neurones throughout the mediobasal hypothalamus (Dreifuss *et al.*, 1968; Murphy and Renaud, 1969; Renaud, 1976*c*,*d*, 1977*a*).

NEUROPHARMACOLOGY OF TUBERO-INFUNDIBULAR NEURONES

In comparison with the number of studies conducted on magnocellular neurones, there has been relatively little neuropharmacological investigation of tubero-infundibular neurones. However, over the years a number of putative transmitter agents have been tested by Renaud and his colleagues on antidromically identified cells of the ventromedial nucleus (table 3.1) (Renaud, 1975, 1976*e*). In the majority of instances the dicarboxylic amino acids (glutamate and aspartate) enhance unit activity, whereas the monocarboxylic amino acids (gamma aminobutyric acid (GABA) and glycine) generally depress either spontaneous or induced (chemical or synaptic) cellular activity. Preliminary results suggest that glutamate is approximately equipotent or more potent than aspartate. GABA on the other hand is generally more potent than glycine. Acetylcholine, noradrenaline, dopamine and histamine usually depress cellular activity. Cells whose activity is enhanced by histamine display one of two excitatory patterns: a brisk onset of excitation lasting only as long as the application, and a delayed excitation which outlasts the application period by several seconds.

Synaptic inhibition is a prominent feature of electrophysiological studies in the ventromedial hypothalamic nucleus. It is of interest to note that most putative neurotransmitter agents applied to these neurones also evoke depression of activity. Picrotoxin and bicuculline, applied by microiontophoresis with currents of 40–200 nA, promote a substantial shift to the right in the GABA log-current response plots, and much smaller shifts in the glycine log-current response plots. In certain instances these agents do produce a decrease in the duration of inhibition or convert an inhibition into a profound excitation. In a similar fashion strychnine iontophoresis (20–60 nA) more effectively antagonises glycine- rather than GABA-evoked depression in excitability, but has no effect on synaptic inhibition. However, intravenous picrotoxin and bicuculline, but not strychnine,

Table 3.1 Microiontophoresis in the Hypothalamic Ventromedial Nucleus*

Agent tested	*Influence on spike discharge pattern (no. of cells)*			*Effective current range* (nA)
	Increase	Decrease	No response	
L-Glutamate	80	2	18	12–80
L-Aspartate	15	0	3	15–50
GABA	0	40	2	0–20
Glycine	0	22	3	4–25
Histamine	12	42	5	3–25
Dopamine	0	23	4	20–50
Noradrenaline	0	15	2	10–40
Acetylcholine	3	10	2	5–30
TRH	0	46	20	10–30
LH-RH	0	13	10	10–30
GH-RIH (somatostatin)	0	28	10	5–80

*Reproduced with permission from Renaud (1978*a*)

decrease the duration of synaptic inhibition evoked from various sources (Pittman *et al.*, 1977), suggesting GABA rather than glycine may be an inhibitory synaptic transmitter in this nucleus.

Table 3.1 also shows that TRH, LH-RH and GH-RIH or somatostatin are also effective depressants of ventromedial nucleus neuronal activity. These responses are not antagonised by picrotoxin or bicuculline, and have often been associated with an increase in size of the extracellular action potential. Several analogues of these peptides have been examined on hypothalamic and other central neurones and the 3-methyl-histidine TRH analogue is usually more potent than native TRH.

Monoamines

The importance of amines has been reviewed by Wurtman (1970, 1971). Moss and his colleagues, reporting on the sensitivity of tubero-infundibular neurones in the arcuate nucleus to monoamines have noted that tubero-infundibular cells whose activity is enhanced during microiontophoretic application of noradrenaline are either unresponsive or decrease their firing frequency during dopamine microiontophoresis (Moss *et al.*, 1975). The reverse pattern applies to tubero-infundibular cells whose activity is enhanced by dopamine. Furthermore most of the tubero-infundibular neurones in the arcuate nucleus appear to be unresponsive to microiontophoretic application of the hypothalamic peptide TRH (Moss, 1976, 1977; Moss *et al.*, 1975). This contrasts rather sharply with the sometimes striking effect these peptides have on other hypothalamic or central neurones (Dyer and Dyball, 1974; Kawakami and Sakuma, 1974, 1976; Renaud *et al.*, 1975, 1976).

Peptides (TRH and LH–RH)

Since Renaud (1977*b*) has recently reviewed the wealth of evidence which shows that TRH and LH–RH are localised in both the median eminence and other regions of the central nervous systems and that they affect animal behaviour, only the results obtained by microiontophoresis will be considered here.

Although a number of studies have shown microiontophoretic application of TRH alters the excitability of a certain percentage of neurones in the cuneate nucleus, cerebral and cerebellar cortices, hypothalamus and preoptic areas (Dyer and Dyball, 1974; Moss, 1976, 1977; Renaud and Martin, 1975; Renaud *et al.*, 1975), Renaud and his colleagues have tended to stress the great frequency with which TRH-sensitive neurones are encountered in the hypothalamus and how its action here is predominantly inhibitory (figure 3.18). This inhibitory action of TRH may well be quite specific since it is unaffected by the amino acid antagonist bicuculline and strychnine (Renaud *et al.*, 1975). TRH, however, has been shown to enhance the excitability of a small percentage of central neurones (Dyer and Dyball, 1974; Moss, 1977) where the observed action is consistent with a direct effect on the neurones under study. It is of interest to note that Moss found TRH to have little effect on identified infundibular neurones and to excite predominantly the neurones of the arcuate and ventromedial region which

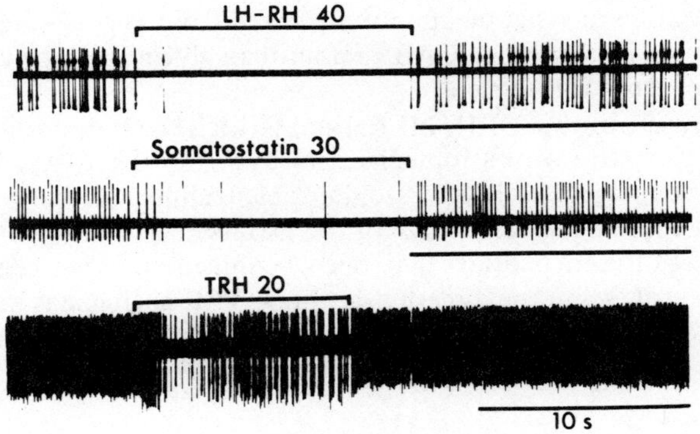

Figure 3.18 Representative oscillograph traces to show the depressant effects of microiontophoretic application of luteinising hormone-releasing hormone (LH-RH), cyclic somatostatin, and thyrotrophin-releasing hormone (TRH) on the frequency of spike discharges recorded from three separate neurones. The cell in the top trace was located in the ventromedial nucleus, the cell in the middle trace was recorded from the parietal cortex, and the cell in the lower trace was recorded from the cerebellar cortex of pentobarbitone-anaesthetised Sprague–Dawley rats. The numbers above each bar indicate the microiontophoretic current applied in nanoamperes. The time calibration (10s) is represented by the horizontal bar below each trace. (Reproduced with permission from Wilber *et al.*, 1976)

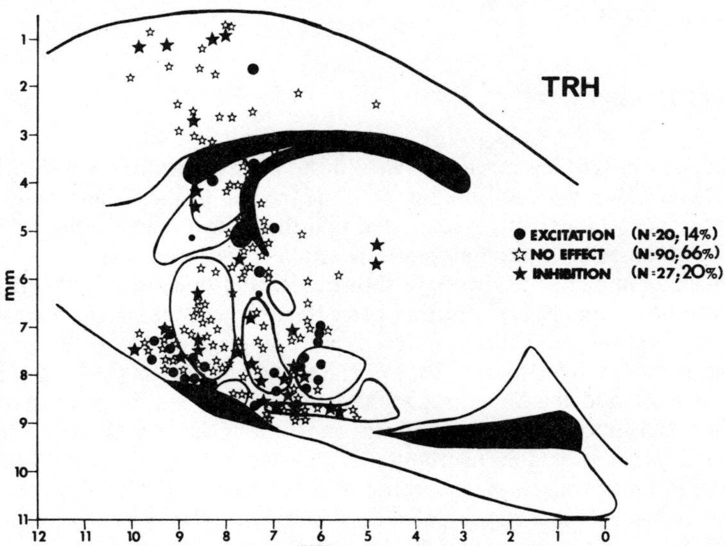

Figure 3.19 Schematic diagram of a sagittal section through the rat brain to show the localisation of excitatory, non-responsive and inhibitory effects of thyrotrophin-releasing hormone (TRH) applied by microiontophoresis to single neurones. (Reproduced with permission from Moss, 1977)

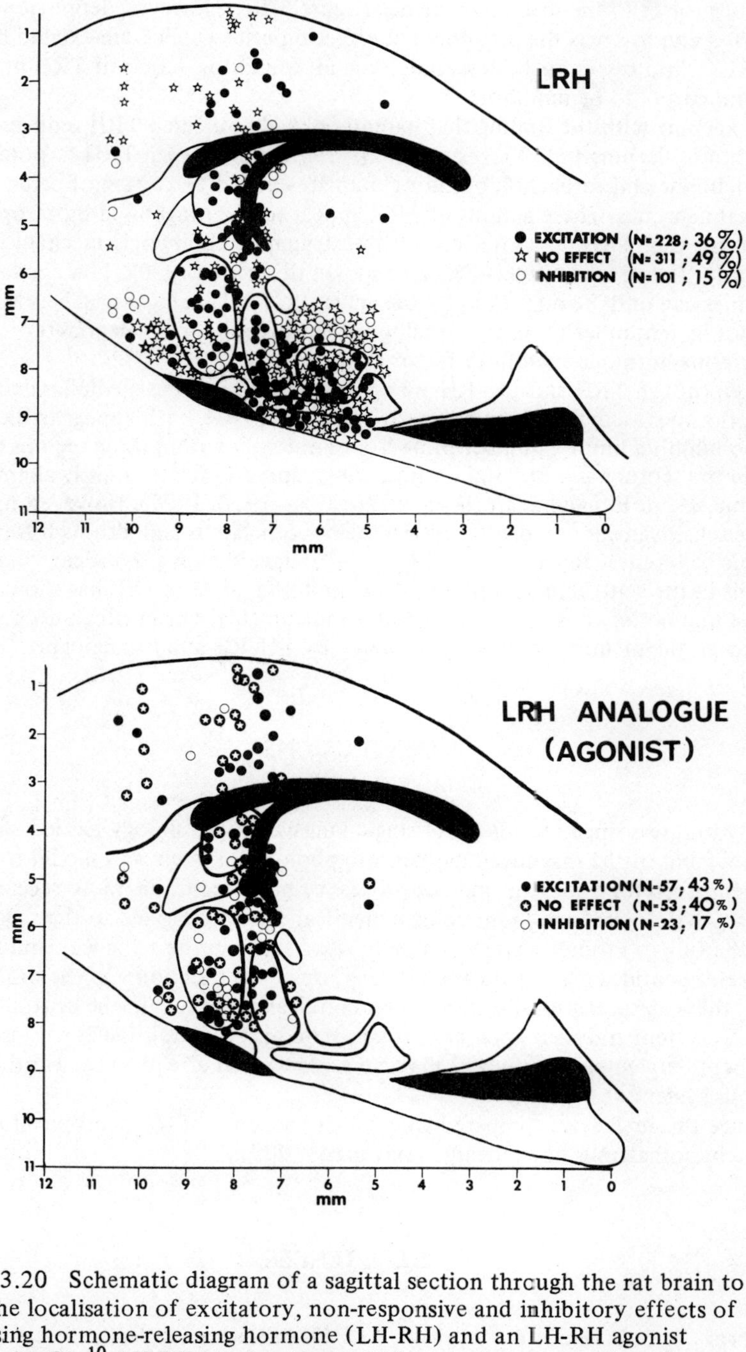

Figure 3.20 Schematic diagram of a sagittal section through the rat brain to show the localisation of excitatory, non-responsive and inhibitory effects of luteinising hormone-releasing hormone (LH-RH) and an LH-RH agonist D-ala^6, des-Gly^{10}-LH-RH ethylamide applied to single neurones by microiontophoresis. (Reproduced with permission from Moss, 1977)

did not project to the median eminence (figure 3.19). However, depressions also occurred and this was the predominant effect on cells which were excited by LH–RH. Curiously enough Moss, like Renaud, found the action of TRH in the cerebral cortex to be inhibitory.

In keeping with the finding that parenterally administered TRH reduces the barbiturate sleeping time Yarbrough *et al.* (1978) have shown TRH to potentiate acetylcholine and carbachol, but not glutamate-evoked excitation of cerebral cortical neurones. These actions of TRH appear to be antagonised by atropine, suggesting that TRH may interfere with postsynaptic cholinergic mechanisms. Recently Yarbrough *et al.* (1978) have shown this action of TRH on cortical neurones can only be mimicked by the analeptic TRH analogues such as MK-771 and not by a number of inactive analogues, somatostatin or melanocyte-stimulating hormone inhibitory factor (MIF).

Even though TRH and LH–RH are confined mainly to the mediobasal and preoptic areas of the hypothalamus (Brownstein, 1977), both appear to excite and to inhibit a limited number of neurones not only within these regions but also in the septum and cerebral cortices (Dyer and Dyball, 1974; Kawakami and Sakuma, 1974; Renaud *et al.*, 1975, 1976; Moss, 1976, 1977). However, most of the sensitive cells are located within the hypothalamus and although Renaud and his colleagues (Renaud *et al.*, 1975, 1976) have shown a significant number of cells in the ventromedial nucleus to be inhibited, Moss (1977) has shown that a large number of cells of the arcuate and ventromedial nuclei which do not project to the median eminence are excited by LH–RH and its analogues (figure 3.20).

CONCLUSIONS

Steady progress since the advent of single-unit electrophysiology has led to the development of the magnocellular neurohypophyseal system as a model for the neurosecretory neurone and the process of neurosecretion. More recent electrophysiological and immunohistochemical evidence has led to the suggestion that the tubero-infundibular system may also behave in the same way and the release of peptides may modulate adenohypophyseal function. On the other hand, the widespread localisation of certain releasing factors in the central nervous system and their apparent influence on many neural tissues suggests that peptidergic neural networks may have widespread effects in the central nervous system.

Since this review was prepared an extensive account of the neurophysiology of the hypothalamus has appeared (Hayward, 1977).

REFERENCES

Abrahams, V. C., Koelle, G. B., and Smart, P. (1957). Histochemical demonstration of cholinesterases in the hypothalamus of the dog. *J. Physiol. (Lond.)*, **139**, 137–144

Arnauld, E., Dufy, B. and Vincent, J. D. (1975). Hypothalamic supraoptic neurones:

rates and patterns of action potential firing during water deprivation in the unanaesthetised monkey. *Brain Res.*, **100**, 315–325

Arnauld, E., Vincent, J. D. and Dreifuss, J. J. (1974). Firing patterns of hypothalamic neurones during water deprivation in monkeys. *Science*, **185**, 535–537

Barker, J. L., Crayton, J. W. and Nicoll, R. A. (1971*a*). Supraoptic neurosecretory cells: adrenergic and cholinergic sensitivity. *Science*, **171**, 208–210

Barker, J. L., Crayton, J. N. and Nicoll, R. A. (1971*b*). Noradrenaline and acetylcholine responses of supraoptic neurosecretory cells. *J. Physiol. (Lond.)*, **218**, 19–32

Barker, J. L., Crayton, J. W. and Nicoll, R. A. (1971*c*). Antidromic and orthodromic responses of paraventricular and supraoptic neurosecretory cells. *Brain Res.*, **33**, 353–366

Barry, J. and Dubois, M. P. (1975). Immunofluorescence study of LRF- producing neurones in the cat and the dog. *Neuroendocrinology*, **18**, 290–298.

Barry, J., Dubois, M. P. and Carette, B. (1974). Immunofluorescence study of the preoptic-infundibular LRF neurosecretory pathway in the normal, castrated or testosterone-treated male guinea pig. *Endocrinology*, **95**, 1416–1423

Bennett, J. P. Jr., Arregui, A. and Snyder, S. H. (1976). Angiotensin II as a possible mammalian central neurotransmitter: synaptic neurochemistry in normal mammalian and Huntington chorea brain tissue. *Neurosci. Abstr.*, **2**, 775.

Bennett, C. T. and Pert, A. (1974). Antidiuresis produced by injections of histamine into the cat supraoptic nucleus. *Brain Res.*, **78**, 151–156

Bisset, G. W. (1968). The milk-ejection reflex and the action of oxytocin, vasopressin and synthetic analogues on the mammary gland. Neurohypophyseal hormones and similar polypeptides. In B. Berde (ed.), *Handbook of Experimental Pharmacology*, Chapt. 23, Springer-verlag, Berlin, 475–544

Bisset, G. W. (1976). 'Neurohypophyseal hormones'. In J. A. Parsons (ed.), *Peptide Hormones,* The Macmillan Press Ltd., London, 145–177

Blackwell, R. E. and Guillemin, R. (1973). Hypothalamic control of adenohypophyseal secretions. *Ann. Rev. Physiol.*, **35**, 357–370

Bloom, F. E., Oliver, A. P. and Salmoiraghi, G. C. (1963). The responsiveness of individual hypothalamic neurons to microelectrophoretically administered endogenous amines. *Int. J. Neuropharmac.*, **2**, 181–193

Brawer, J. (1972). The fine structure of the ependymal tanycytes at the level of the arcuate nucleus. *J. Comp. Neurol.*, **145**, 25–42

Brawer, J. and Sonnenschein, C. (1975). Cytopathological effects of oestradiol on the arcuate nucleus of the female rat. A possible mechanism for pituitary tumorigenesis. *Am. J. Anat.*, **144**, 57–88

Brimble, M. J. and Dyball, R. E. J. (1977). Characterisation of the responses of oxytocin and vasopressin-secreting neurones in the supraoptic nucleus to osmotic stimulation. *J. Physiol. (Lond.)*, **271**, 253–271

Brownstein, M. J. (1977). Neurotransmitters and hypothalamic hormones in the central nervous system. *Fed. Proc.*, **36**, 1960–1963

Brownstein, M. J., Palkovits, M., Saavedra, J. M. and Kizer, J. S. (1976). Distribution of hypothalamic hormones and neurotransmitters within the diencephalon. In L. Martini and W. F. Ganong (eds.), *Frontiers in Neuroendocrinology*, Raven Press, New York 1–23

Brownstein, M. J., Saavedra, J. M., Axelrod, J., Zeman, G. H. and Carpenter, D. O. (1974*a*). Coexistence of several putative neurotransmitters in single identified neurones of aplysia. *Proc. Nat. Acad. Sci.*, **71**, 4662–4655

Brownstein, M. J., Saavedra, J. M., Palkovits, M. and Axelrod, J. (1974*b*). Histamine content of hypothalamic nuclei of the rat. *Brain Res.*, **77**, 151–156

Buggy, J., Fisher, A. E., Hoffman, W. E., Johnson, A. K. and Phillips, M. I. (1975). Ventricular obstruction: effect of drinking induced by intracranial angiotensin. *Science*, **190**, 72–74

Burnstock, G. (1976). Do some nerve cells release more than one transmitter? *Neuroscience*, **1**, 239–248.
Burt, D. R., and Snyder, S. H. (1975). Thyrotrophin releasing hormone (TRH)–apparent receptor binding in rat brain membranes. *Brain Res.*, **93**, 309–328
Cajal, S. R. (1911). Histologie du Système Nerveux de l'Homme et des Vertèbres, Vol. 2, Naloine, Paris
Christ, J. F. (1966). Nerve supply, blood supply and cytology of the neurohypophysis. In G. W. Harris and B. T. Donovan (eds.), *The Pituitary Gland*, University of California Press, Berkeley, Calif., 62–130
Crawford, J. M. and Curtis, D. R. (1966). Pharmacological studies on feline Betz cells. *J. Physiol. (Lond.)*, **186**, 121–138
Cross, B. A., Dyball, R. E. J., Dyer, R. G., Jones, C. W., Lincoln, D. W., Morris, J. F. and Pickering, B. T. (1975). Endocrine neurones. *Recent Prog. Hormone Res.*, **31**, 243–294
Cross, B. A. and Green, J. D. (1959). Activity of single neurones in the hypothalamus: effect of osmotic and other stimuli. *J. Physiol. (Lond.)*, **148**, 554–569
Cross, B. A., Moss, R. L. and Urban, I. (1971). Effect of iontophoretic application of acetylcholine and noradrenaline to antidromically identified paraventricular neurones. *J. Physiol. (Lond.)*, **214**, 28–30P
Cross, B. A. and Silver, I. A. (1966). Electrophysiological studies on the hypothalamus. *Brit. Med. Bull.*, **22**, 254–260
Curtis, D. R. and Eccles, R. M. (1958). The excitation of Renshaw cells by pharmacological agents applied electrophoretically. *J. Neurophysiol.*, **141**, 435–445
Curtis, D. R. and Ryall, R. W. (1966*a*). The excitation of Renshaw cells by cholinomimetics. *Expl. Brain Res.*, **2**, 49–65
Curtis, D. R. and Ryall, R. W. (1966*b*). The acetylcholine receptors of Renshaw cells. *Expl. Brain Res.*, **2**, 66–80
Dale, H. A. (1935). Pharmacology and nerve endings. *Proc. R. Soc. Med.*, **28**, 319–332
De Wied, D., Witter, A. and Greven, H. M. (1975). Behaviourally active ACTH analogues. *Biochemical Pharmacology*, **24**, 1463–1468
Dingledine, R. and Kelly, J. S. (1978). Cholinergic processes at synaptic junctions. In J. D. Feldman, N. B. Gilula and J. D. Pitts (eds.), *Intercellular Junctions and Synapses in Development*, Chapman and Hall, London, 141–179
Dreifuss, J. J., Harris, M. C. and Tribollett, E. (1976*a*). Excitation of phasically firing hypothalamic supraoptic neurones by carotid occlusion in rats. *J. Physiol. (Lond.)*, **257**, 337–354
Dreifuss, J. J., Kalnins, I., Kelly, J. S. and Ruf, K. B. (1971). Action potentials and release of neurohypophyseal hormones *in vitro*. *J. Physiol. (Lond.)*, **215**, 805–817
Dreifuss, J. J. and Kelly, J. S. (1970). Excitation of identified supraoptic neurones by the iontophoretic application of acetylcholine. *J. Physiol. (Lond.)*, **210**, 170–172P
Dreifuss, J. J. and Kelly, J. S. (1972*a*). Recurrent inhibition of antidromically identified rat supraoptic neurones. *J. Physiol. (Lond.)*, **220**, 87–103
Dreifuss, J. J. and Kelly, J. S. (1972*b*). The activity of identified supraoptic neurones and their responses to acetylcholine applied by iontophoresis. *J. Physiol. (Lond.)*, **220**, 105–118
Dreifuss, J. J., Murphy, J. T. and Gloor, P. (1968). Contrasting effects of two identified amygdaloid efferent pathways on single hypothalamic neurones. *J. Neurophysiol.*, **31**, 237-248
Dreifuss, J. J., Nordmann, J. J. and Vincent, J.-D. (1973). Recurrent inhibition of supraoptic neurosecretory cells in Brattleboro rats. *J. Physiol. (Lond.)*, **237**, 25–27P
Dreifuss, J. J., Tribollet, E. and Baertschi, A. J. (1976*b*). Excitation of supraoptic neurones by vaginal distention in lactating rats: correlation with neurohypophyseal hormone release. *Brain Res.*, **113**, 600–605
Dubois, M. P. and Kolodziejczyk, E. (1975). Centre hypothalamiques due rat secretant

la somatostatine: reparition des pericaryons en 2 systems magno et parvocellulaires (études immunocytologiques). *C. R. Acad. Sci. Paris*, **281**, 1737–1740

Dufy, B., Dufy-Barbe, L. and Poulain, D. (1974). Gonadotropin release in relation to electrical activity in hypothalamic neurones. *J. Neural Transmission*, **35**, 47–52

Dyball, R. E. J. (1971). Oxytocin and ADH secretion in relation to electrical activity on antidromically identified supraoptic and paraventricular units. *J. Physiol. (Lond.)*, **214**, 245–256

Dyball, R. E. J. (1974). Single unit activity in the hypothalamo-neurohypophyseal system system of Brattleboro rats. *J. Endocrinol.*, **60**, 135–143

Dyball, R. E. J. and Koizumi, K. (1969). Electrical activity in the supraoptic and paraventricular nuclei associated with neurohypophyseal hormone release. *J. Physiol. (Lond.)*, **201**, 711–722

Dyer, R. G. and Dyball, R. E. J. (1974). Evidence for a direct effect of LRF and TRF on single unit activity in the rostral hypothalamus. *Nature*, **252**, 486–488

Dyer, R. G., Macleod, D. N. K. and Ellendorf, F. (1976). Electrophysiological evidence for sexual dimorphism and synaptic convergence in the preoptic and anterior hypothalamic areas of the rat. *Proc. R. Soc. Lond. B.*, **193**, 421–440

Eccles, J. C., Eccles, R. M. and Fatt, P. (1956). Pharmacological investigations on a central synapse operated by acetylcholine. *J. Physiol. (Lond.)*, **131**, 154–159

Eccles, J. C., Fatt, P. and Koketsu, K. (1954). Cholinergic and inhibitory synapses in a pathway from motor-axon collaterals to motoneurones. *J. Physiol. (Lond.)*, **126**, 524–562

Epelbaum, J., Brazeau, P., Tsang, D. Brawer, J. and Martin, J. B. (1977). Subcellular distribution of radioimmunoassayable somatostatin in rat brain. *Brain Res.*, **126**, 309–324

Epstein, A. N., Fitzsimons, J. T. and Rolls, B. J. (1970). Drinking induced by injection of angiotensin into the brain of the rat. *J. Physiol. (Lond.)*. **210**, 457–474

Feldberg, W. and Vogt, M. (1948). Acetylcholine synthesis in different regions of the central nervous system. *J. Physiol. (Lond.)*, **107**, 372–381

Felix, D. (1976). Peptide and acetylcholine action of neurones of the cat subfornical organ. *Naunyn-Schmiedeberg's Arch. Pharmacol.*, **292**, 15–20

Felix, D. and Akert, K. (1974). The effect of angiotensin II on neurones of the cat subfornical organ. *Brain Res.*, **76**, 350–353

Felix, D. and Phillips, M. I. (1978). Effects of angiotensin II on central neurones. In R. W. Ryall and J. S. Kelly (eds.). *Iontophoresis and Transmitter Mechanisms in the Mammalian Central Nervous System,* Elsevier/North Holland, Amsterdam, 104–106

Freund-Mercier, M. J. and Richard, P. H. (1977). Spontaneous and reflex activity of paraventricular nucleus units in cycling and lactating rats. *Brain Res.*, **130**, 505–520

Fuxe, K., Ganten, D., Hökfelt, T. and Bome, P. (1976). Immunohistochemical evidence for the existence of angiotensin II containing nerve terminals in the brain and spinal cord of the rat. *Neurosci. Lett.*, **2**, 229–234

Fuxe, K. and Hökfelt, T. (1967). The influence of central catecholamine neurones in the hormone secretion from the anterior and posterior pituitary. In F. Stutinsky (ed.). *Neurosecretion,* Springer, Berlin

Ganten, D., Hutchinson, J. S., Schelling, J. P., Ganten, U. and Fischer, H. (1976). The iso-renin angiotensin systems in extra-renal tissue. *Clin. Exp. Pharm. Physiol.*, **2**, 103–126

Garbarg, M., Barbin, G., Feger, J. and Schwartz, J.-C. (1974). Histaminergic pathway in rat brain evidenced by lesions of the medial forebrain bundle. *Science*, **186**, 833–835

Ginsburg, M. (1968). Production, release, transportation and elimination of the neurohypophyseal hormones. In B. Berde (ed.), *Handbook of Experimental Pharmacology*, Springer-Verlag, Berlin, 286–371

Ginsburg, M. and Brown, L. M. (1956). Effect of anaesthetics and haemorrhage on the release of neurohypophyseal antidiuretic hormone. *Brit. J. Pharmacol. Chemother.*, **11**, 236–244

Gordon, G. and Jukes, M. G. M. (1964). Descending influences on the exteroceptive organizations of the cat's gracile nucleus. *J. Physiol. (Lond.)*, **173**, 291–319

Green, M. D., Simon, M. L. and Lomax, P. (1975). Histamine as a neurotransmitter in the central thermoregulatory pathways of the rat. *Proc. West. Pharmacol. Soc.*, **18**, 110–113

Gronan, R. J. and York, D. H. (1976). Effect of angiotensin on cells in the preoptic area of rats. *Neurosci Abst.*, **2**, 426

Haas, H. L. (1974). Histamine: action on single hypothalamic neurones. *Brain Res.*, **76**, 363–366

Haas, H. L., and Wolf, P. (1977). Central actions of histamine: microelectrophoretic studies. *Brain Res.*, **122**, 269–279

Haas, H. L., Wolf, P. and Nussbaumer, J. C. (1975). Histamine: action on supraoptic and other hypothalamic neurones of the cat. *Brain Res.*, **88**, 166–170

Halász, B. (1969). The endocrine effects of isolation of the hypothalamus from the rest of the brain. In W. F. Ganong and L. Martini (eds.), *Frontiers in Neuroendocrinology*, Oxford University Press, New York, 307–342

Harris, G. W. (1955). *Neural Control of Pituitary Gland.* Edward Arnold, London

Harris, G. W., Manabe, Y. and Ruf, K. B. (1969). A study of the parameters of electrical stimulation of unmyelinated fibers in the pituitary stalk. *J. Physiol. (Lond.)*, **203**, 67–81

Harris, M. C. and Sanghera, M. (1974). Projection of medial basal hypothalamic neurones to the preoptic anterior hypothalamic areas and the paraventricular nucleus in the rat. *Brain Res.*, **81**, 401–411

Haymaker, W., Anderson, E. and Nauta, W. J. H. (1969). *The Hypothalamus*, Thomas, Springfield

Hayward, J. N. (1975). Neural control of the posterior pituitary. *Ann. Rev. Physiol.*, **37**, 191–210

Hayward, J. (1977). Functional and morphological aspects of hypothalamic neurones. *Physiol. Rev.*, **57**, 574–658

Hayward, J. N. and Jennings, D. P. (1973). Activity of magnocellular neuroendocrine cells in the hypothalamus of unanaesthetised monkeys. Part II: Osmosensitivity of functional cell types in the supraoptic nucleus and the internuclear zone. *J. Physiol. (Lond.)*, **232**, 545–572

Ishida, A. (1970). The oxytocin and the compound action potential evoked by electrical stimulation of the isolated neurohypophysis of the rat. *Jap. J. Physiol.*, **20**, 84–96

Kandel, E. R. (1964). Electrical properties of hypothalamic neuroendocrine cells. *J. gen. Physiol.*, **47**, 691–717

Kawakami, M. and Sakuma, Y. (1974). Responses of hypothalamic neurons to the micro-iontophoresis of LH-RH, LH and FSH under various levels of circulating ovarian hormones. *Neuroendocrinology*, **15**, 290–307

Kawakami, M. and Sakuma, Y. (1976). Electrophysiological evidence for possible participation of periventricular neurons in anterior pituitary regulation. *Brain Res.*, **101**, 79–94

Kelly, J. S. (1975). Microiontophoretic application of drugs onto single neurons. In L. L. Iversen, S. D. Iversen and S. H. Snyder (eds.), *Handbook of Psychopharmacology* Vol 2, Plenum Press, New York and London, 29–67

Kelly, J. S. and Dreifuss, J. J. (1970). Antidromic inhibition of identified rat supraoptic neurones. *Brain Res.*, **22**, 406–409

Kelly, J. S., Simmonds, M. A. and Straughan, D. W. (1975). Microelectrode techniques. In P. B. Bradley (ed.), *Methods in Brain Research*, Wiley, New York, 333–337

Koizumi, K., Ishikawa, T. and McC. Brooks, C. (1973). The existence of facilitatory axon collaterals in neurosecretory cells of the hypothalamus. *Brain. Res.*, **63**, 408–413

Koizumi, K. and Yamashita, H. (1972). Studies of antidromically identified neurosecretory cells of the hypothalamus by intracellular and extracellular recordings. *J. Physiol. (Lond.)*, **221**, 683–705

Krnjevic, K. and Phillis, J. W. (1963*a*). Acetylcholine-sensitive cells in the cerebral cortex *J. Physiol. (Lond.)*, **166**, 296–327

Krnjevic, K. and Phillis, J. W. (1963*b*). Pharmacological properties of acetylcholine-sensitive cells in the cerebral cortex. *J. Physiol. (Lond.)*, **166**, 328–350

Krnjevic, K., Pumain, R. and Renaud, L. (1971). The mechanism of excitation by acetylcholine in the cerebral cortex. *J. Physiol. (Lond.)*, **215**, 247–268

Leontovich, T. A. (1970). The neurons of the magnocellular neurosecretory nuclei of the dog's hypothalamus. *J. Hirnforsch.*, **II**, 499–517

Lěránth, Cs., Zaborszky, L., Marton, J. and Palkovits, M. (1975). Quantitative studies on the supraoptic nucleus in the rat. I. Synaptic organization. *Exp. Brain Res.*, **22**, 509–523

Lichtensteiger, W. and Keller, P. J. (1974). Tubero-infundibular dopamine neurons and the secretion of luteinizing hormone and prolactin: extrahypothalamic influences, interaction with cholinergic systems and the effect of urethane anesthesia, *Brain Res.*, **74**, 279–303

Lincoln, D. W. and Wakerley, J. B. (1974). Electrophysiological evidence for the activation of supraoptic neurones during the release of oxytocin. *J. Physiol. (Lond.)*, **242**, 533–554

Lincoln, D. W. and Wakerley, J. B. (1975). Factors governing the periodic activation of supraoptic and paraventricular neurosecretory cells during suckling in the rat. *J. Physiol. (Lond.)*, **250**, 443–461

Makara, G. B., Harris, M. C. and Spyer, K. M. (1972). Identification and distribution of tubero-infundibular neurones. *Brain Res.*, **40**, 283–290

Moss, R. L. (1976). Unit responses in preoptic and arcuate neurons related to anterior pituitary function. In: L. Martini and W. F. Ganong (eds.), *Frontiers in Neuroendocrinology*, Vol 4, Raven Press, New York, 95–128

Moss, R. L. (1977). Role of hypophysiotropic neurohormone in mediating neural and behavioural events. *Fed. Proc.*, **36**, 1978–1983

Moss, R. L., Dyball, R. E. J. and Cross, B. A. (1972*a*). Excitation of antidromically identified neurosecretory cells in the paraventricular nucleus by oxytocin applied iontophoretically. *Exp. Neurol.*, **34**, 95–102

Moss, R. L., Kelly, M. and Riskind, P. (1975). Tubero-infundibular neurons: dopaminergic and norepinephrinergic sensitivity. *Brain Res.*, **89**, 265–277

Moss, R. L., Urban, I. and Cross, B. A. (1972*b*). Microelectrophoresis of cholinergic and aminergic drugs on paraventricular neurons. *Am. J. Physiol.*, **223**, 310–318

Mroz, E. A., Brownstein, M. J. and Leeman, S. E. (1976). Evidence for substance P in the habenulo-interpeduncular tract. *Brain Res.*, **113**, 597–599

Murphy, J. T. and Renaud, L. P. (1969). Mechanisms of inhibition in the ventromedial nucleus of the hypothalamus. *J. Neurophysiol.*, **32**, 85–102

Negoro, H. and Holland, R. C. (1972). Inhibition of unit activity in the hypothalamic paraventricular nucleus following antidromic activation. *Brain Res.*, **42**, 385–402

Negoro, H., Visessuwan, S. and Holland, R. C. (1973). Inhibition and excitation of units in paraventricular nucleus after stimulation of the septum, amygdala and neurohypophysis. *Brain Res.*, **57**, 479–483

Nicoll, R. A. and Barker, J. L. (1971*a*). Excitation of supraoptic neurosecretory cells by angiotensin II. *Nature New Biol.*, **233**, 172–174

Nicoll, R. A. and Barker, J. L. (1971*b*). The pharmacology of recurrent inhibition in the supraoptic neurosecretory system. *Brain Res.*, **35**, 501–511

Novin, D., Sundsten, J. W., and Cross, B. A. (1970). Some properties of antidromically activated units in the paraventricular nucleus of the hypothalamus. *Expl. Neurol.*, **26**, 330–341

Olivecrona, H. (1957). Paraventricular nucleus and pituitary gland. *Acta Physiol. Scand.*, **40**, Suppl. 136, 1–178

Oomura, Y., Ono, T. and Ooyama, H. (1970). Inhibitory action of the amygdala on the central hypothalamic area in rats. *Nature*, **228**, 1108–1110

Oomura, Y., Ono, T., Sugimori, M. and Wayner, M. J. (1976). Acetylcholine, an inhibitory transmitter on the rat lateral hypothalamus. *Brain Res Bull.*, **1**, 151–153

Parry, H. B. and Livett, B. G. (1976). Neurophysin in the brain and pituitary gland of normal and scrapie-affected sheep. Part I: Its localization in the hypothalamus and neurohypophysis with particular reference to a new hypothalamic neurosecretory pathway to the median eminence. *Neuroscience,* **1**, 275–299

Phillips, M. I. and Felix, D. (1976). Specific angiotensin II receptive neurons in the cat subfornical organ. *Brain Res.,* **109**, 531–540

Phillis, J. W., Kostopoulos, G. K. and Odutola, A. (1975). On the specificity of histamine H_2-receptor antagonists in the rat cerebral cortex. *Canad. Physiol. Pharmacol.,* **53**, 1205–1209

Phillis, J. W. and Limacher, J. J. (1974). Excitation of cerebral cortical neurons by various polypeptides. *Exp. Neurol.*, **43**, 414–423

Phillis, J. W., Tebecis, A. K. and York, D. H. (1968). Histamine and some antihistamines: their actions on cerebral cortical neurones. *Brit. J. Pharmacol.*, **33**, 426–440

Pickford, M. (1939). The inhibitory effect of acetylcholine on diuresis in the dog and its pituitary transmission. *J. Physiol. (Lond.)*, **95**, 226–238

Pittman, Q. J., Blume, H. W., MacKenzie, B. W. and Renaud, L. P. (1977). GABA and glycine and synaptic inhibition in the hypothalamic ventromedial nucleus of the rat. *Canad. Physiol.*, 8, 56

Plotnikoff, N. P. and Kastin, A. J. (1976). Neuropharmacology of hypothalamic releasing factors. *Biochemical Pharmacology,* **25**, 363–365

Porter, J. C., Kameri, I. A. and Grazia, Y. R. (1971). Pituitary blood flow and portal vessels. In L. Martini and W. F. Ganong (eds.), *Frontiers in Neuroendocrinology*, Oxford University Press, Oxford, 145–175

Poulain, D. A., Wakerley, J. B. and Dyball, R. E. J. (1977). Electrophysiological differentiation of oxytocin- and vasopressin-secreting neurones. *Proc. Roy. Soc. Lond. B.,* **196**, 367–384

Prange, A. J. Jr., Nemeroff, C. B., Lipton, M. A., Breese, G. R. and Wilson, I. C. (1978). Peptides and the central nervous system. In L. L. Iversen, S. D. Iversen and S. H. Snyder (eds.), *Handbook of Psychopharmacology*, Vol. 13, Plenum Press, New York, to be published

Ramirez, V. D., Gautron, J. P., Epelbaum, J., Pattou, E., Zamura, A. and Kordon, C. (1975). Distribution of LH-RH in subcellular fractions of the basomedial hypothalamus. *Mol. Cell. Endocrinol.*, **3**, 339–350

Renaud, L. P. (1975). Electrophysiological evidence to suggest that hypothalamic releasing (inhibiting) peptides may be liberated from nerve terminals in the CNS. *Neurosci. Abstr.*, **1**, 441

Renaud, L. P. (1976*a*). Tuberoinfundibular neurons in the basomedial hypothalamus of the rat: electrophysiological evidence for axon collaterals to hypothalamic and extrahypothalamic areas. *Brain Res.,* **105**, 59–72

Renaud, L. P. (1976*b*). Tuberoinfundibular neurons: electrophysiological studies on afferent and efferent connections. *The Physiologist,* **19**, 388

Renaud, L. P. (1976*c*). Influence of amygdala stimulation on the activity of identified tuberoinfundibular neurons in the rat hypothalamus. *J. Physiol. (Lond.).,* **260** 237–252

Renaud, L. P. (1976*d*). An electrophysiological study of amygdalo-hypothalamic projections to the ventromedial nucleus of the rat. *Brain Res.,* **105**, 45–58

Renaud, L. P. (1976*e*). Response of identified ventromedial hypothalamic nucleus neurons to putative neurotransmitters applied by microiontophoresis. *Brit. J. Pharmacol.,* **55**, 277–278P

Renaud, L. P. (1977*a*). Influence of medial preoptic-anterior hypothalamic area stimulation on the excitability of mediobasal hypothalamic neurones in the rat. *J. Physiol. (Lond.)*, **264**, 541–564

Renaud, L. P. (1977*b*). TRH, LH-RH and somatostatin: distribution and physiological action in neural tissue. In W. M. Cowan and J. A. Ferendelli (eds.), *Neuroscience Symposia*, Vol. 2, Raven Press, New York, 265–290

Renaud, L. P. (1978*a*). Influence of peptides and putative neurotransmitters on the excitability of identified hypothalamic neurons. In R. W. Ryall and J. S. Kelly (eds.), *Iontophoresis and Transmitter Mechanisms in the Mammalian Central Nervous System.* Elsevier, Amsterdam, 127–129

Renaud, L. P. (1978*b*). Neurophysiological organization of the Endocrine Hypothalamus. Association for Research in Nervous and Mental Disease Research Publication, Vol. 56, in S. Reichlin, R. J. Baldessarini and J. B. Martin (eds.), *The Hypothalamus,* Raven Press, New York, 269–301 (in press)

Renaud, L. P. and Martin, J. B. (1975). Electrophysiological studies of connections of hypothalamic ventromedial nucleus neurons in rat–evidence for a role in neuroendocrine regulation. *Brain Res.*, **93**, 145–151

Renaud, L. P., Martin, J. B. and Brazeau, P. (1975). Depressant action of TRH, LH-RH and somatostatin on activity of central neurons. *Nature*, **255**, 233–235

Renaud, L. P., Martin, J. B. and Brazeau, P. (1976). Hypothalamic releasing factors: physiological evidence for a regulatory action on central neurons and pathways for their distribution in brain. *Pharmacol. Biochem. Behav.*, **5**, Suppl. 1, 171–178

Roberts, J. S. (1971). Progesterone-inhibition of oxytocin release during vaginal distention: evidence for a central site of action. *Endocrinology,* **89**, 1137–1141

Roberts, J. S. (1973). Functional integrity of the oxytocin-release reflex in goats: dependence on oestrogen. *Endocrinology*, **93**, 1309–1314

Roberts, J. S. (1975). Cyclical fluctuations in reflexive oxytocin release during the oestrous cycle of the goat. *Biol. Reproduct.*, **13**, 314–317

Roberts, J. S. and Share, L. (1969). Effects of progesterone and oestrogen on blood levels of oxytocin during vaginal distention. *Endocrinology*, **84**, 1076–1081

Roberts, J. S. and Share, L. (1970). Inhibition by progesterone of oxytocin secretion during vaginal stimulation. *Endocrinology,* **87**, 812–815

Sakai, K. K., Marks, B. H., George, J. M. and Koestner, A. (1974). Specific angiotensin II receptors in organ-cultured canine supraoptic nucleus cells. *Life Sci.*, **14**, 1337–1344

Sawaki, Y. and Yagi, K. (1973). Electrophysiological identification of cell bodies of the tuberoinfundibular neurons in the rat. *J. Physiol. (Lond.)*, **230**, 75–85

Sawaki, Y. and Yagi, K. (1976). Inhibition and facilitation of antidromically identified tuberoinfundibular neurons following stimulation of the median eminence in the rat. *J. Physiol. (Lond.)*, **260**, 447–460

Schally, A. V., Arimura, A. and Kastin, A. J. (1973). Hypothalamic regulatory hormones. *Science,* **179**, 341–350

Schwartz, J.-C. (1975). Histamine as a transmitter in brain. *Life Sci.*, **17**, 503–528

Severs, W. B. and Daniels-Severs, A. E. (1973). Effects of angiotensin on the central nervous system. *Pharmacol. Rev.*, **25**, 415–449

Shute, C. C. D. (1970). Distribution of cholinesterase and cholinergic pathways. In L. Martini, M. Motta and F. Fraschini (eds.), *The Hypothalamus*, Academic Press, New York, 167–179

Silverman, A. J. (1976). Ultrastructural studies on the localization of neurohypophyseal hormones and their carrier proteins. *J. Histochem. Cytochem.*, **24**, 816–827

Simpson, J. B. and Routtenberg, A. (1973). Subfornical organ: site of drinking elicitation by angiotensin II. *Science,* **181**, 1172–1175

Sirett, N. E., McLean, A. S., Bray, J. J. and Hubbard, J. I. (1977). Distribution of angiotensin II receptors in rat brain. *Brain Res.*, **122**, 299–312

Spehlmann, R. (1963). Acetylcholine and prostigmine electrophoresis at visual cortex neurones. *J. Physiol. (Lond.)*, **26**, 127–139

Swanson, L. W. (1977). Immunohistochemical evidence for a neurophysin-containing autonomic pathway arising in the paraventricular nucleus of the hypothalamus. *Brain Res.*, **128**, 346–353

Szentágothai, J., Flerko, B., Mess, B. and Halász, B. (1968). Hypothalamic control of the anterior pituitary, Akademiai Kiado, Budapest

Wakerley, J. B. and Lincoln, D. W. (1973). The milk ejection reflex of the rat: a 20- to 40-fold acceleration in the firing of paraventricular neurones during oxytocin release. *J. Endocrinol.*, **57**, 477–493

Wakerley, J. B., Poulain, D. A., Dyball, R. E. J. and Cross, B. A. (1975). Activity of phasic neurosecretory cells during haemorrhage. *Nature*, **258** 82–84

Wayner, M. J., Ono, T. and Nolley, D. (1973). Effects of angiotensin II in central neurones. *Pharmac. Biochem. Behav.*, **1**, 679–691

Wilber, J. F., Montoya, E., Plotnikoff, N., White, W. F., Gendrich, N. P., Renaud, L. and Martin, J. B. (1976). Gonadotrophin-releasing hormone and thyrotrophin-releasing hormone: distribution and effects in the central nervous system. *Recent Prog. Hormone Res.*, **32**, 117–153

Wimersma Greidanus, Tj. B., Bohus, B. and de Wied, D. (1975). The role of vasopressin in memory processes. In W. H. Gispen, Tj. B. Van Wimersma Greidanus, B. Bohus and D. de Wied (eds.), *Progress in Brain Research*, Vol. 42, Elsevier, Amsterdam, 135–141

Wolf, P. and Monnier, M. (1973). Electroencephalographic, behavioural and visceral effects of intraventricular infusion of histamine in the rabbit. *Agents and Actions*, **3**, 196

Wurtmann, R. J. (1970). Neuroendocrine transducer cells in mammals. In F. O. Schmitt (ed.), *The Neurosciences Second Study Program*, Rockefeller University Press, New York, 530–538

Wurtman, R. J. (1971). Brain monoamines and endocrine function. *Neuroscience Res. Prog. Bull.*, 9, 171–297

Wuttke, W. (1974). Preoptic unit activity and gonadotropin release. *Exp. Brain Res.*, **19**, 205–216

Yagi, K., Azuma, T. and Matsuda, K. (1966). Neurosecretory cell: capable of conducting impulse in rats. *Science*, **154**, 778–779

Yagi, K. and Sawaki, Y. (1974). Recurrent inhibition and facilitation demonstration in the tuberoinfundibular system and effects of strychnine and picrotoxin. *Brain Res.*, **84**, 155–159

Yamashita, H., Koizumi, K. and Brooks, C. McC. (1970). Electrophysiological studies of neurosecretory cells in the cat hypothalamus. *Brain Res.*, **20**, 462–466

Yarbrough, G. G., Haubrich, D. R. and Schmidt, D. E. (1978). Thyrotrophin releasing hormone (TRH) and MK-771 interaction with CNS cholinergic mechanisms. In R. W. Ryall and J. S. Kelly (eds.), *Iontophoresis and Transmitter Mechanisms in the Mammalian Central Nervous System.* Elsevier, Amsterdam, 136–138

Zaborszky, L., Leranth, Cs., Makara, G. B. and Palkovits, M. (1975). Quantitative studies in the supraoptic nucleus in the rat. II. Afferent fibre connections. *Exp. Brain Res.*, **22**, 525–540

Zimmerman, E. A. (1976). Localization of neurosecretory peptides in neuroendocrine tissues. In F. Naftolin, K. J. Ryan and J. Davies (eds.), *Subcellular Mechanisms in Reproductive Neuroendocrinology*, Elsevier, Amsterdam, 81–108

Zimmerman, E. A. and Antunes, J. L. (1976). Organization of the hypothalamic-pituitary system: current concepts from immunohistochemical studies. *J. Histochem. Cytochem.*, **24**, 807–815

Zolovick, A. J. (1972). Effects of lesions and electrical stimulation of the amygdala on hypothalamic-hypophyseal regulation. In B. E. Eleftheriou (ed.), *The Neurobiology of the Amygdala*, Plenum Press, New York, 643–684

Discussion

Bisset (London)
During continuous suckling in the anaesthetised rat, milk ejection occurs only intermittently so that there is a pulsatile release of hormone apparently correlated with bursts of activity within the nerve cells. You suggested that there is a pattern of firing imposed on the cells. Does this imply some kind of gating mechanism?

Kelly
I think that this modulation occurs outside the nucleus. The firing of the cells comes in bursts and the hormone is not released in response to a single burst but requires a train in excess of 30 Hz. The nucleus may be modulated by non-neuronal mechanisms—that is, the effect is easier to produce after priming with oestrogen. However, the pulsatile release of the oxytocin is not related to the neuronal bursting. The bursting cell seems to be related to vasopressin not oxytocin release and the cycle is approximately once every five minutes, not once every twenty minutes as in the pulsatile milk release.

Morley (Alderley Park)
There are interesting cases where peptides are stimulatory in certain cell populations and inhibitory in others. Could peptides have biphasic dose-response curves and elicit stimulation followed by inhibition? Furthermore is it possible that either certain cells are more sensitive so that a given concentration of peptide is inhibitory or that the peptide is more accessible to certain cells?

Kelly
This cannot be answered until the problem is investigated using intracellular recordings. By analogy, cells inhibited by acetylcholine are all spontaneously active, firing at 20 to 50 per second. Whereas the cells which are excited (by acetylcholine) are usually slow firing. I think this also holds true for the catecholamines. Until we know more about the sodium and potassium channels it is difficult to be sure. Concerning the peptides, I think the slow action of substance P is a technical manifestation. The transport number of substance P is between 5000 to 50,000 times less than glutamate. Thus the concentration of substance P reaching the cell is not only low but it arrives very slowly and I think the slow effects are due to this rather than any other reason.

Wilson (Dublin)
Could you comment on the observation that the hypothalamus of the guinea-pig has the highest concentration of ascorbic acid of the whole brain, yet when the animal dies of scurvy there is no ascorbic acid in the adrenal cortex but 50 per cent remains in the hypothalamus. When guinea-pigs are given metrazole just prior to death the ascorbic acid falls a further 10 per cent. Could ascorbic

acid act as a hypothalamic modulator, particularly as it is known to affect both prostaglandin and adrenaline synthesis, and furthermore could its action as a modulator be affected by the diet?

Kelly

There are so many things concentrated within the hypothalamus and I haven't mentioned them. I can add very little to this topic except that taurine and particularly imidazole acetic acid are found in high concentrations in this region and when applied to neurones are extremely inhibitory. Whether or not this has relevance to an action for ascorbic acid is uncertain.

Halász

You mentioned that if you stimulated the median eminence antidromically you could find cell bodies in the hypothalamus. How do you know the stimulated area is related to the median eminence? Also there are some nerve cell bodies in the median eminence and it could be these that you are stimulating.

Kelly

What you say is true, but when the median eminence is exposed and the electrodes placed very carefully on it, it seems that the cells thought to project to the median eminence actually do so. This technique is preferable to placing the electrodes from above, thus relying on stereotaxic co-ordinates.

Pickford (Nottingham)

Is anything known about vasopressin output during the menstrual cycle in humans? I wonder whether vasopressin as well as oestrogens are responsible for pre-menstrual tension and for the relief when the diuresis is over?

Kelly

I don't know about humans but in rats the output of these hormones does follow the sexual cycle. Oestrogens will increase the number of action potentials seen in the supraoptic nucleus and also in the preoptic area. It has been shown that females have a higher spontaneous firing rate than males in the preoptic area, which varies during the sexual cycle. Amygdaloid stimulation is also different in males and females, males being more sensitive. This also seems to be a hormonal effect as the hormonal stimulation during the first four days of life determines what the input of the amygdala to these preoptic cells will be in the adult.

4
The hypothalamus and the pharmacology of thermoregulation

A. S. Milton*

INTRODUCTION

Homeothermic animals maintain a constant deep body temperature despite considerable variations in both their external and internal environments. To maintain deep body temperature, or perhaps more accurately body heat content, the heat production of the body must equal heat-loss. Thermoregulation is therefore simply the mechanism by which these parameters are regulated. The hypothalamus is normally regarded as being the centre for thermoregulation.

What then is the evidence that the hypothalamus is involved in thermoregulation? During the last century experiments were carried out in which various areas of the brain tissue were destroyed and, following destruction of the hypothalamic area, it was found that animals had impaired thermoregulation.

Similar experiments have been carried out in more recent years and at more sophisticated levels (see Cooper, 1966; Bligh, 1973) and the results also indicate that if the hypothalamus is damaged, thermoregulation is impaired. One must stress here the word 'impaired' because animals, including man, are able to thermoregulate to some extent in the absence of the hypothalamus. Perhaps the simplest way to look at central thermoregulation is to consider that there is a chain of centres which can affect thermoregulation and that normally the highest of these centres functionally is the determinate one. If the highest centre is destroyed then the next centre down the chain of command appears to be able to take over. Whether in fact these other centres play a part in normal physiological thermoregulation is difficult to decide because the only way to show that they can affect body temperature is to isolate them from each other, an unnatural state of affairs.

Further evidence for the involvement of the hypothalamus in thermoregulation has been derived from studies on cooling or heating brain structures. Warming or cooling the internal carotid artery elicits appropriate thermoregulatory mechanisms, thereby indicating the presence of thermo-sensitive structures in the vascular

*Department of Pharmacology, University Medical Buildings, Foresterhill, Aberdeen, AB9 2ZD, Scotland.

territory of this artery (Kahn, 1904; Downey *et al.*, 1964). Diathermic warming of the region between the optic chiasma and anterior commissure in anaesthetised cats evokes panting, sweating and an increase in the respiratory rate (Magoun *et al.*, 1938). Similarly, heating the anterior hypothalamus in the dog and monkey has been shown to inhibit shivering and produce peripheral vasodilation (Hemingway *et al.*, 1940; Eliasson and Ström, 1950). Conversely, cooling of the anterior hypothalamus in a variety of laboratory species elicits heat-gain effector responses such as shivering and cutaneous vasoconstriction (Hammel *et al.*, 1960; Satinoff, 1964).

Finally, unit activity studies have confirmed the presence of thermo-sensitive cells in the preoptic anterior hypothalamus. Nakayama *et al.* (1961, 1963) demonstrated that an increase in anterior hypothalamic temperature in the anaesthetised cat was associated with an increase in discharge frequency of a proportion of preoptic cells. Cold-sensitive units were subsequently identified by other workers (Eisenman, 1969; Hellon, 1967; Cabanac *et al.*, 1968).

Various temperature response patterns have been observed, and on the basis of these Eisenman and Jackson (1967) classified thermo-sensitive cells of the preoptic anterior hypothalamus into two categories: thermodetectors–those cells in which the firing rate changes as a function of local temperature–and thermoregulatory interneurones. Thermo-sensitive cells also occur in the posterior hypothalamus. Some of these cells are responsive to local temperature changes whereas others receive input from the skin, preoptic anterior hypothalamus and spinal areas (Stuart *et al.*, 1963). Inputs from skin thermodetectors additionally converge in the preoptic anterior hypothalamus (Wit and Wang, 1968; Hellon, 1970).

Although the hypothalamus is considered to be the principal thermoregulatory site in the brain, it is clear that additional structures such as the medulla oblongata and spinal cord are thermo-sensitive. Behavioural thermo-regulation is probably under the control of higher centres in the brain, although the hypothalamus would appear to be the most important. Interpretation of the physiological responses in terms of engineering concepts has created difficulty in understanding the role of the hypothalamus. Engineers and physicists have designed models of thermoregulation which have become so complicated that they are difficult to understand. However, one of the simpler concepts is that of the hypothalamus as a thermostat, that this thermostat is set at 37.5°C and that any deviation from this temperature is immediately corrected by appropriate changes in either heat-gain or heat-loss mechanisms.

Using this concept of a thermostat, it is often said, for example, that fever is the result of bacterial pyrogen resetting the thermostat at a higher temperature. At its simplest, the hypothalamus may be looked upon as a black box into which afferent impulses from the temperature-sensitive nerve endings in the skin, deep body structures, and also from within the central nervous system itself are continually flowing. The efferent impulses which emerge from the hypothalamus are then responsible for changes in various physiological processes resulting in either heat-gain or heat-loss–that is, if the incoming information indicates that body temperature is high, then the efferent signals going out will be concerned with lowering deep body temperature, whereas if the input to the hypothalamus suggests that deep body temperature is low, then the efferent signal will be concerned with raising deep body temperature. Thus at thermal neutrality the total sum of input from cold sensors and heat sensors would be zero, and there would be zero efferent activity.

Another difficulty which arises from earlier experimentation is the idea that the hypothalamus itself is heat-sensitive. This is probably a misconception. If one considers the black-box hypothesis, it would seem unnecessary to regard the hypothalamus itself as being temperature-sensitive. In fact it would probably be a considerable disadvantage as this is an area concerned solely with the processing of information. This does not exclude the possibility that parts of the hypothalamus or surrounding structures may demonstrate thermo-sensitivity and experiments have been carried out in which the hypothalamic area has been heated or cooled, resulting in changes in deep body temperature irrespective of the thermal state of the animal. Temperature sensors are to be found in various parts of the body, and areas which have been implicated are the skin, deep body structures and possibly the hypothalamus itself (see figure 4.1).

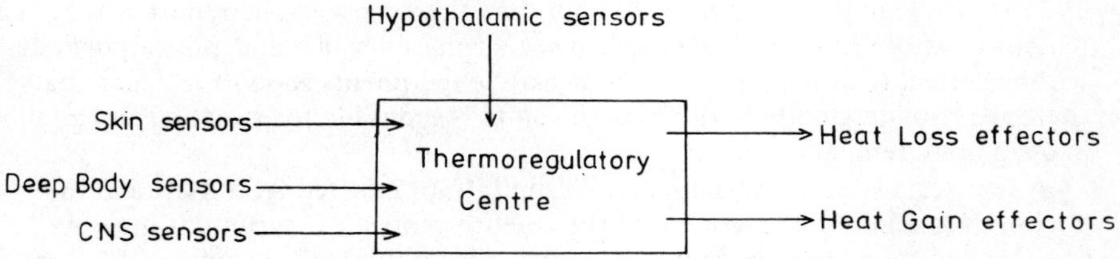

Figure 4.1 Diagrammatic representation of hypothalamic thermoregulation

When considering various transmitter substances which are thought to be involved in thermoregulation, it is apparent that a large majority produce more than one effect on deep body temperature. They may, for example, cause either a fall or a rise in deep body temperature in the same species depending on experimental conditions, or they may produce different effects in different species—causing a rise in one species and a fall in deep body temperature in another species. There is only a limited number of neurotransmitters and therefore it is not surprising that the same transmitter should be implicated in more than one pathway. This is no different from the fact that acetylcholine is the transmitter at all autonomic ganglia, both sympathetic and parasympathetic, and yet in many cases the parasympathetic and sympathetic responses are working in opposite directions. Confusion arises following the injection of relatively large amounts of a substance which reaches more than one pathway at the same time. The effects seen will depend upon the relative sensitivity of these pathways to the substance in question and also variations in experimental techniques, environmental conditions and anatomical differences between species.

NEUROTRANSMITTERS AND THERMOREGULATION

All nervous activity involves synapses and hence neurotransmitter substances and considerable attention has therefore focussed on the possible role of these transmitters in the hypothalamus and related structures.

Acetylcholine

Since acetylcholine (ACh) is the archetypal transmitter substance, it is not surprising to find that cholinergic substances were among the very first to be injected into the cerebro-ventricular system and their effects on deep body temperature measured. What is surprising is that the earliest experiments were performed over 45 years ago when Cushing (1931*a*) injected pilocarpine in doses of 5 and 2.5 mg into the lateral cerebral ventricles of seven patients. The injections produced nausea and vomiting, a marked flush and very heavy sweating. The temperature began to fall after about 30 min. and remained depressed for about 3 h (figure 4.2). Cushing concluded that these effects were due to a central autonomic stimulation, predominantly of the parasympathetic division. In a subsequent paper, he found that an intracranial injection of atropine abolished the effect of pilocarpine (Cushing, 1931*b*), whilst in yet another paper published the same year, he reported that in a patient in whom the hypothalamus had been damaged by a lesion, pilocarpine was without effect (Cushing, 1931*c*). These early experiments support the idea that there is a cholinergic site in the hypothalamus responsible for sweating and a fall in deep body temperature.

A few years later, Henderson and Wilson (1936) injected ACh itself and the cholinesterase inhibitor eserine into the cerebro-ventricular system of man. As Cushing had found with pilocarpine, so they found that ACh in a dose of 2.5 mg produced vomiting and retching and sweating, and this was accompanied by a

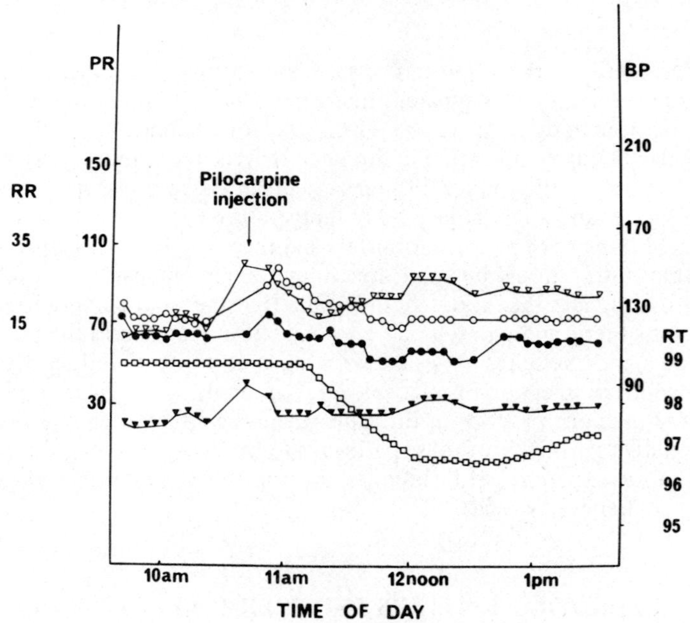

Figure 4.2 The effect of an intraventricular injection of 2.5 mg pilocarpine into a human subject on pulse rate (PR, min^{-1}, ●–●), respiratory rate (RR, min^{-1}, ○–○), blood pressure (BP, mm Hg, systolic ▽–▽, diastolic ▼–▼) and rectal temperature (RT, °F, □–□). Redrawn from Cushing (1931*a*)

fall in deep body temperature. They concluded that the fall in temperature could be entirely accounted for by the degree of sweating.

Eserine in a dose of 50 μg produced similar effects to ACh and also potentiated the response to ACh which was abolished by atropine (0.5 mg). Similar doses given parenterally were without effect and therefore the effects seen were of central origin.

Unfortunately, in spite of this promising start, our knowledge of the role of ACh in thermoregulation is still confusing as many conflicting papers bear witness. For example, Findlay and Thompson (1968) found that ACh was without effect when injected into the lateral ventricle of the ox, whereas Beckman and Carlisle (1969) found that in rats ACh produced a fall in deep body temperature and Myers and Yaksh (1969) noted a rise in temperature in the same species. The problem of using ACh itself is that it is readily hydrolysed by cholinesterase and may be destroyed before it has time to reach its site of action. One way to try to prevent this is to inject a cholinesterase inhibitor at the same time. Though this may prevent the destruction of the ACh, it may complicate the picture by having an action of its own. It is for this reason that many authors have used stable cholinergic compounds in place of ACh, although here again one must raise the criticism that in the CNS their effects may differ considerably from their peripheral effects. In particular, one cannot assume that there are nicotinic and muscarinic receptors for ACh in the CNS identical to their peripheral counterparts.

In 1972 Hall put forward the view that both cholinergic heat-loss and heat-gain mechanisms are implicated in the hypothalamic pathways involved in thermoregulation. He used the cat as his experimental animal and found that nicotine produced a fall in deep body temperature which was associated with skin vasodilation, tachypnoea and panting. These responses could be prevented by the intraventricular injection of the ganglion-blocking drugs hexamethonium and mecamylamine, but not by the muscarinic-blocking drug atropine. Conversely, Hall found that carbachol produced a rise in deep body temperature which was associated with skin vasoconstriction, piloerection and severe shivering. The effects of carbachol were abolished by atropine but not by hexamethonium or mecamylamine. Hexamethonium and mecamylamine were found to produce a prolonged rise in temperature, together with skin vasoconstriction, piloerection and intense shivering. Atropine, on the other hand, had no effect. Interestingly, a mixture of ACh and eserine was also without effect on body temperature. Hall proposed that the cholinergic heat-gain pathways were muscarinic whereas the heat-loss pathways were nicotinic. These observations in the cat differ from those previously mentioned in man in that the heat-loss mechanisms would appear to involve muscarinic receptors.

It is rather surprising that Hall (1972) was unable to find any effect with an ACh-eserine mixture or with atropine itself in the cat. In recent experiments Dascombe and Milton, unpublished) ACh, eserine and ACh eserine mixtures have been injected into the third ventricle of the conscious cat at an ambient temperature of 22°C. ACh was found to have no significant effect on deep body temperature, whereas eserine itself produced a slight rise. The mixture of ACh and eserine also produced an incease in deep body temperature which was considerably greater than that produced by eserine alone. The effects of eserine and of ACh-eserine mixtures were abolished by atropine which itself was found to decrease profoundly deep body temperature, producing rapid ventilation and ear skin vasodilatation.

The abolition of the rise in deep body temperature appeared to be a true antagonism, as all the observable effects of the ACh-eserine mixture were abolished. This was in contrast to the reduction in the hyperthermic effect of prostaglandin E_2, where the total effect appeared to be a combination of the increase in heat-gain produced by the prostaglandin and a decrease in body heat produced by the atropine. In these experiments there was no evidence of a hypothermic response to either eserine or to the ACh-eserine mixture.

The situation at the present time would suggest that there are cholinergic synapses within the central nervous system which form part of the pathways involved in both heat-loss and heat-gain mechanisms, but we still have a long way to go before we can define the role which ACh plays in the physiological control of body temperature.

5-Hydroxytryptamine, Adrenaline and Noradrenaline

One of the most significant papers in the field of thermoregulation is that by Feldberg and Myers (1964) which has the title 'Effects on temperature of amines injected into the cerebral ventricle: a new concept of temperature regulation'. In this paper, Feldberg and Myers describe the effects of injections of 5-hydroxytryptamine (5-HT) and catecholamines into the cerebral ventricles of the conscious cat. 5-HT produced a rise in deep body temperature whereas adrenaline (A) and noradrenaline (NA) produced a fall. The effects of these substances were attributed to an action of the hypothalamus and the authors suggested that

> '... there are two possible ways in which the amines may affect body temperature. They may be continuously released and the normal temperature may be the outcome of a fine balance between the release of 5-HT and of the catecholamines. Changes in temperature would then be brought about by a disturbance of this balance. Or, normal temperature is maintained independent of the release of the three amines, their release being the mechanism by which changes in temperature are effected.'

Feldberg and Myers were not the first to notice the effect of monoamines on thermoregulation. In 1943, von Euler *et al.* reported that intra-cranial injection of small doses of A modified rectal temperature in rabbits, and later, in 1957, Brodie and Shore were the first to suggest a role for the hypothalamic monoamines in temperature regulation. Indirect evidence in support of this idea had first come from the work of Horita and Gogerty (1958) and Canal and Ornesi (1961) who showed that peripheral or central administration of 5-hydroxytryptophan produced dose-related hyperthermia in rabbits. The febrile response in each case was potentiated by prior treatment with a monoamine oxidase inhibitor and attenuated by a 5-HT receptor antagonist.

Since the paper by Feldberg and Myers in 1964, many other workers have attempted to confirm and extend these studies, and there is a wide variation in the results obtained. Initially inter-species differences were observed. Dogs and monkeys responded to the monoamines in the same way as the cat (Feldberg *et al.*, 1966, 1967). The rabbit and sheep, however, responded in an exactly opposite manner. Intraventricular injection of 5-HT usually produced hypothermia and NA was invariably hyperthermic in these animals (Bligh, 1966; Cooper *et al.*, 1964,

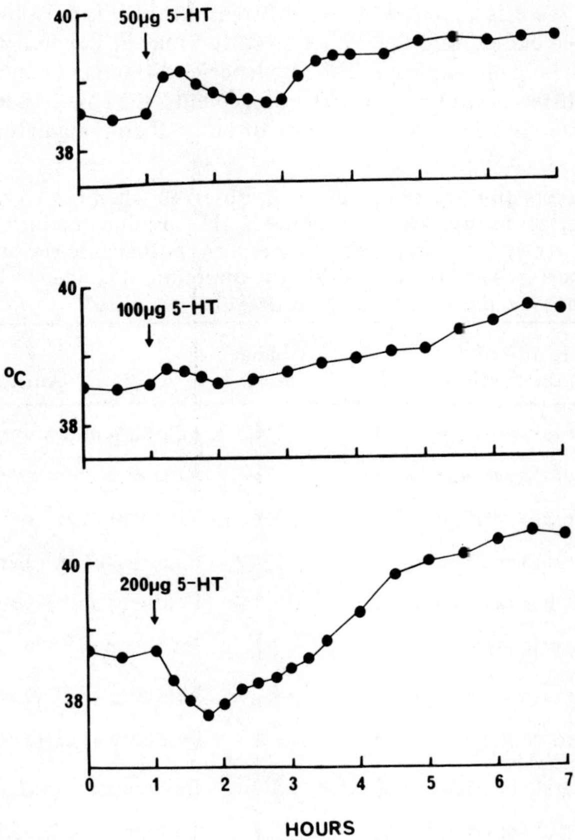

Figure 4.3 Rectal temperature records of an unanaesthetised cat. At the arrows 5-hydroxytryptamine (5-HT) was injected into the right lateral ventricle. Records from the same cat with 7 days between each experiment. (After Milton and Harvey, 1975, with permission).

1965). Other species, including the mouse, rat, goat and ox, showed further response patterns (Brittain and Handley, 1967; Andersson *et al.*, 1966; Findlay and Robertshaw, 1967). Various responses were also seen in the same species in different studies—for example the contrasting reports on the response of the cat to intraventricular 5-HT injection (Feldberg and Myers, 1964; Kulkarni, 1967). Figure 4.3 from Milton and Harvey (1975) shows the effect of injecting increasing doses of 5-HT in the same animal and it will be seen that both a fall and a rise in deep body temperature occur depending upon the dose. More recently, Komiskey and Rudy (1977) using microinjections, have elegantly mapped out two discrete sites in the cat hypothalamus, one responsible for a rise in deep body temperature and one responsible for a fall in deep body temperature.

The inter-species differences are the most puzzling, as it is difficult to conceive that the same neurotransmitter has an entirely different function in different species. Since there is a limited number of transmitters it is most likely, as has been mentioned earlier, that the same transmitter may be concerned with both heat-loss and heat-gain pathways. The inter-species differences may therefore be due to slight differences in anatomical arrangements, the ease of penetration of neurotransmitter and drugs to active sites, or in the thermoregulatory state of the

Table 4.1 Effects on body temperature T_b observed when the monoamines noradrenaline (NA) and 5-hydroxytryptamine (5-HT) are injected into the cerebral ventricles or directly into the preoptic/anterior hypothalamic region of the brains of different species. The arrows indicate the direction of change in body temperature. A light arrow represents a feeble or irregular response*

Species	Route of administration	T_b NA	Change 5-HT	Authors
Cat	Lateral ventricle	↓	↑	Feldberg and Myers (1963, 1964)
	Lateral ventricle	↓	↑	Ruckebusch *et al.* (1965)
	Lateral ventricle		↓	Kulkarni (1967)
	Lateral ventricle		↑↓	Banerjee *et al.* (1968)
	Ant. hypothalamus	↑	↑	Feldberg and Myers (1965)
	Preoptic area		↓	Jacobson (1967)
Dog	Lateral ventricle	$\downarrow_1$	$\uparrow_1$	Feldberg *et al.* (1966)
	Third ventricle	↓	↑	Feldberg *et al.* (1967)
Rabbit	Lateral ventricle	↑	↑↓	Ruckebusch *et al.* (1965)
	Lateral ventricle	↑	↓	Cooper *et al.* (1965)
Sheep	Lateral ventricle	↑	↓	Ruckebusch *et al.* (1965)
	Lateral ventricle	↑	↓	Bligh (1966)
Goat	Third ventricle		↓	Andersson *et al.* (1966)
Ox	Lateral ventricle	$\downarrow_4$	↓	Findlay and Thompson (1968)
Rat	Lateral ventricle	$\uparrow_2$ $\downarrow_3$	↓	Feldberg and Lotti (1967)
	Lateral ventricle	↑	↓	Myers and Yaksh (1968)
Mouse	Lateral ventricle	↓	↓	Brittain and Handley (1967)
Monkey	Lateral ventricle	↓	$\uparrow_2$ $\downarrow_3$	Myers and Sharpe (1967)
	Ant. hypothalamus	↓	↑	Myers and Yaksh (1969)
	Third ventricle	$\downarrow_1$	$\uparrow_1$	Feldberg *et al.* (1967)

*Reproduced from Bligh *et al.* (1971) with permission
Subscripts: 1 = anaesthetised, 2 = low dose level, 3 = high dose level and 4 = at low or moderate ambient temperatures only

animal at the time of the experiment. For example, in the rabbit, goat or sheep, NA appears to act in a non-specific inhibitory manner on whichever effector pathway is dominant at the time. NA therefore produces a rise in body temperature in a hot environment and a fall in body temperature in a cold environment (Bligh and Cottle, 1969; Bligh *et al.*, 1971). Table 4.1 by Bligh *et al.* (1971) gives a summary of the effects of the monoamines in different species. However, more research is necessary before a definitive answer to the question of inter-species differences can be given. In addition to the injection of amines themselves into the hypothalamus, pharmacological manipulation of brain amines has also been attempted in a wide variety of studies to assess their possible thermoregulatory transmitter function.

Intraventricular injections of α-adrenoceptor-blocking agents produce hypothermia in rabbits and rats and hyperthermia in cats (Feldberg and Saxena, 1971*a*; Uchimura and Murakami, 1973). Milton and Harvey (1975) showed that in the cat the 5-HT-blocking drug methysergide abolished the initial shivering response following the injection of 5-HT. Depletion of central monoamines by 6-hydroxydopamine or parachlorphenylalanine (*p*CPA), though having no direct effect on normal temperature regulation, affected the ability of cats to thermoregulate under conditions of heat and cold stress (Milton and Harvey 1975; Harvey and Milton. 1974, 1976). 6-hydroxydopamine treatment was also found to potentiate pyrogen fever, whereas *p*CPA treatment reduced the response. There are several reports of a rise in cerebral 5-HT turnover in heat-stressed rats (Tagliamonte *et al.*, 1971; Simmonds, 1970). Cerebral 5-HT turnover is also increased in rats following endotoxin administration.

Further support for the idea of the monoamines being involved in thermoregulation comes from studies on release of monoamines into cerebrospinal fluid. Both 5-HT and the catecholamines appear in the CSF of conscious monkeys during pyrogen fever (Myers, 1971). In addition, 5-HT is released from the preoptic anterior hypothalamus of this species during peripheral cooling. Conversely, in the cat NA release from the preoptic anterior hypothalamus coincides with activation of heat-loss effector mechanisms (Myers and Chinn, 1973). Finally, the effects of the monoamines have been studied on hypothalamic thermo-sensitive neurones. When the amines were applied in the rat the results obtained by microiontophoresis appeared to correlate with those from microinjection studies. Thus NA microinjected into the rat elicited a temperature rise (Myers and Yaksh, 1968), which is consistent with the finding that warm-sensitive cells are depressed and cold-sensitive cells activated by iontophoretic NA (Murakami, 1973). Conversely, 5-HT exerts its hyperthermic action via depression of cold-sensitive cells and activation of warm-sensitive ones.

Dopamine

Dopamine (DA) has recently attracted the attention of pharmacologists working in the field of thermoregulation, and the effects described are less confusing than those found with the other monoamines. In the majority of species examined DA produces a fall in deep body temperature when injected into the central nervous system—for example in the cat (Kennedy and Burks, 1974), the rat (Cox and Lee, 1977*a*; *b*) and the baboon (Toivola and Gale, 1970). However, in the rabbit Hill and Horita (1971, 1972) have reported that DA produces a rise in deep body temperature. In contrast, it has recently been observed that N_6-adenosine dopamine produces a very marked fall in deep body temperature when injected into the hypothalamus of both the rabbit and the cat, and interestingly also when injected intravenously (Milton, unpublished). The majority of studies using DA have shown that its effects can be blocked by DA antagonists such as haloperidol and pimozide. In addition Kruk (1972) and Cox and Lee (1977*a, b*) have shown that apomorphine, which is said to be a specific DA agonist, produced hypothermia in the rat and this effect is also blocked by the DA antagonist pimozide.

If DA does have a role in thermoregulation, then it would appear from the available evidence to be acting as a neurotransmitter in heat-loss pathways.

Prostaglandins

Prostaglandins (PG) were first reported to increase body temperature by Milton and Wendlandt in 1970. When PGE_1 was injected into the third cerebral ventricle of the conscious cat it produced vigorous shivering, ear vasoconstriction and a rapid rise in deep body temperature. The threshold dose required to produce a rise in deep body temperature was extremely small, in the order of 3×10^{-11} mol, and the duration of the hyperthermia was short, particularly when compared with the long-lasting fever produced by intraventricular injection of bacterial pyrogen (figure 4.4). In all other respects the hyperthermia produced by PGE_1 resembled that produced by bacterial pyrogen, both behaviourally and in its effects upon the autonomic nervous system. At the same dose levels as PGE_1, PGs, A_1, $F_{1\alpha}$ and $F_{2\alpha}$ were without significant effect on deep body temperature. Recently, Ewen *et al.* (1976) have reinvestigated the effects of $PGF_{2\alpha}$ and found that in the dose range of 800 ng to 12.8 mg $PGF_{2\alpha}$ produced a dose-related hyperthermia, vasoconstriction but no shivering. In addition they found that prostaglandin D_2 had no

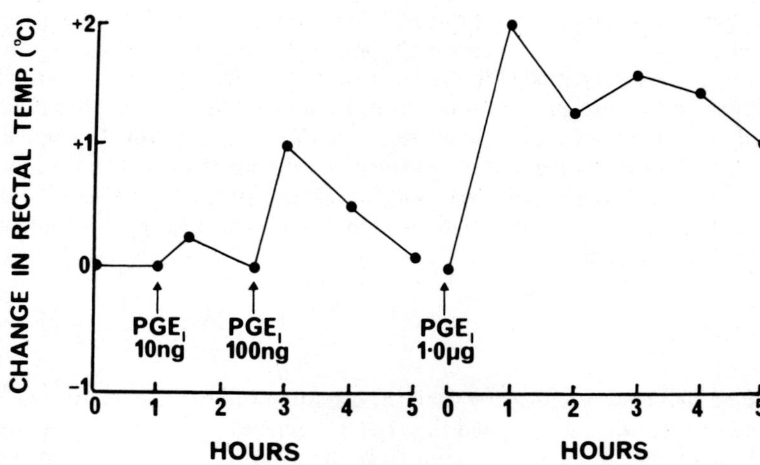

Figure 4.4 Rectal temperature of an unanaesthetised cat. At the arrows injections into the third ventricle of PGE_1 in the doses indicated. (After Milton and Wendlandt, 1971*b*, with permission)

effect on body temperature. PGE_1 has also been shown to produce hyperthermia in the rat and rabbit (Milton and Wendlandt, 1971*a, b*). These workers also showed that PGE_2 had almost identical actions to PGE_1.

Feldberg and Saxena (1971*b*) located the site of action of PGE_1 to the preoptic area of the anterior hypothalamus. Also in 1971, Feldberg and Saxena (1971*c*) showed that when PGE_1 was infused into the cerebral ventricles of the cat the hyperthermia produced was sustained for only as long as the infusion lasted. Thereafter the deep body temperature rapidly returned to the pre-infusion level (figure 4.5).

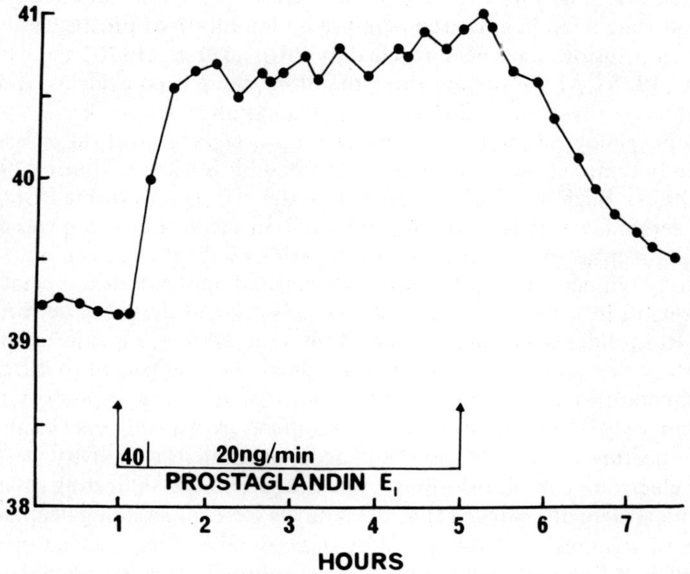

Figure 4.5 Record of rectal temperature obtained from an unanaesthetised cat. ↑–↑ indicates 4 h period of infusion of PGE_1 into the cannulated left ventricle in the doses indicated. (After Feldberg and Saxena, 1971*c*, with permission)

As a consequence of the reported actions of the PGs of the E series in raising deep body temperature by an action on the hypothalamus, the question arises as to whether the prostaglandins are involved in normal thermoregulation as opposed to pathological conditions, and this will be discussed later. In 1975 Cammock *et al.* (1976) measured the PGE levels in cerebrospinal fluid collected from the cisterna magna of cats subjected to heat and cold exposure. During exposure to cold (0°C) the animals exhibited autonomic and behavioural thermoregulatory responses including shivering and ear skin vasoconstriction and the rectal temperature was found to increase slightly. When the animals were subjected to heat (45°C), they were unable to control their deep body temperature and, although panting, sweating from the paw pads and ear skin vasodilatation occurred, the deep body temperature rose continuously until the experiments were terminated. However, neither during heat nor cold exposure were any differences in the PGE content of the CSF found from that collected from the same animals at room temperature (25°C). It is of course possible that any changes in PGE synthesis or release would occur in the region of the thermoregulatory area of the anterior hypothalamus and such changes might be too small to produce any significant changes in PGE levels in cisternal CSF.

Further evidence against involvement of prostaglandins in normal thermoregulation comes from the observation (Milton, unpublished) that prostaglandin synthetase inhibitors do not interfere with the ability of cats to thermoregulate when subjected to heat and cold stress. The fall in deep body temperature in the absence of fever seen with antipyretic drugs was investigated by Milton (1973), who suggested that the drugs were activating a heat-loss mechanism directly and were not producing a fall in body temperature by inhibition of prostaglandin synthesis. Similar conclusions have been reached by Pittman *et al.* (1976) and Clark and Cumby (1975). At the present time, therefore, there is no evidence that the prostaglandins are involved in normal thermoregulation.

The physiological mechanisms by which prostaglandins of the E series increase deep body temperature have been investigated by Bligh and Milton (1973) and Stitt (1973). Bligh and Milton reported on the effects of infusing PGE_1 into a lateral cerebral ventricle of the Welsh mountain sheep. The sheep was chosen as the experimental animal for these investigations since it can maintain a constant deep body temperature over a very wide range of ambient air temperatures by regulating its heat-loss and heat-gain mechanisms, and these can be readily monitored. Recordings were made of deep body temperature, ear skin temperature, respiratory rate and shivering while the animals were subjected to different ambient temperatures. At an ambient temperature of 10°C the respiratory rate was low, approximately 35 per minute, ear skin temperature was the same as air temperature (indicating vasoconstriction) and occasional bursts of activity were observed on the electromyograph recording from a thigh muscle, indicating shivering. These measurements showed that the animals were maintaining deep body temperature by minimising heat-loss and by occasionally increasing heat production. When PGE_1 (1 μg/min for 45 minutes) was infused into a lateral ventricle the respiratory rate dropped slightly to approximately 30 per minute. There was no change in ear skin temperature but violent shivering was recorded and deep body temperature rose. As soon as the infusion was stopped, shivering ceased and the animals started to pant (respiratory rate approximately 100 per minute), continuing to do so until the deep body temperature had returned to the pre-infusion level.

When the ambient air temperature was raised to 45°C, that is above the deep body temperature, the animals panted vigorously (respiratory rate approximately 200 per minute), the ear vessels were dilated and no shivering was recorded, indicating that the animals were actively preventing body temperature from rising primarily by evaporative heat loss. Under these conditions, when the PGE_1 infusion was started ear skin temperature did not change and no shivering was observed; however, the respiratory rate dropped dramatically to about 30 per minute. Consequently evaporative heat loss by panting was suppressed and deep body temperature rose rapidly. On cessation of the PGE_1 infusion, panting was resumed and the respiratory rate rose well above the pre-infusion level to over 300 per minute. This elevated panting continued until the body temperature had returned to normal.

When the animals were at room temperature (18°C) the respiratory rate was about 150 per minute, ear skin temperature being between ambient air temperature and deep body temperature. No shivering was observed. Under these conditions the animals maintained deep body temperature by changes in vasomotor tone with some evaporative heat loss. At this ambient temperature, PGE_1 infusion produced a fall in respiratory rate to approximately 30 per minute, ear skin temperature dropped, indicating vasoconstriction, and occasional bursts of shivering were recorded. After infusion of PGE_1 was stopped, shivering ceased, ear skin temperature increased, indicating vasodilatation, respiratory rate rose to above 250 per minute, indicating panting and evaporative heat loss, and deep body temperature quickly fell to normal.

These experiments on the Welsh mountain sheep showed that PGE_1 increased deep body temperature by inhibiting both heat-loss mechanisms including evaporative heat loss and surface heat loss and by stimulating heat-gain mechanisms such as shivering (metabolic heat production). The predominant effect depended upon the ambient temperature and therefore upon the thermoregulatory pathways being driven at any particular time. Of particular interest were the observations that as soon as the PGE_1 infusions were stopped, the animals actively lost the heat they had gained during the infusion and deep body temperature was quickly restored to normal. This is reminiscent of the effects of antipyretic drugs in reducing fever produced by bacterial pyrogens. Stitt, also in 1973, made similar findings in the rabbit. In addition he compared the effects of PGE_1 with bacterial pyrogens.

In the cat at an ambient temperature of 18–20°C, PGE_1 and PGE_2 both increase deep body temperature by producing shivering and ear skin vasoconstriction and the animals take up the curled-up position to reduce heat loss. Body temperature returns to normal after the PGE_1 hyperthermia by active heat loss including ear vasodilatation, occasional sweating from the paw pads and occasional panting and the animals resume their normal body position (Milton and Wendlandt, 1971*a*; Feldberg and Saxena, 1971*b, c*).

Cyclic Nucleotides

Many of the neurotransmitters which are thought to be involved in thermoregulatory pathways, such as 5-HT, catecholamines and prostaglandins have been shown to stimulate the formation of cyclic adenosine 3′, 5′-monophosphate (cAMP) in brain tissue, and in addition the muscarinic actions of ACh are thought to be mediated by cyclic guanosine 3′, 5′-monophosphate. In view of these observations it is not surprising that in recent years attempts have been made to find a role for the cyclic nucleotides in hypothalamic temperature regulation.

Injections of cAMP into the brain were reported to produce hyperthermia in both rats (Breckenridge and Lisk, 1969) and cats (Gessa *et al.*, 1970). Varagić and Beleslin (1973) and Clark *et al.* (1974) have both observed a period of hypothermia or a transient hyperthermia before the onset of a long-lasting rise in the body temperature of some cats following administration of the N^6-2-O-dibutyryl derivative of cAMP (Db-cAMP) into a lateral cerebral ventricle. However, numerous non-thermoregulatory effects are associated with these changes in body temperature, including lengthening of the oestrous cycle, hyperphagia, aggression, catatonia, convulsions and stimulation of other motor and autonomic activities. Consequently, it was not clear whether these thermoregulatory effects were due specifically to the cyclic nucleotides or were of a non-specific nature. Recently, however, Dascombe and Milton (1975, 1976) found that microinjection of cAMP and Db-cAMP into the region of the preoptic anterior hypothalamus of the unanaesthetised cat at an ambient temperature of 22 ± 2°C both produced hypothermia associated with ear skin vasodilatation, polypnoea, panting and occasionally sweating from the paw pads. With Db-cAMP the hypothermia was shown to be dose-dependent between 50–500 μg. Microinjection of AMP, ADP and ATP into the preoptic anterior hypothalamus also produced hypothermia in the cat. The order of relative potencies of the adenine nucleotides with respect to the hypothermia produced and to the autonomic thermoregulatory effects observed is similar; Db-cAMP was most potent and cAMP least (figure 4.6).

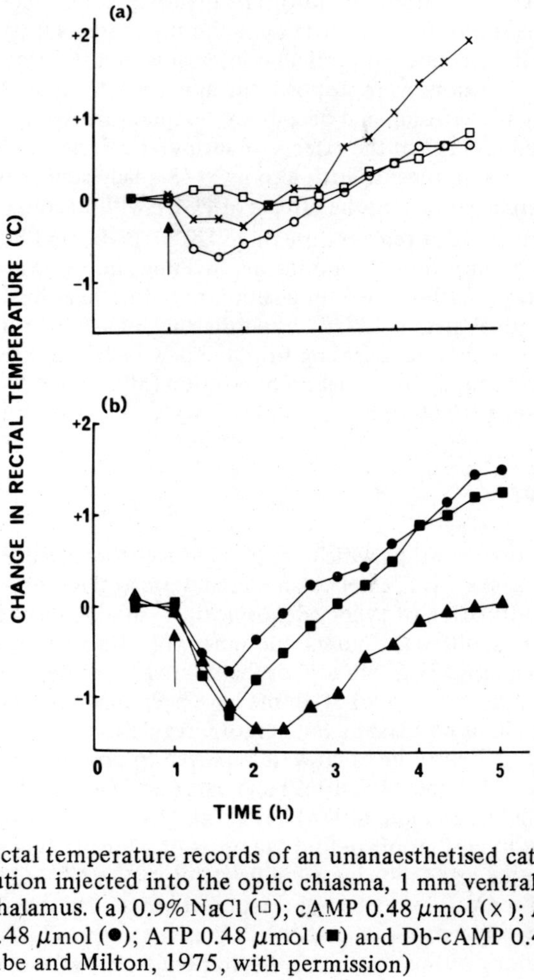

Figure 4.6 Rectal temperature records of an unanaesthetised cat. At the arrow 2.5 μl drug solution injected into the optic chiasma, 1 mm ventral to the preoptic anterior hypothalamus. (a) 0.9% NaCl (□); cAMP 0.48 μmol (×); AMP 0.48 μmol (○). (b) ADP 0.48 μmol (●); ATP 0.48 μmol (■) and Db-cAMP 0.48 μmol (▲). (After Dascombe and Milton, 1975, with permission)

Microinjection into the preoptic anterior hypothalamus of many substances including isotonic saline produces in most cats a non-specific rise in body temperature, which tends to mask the hypothermic response to drugs. A similar non-specific long-lasting rise in temperature has been observed by Feldberg *et al.* (1970) and by Feldberg and Saxena (1970) following ventricular perfusion and by Varagić and Beleslin (1973) in response to intraventricular injections. The tissue damage commonly seen at the site of injection on histological study indicated that the hyperthermia could be due to tissue damage, associated with increased prostaglandin-like activity in the CSF (Dey *et al.*, 1974). 4-Acetamidophenol (paracetamol), an antipyretic which can inhibit prostaglandin synthesis (Flower and Vane, 1972), reduced or abolished the hyperthermic response to the adenine nucleotides when administered either before or after them. The transient rise in temperature seen initially after administration of Db-cAMP in some cats may be due in part to general stimulation of the central nervous system, as suggested by Clark *et al.* (1974). But in some of the experiments of Dascombe and Milton this effect was attributable to the rapid development of pyrexia following microinjection into brain tissue, before the onset of adenine nucleotide-induced hypothermia. The longer latency of the Db-cAMP effect in these animals possibly represented the time for diffusion of the drug to the preoptic anterior hypothalamus from injection sites remote from this region (Myers, 1971).

Dascombe and Milton (1975, 1976) compared the hypothermic effect of the adenine nucleotides with the effects produced in the same cats by microinjection of several putative neurotransmitter substances and various ions. The responses to NA and 5-HT were in agreement with the central effects documented for these monoamines in the cat (Feldberg and Myers, 1965; Kulkarni, 1967). The hypothermic response to both cAMP and to NA in these studies is consistent with the proposal that cAMP mediates many adrenergic responses (Robison *et al.*, 1971).

In the rabbit, by contrast, both intrahypothalamic NA (Cooper *et al.*, 1965) and Db-cAMP (Laburn *et al.*, 1974) cause hyperthermia. If it is shown that the response to Db-cAMP in the rabbit is not a non-specific effect, the apparent species difference in the responses to NA and Db-cAMP in the cat and rabbit would support the involvement of cAMP (Robison *et al.*, 1971). However, hypothermia and associated heat-loss activity in response to equimolar quantities of AMP, ADP and ATP as well as cAMP and Db-cAMP, which differ only in their relative potency, indicates that this effect is a common property of these compounds and is independent of the cyclic structure of cAMP. It is not possible to determine from these experiments whether AMP, ADP and/or ATP modify endogenous cAMP (Rall and Sattin, 1970). The hypothermic response to the adenine nucleotides could be due to a localised increase in free Ca^{2+}, since Duffy and Schwarz (1974) have reported the displacement of Ca^{2+} ions from isolated erythrocyte membranes in response to cAMP and ATP. This hypothesis is strengthened by the observation that when Dascombe and Milton (1975, 1976) injected Ca^{2+} ions in quantities equimolar with the adenine nucleotides, hypothermia was produced.

These workers also found that intravenous injection of 5 and 10 mg cAMP/kg into the unanaesthetised cat, at an ambient temperature of 22 ± 2°C, produced a fall in body temperature. Hypothermia, which began two to three minutes after injection of the nucleotide, was associated with ear skin vasodilatation and occasionally with polypnoea and sweating. Lower doses of cAMP were found to be without effect on body temperature in cats. Although Gessa *et al.* (1970) have reported that cAMP does not pass the blood-brain barrier, Dascombe and Milton have found that cAMP injected intravenously in cats is associated with a dose-dependent increase in the concentration of the nucleotide in cisternal CSF. This suggests that the hypothermic effect of cAMP injected intravenously may be exerted in the hypothalamus, although a possible direct vasodilatory effect of cAMP cannot be discounted (Pettinger *et al.*, 1970).

Effects of phosphodiesterase inhibitors on body temperature

The amount of cAMP in any tissue is the result of a balance between its rate of synthesis, catalysed by adenyl cyclase, and its rate of hydrolysis, catalysed by phosphodiesterase (Cheung, 1970). The highest concentration of phosphodiesterase activity in mammals is in the brain (Sutherland and Rall, 1958), and the methylxanthines, caffeine and theophylline, have been shown to inhibit the activity of this enzyme (Sutherland and Rall, 1958; Butcher and Sutherland, 1962). Administration of large doses of these drugs can produce a low-grade rise in body temperature in man (Soifer, 1957; Reimann, 1967). Dascombe and Milton (1972) and Dascombe (1977) observed a thermogenic effect with caffeine, but not theophylline, when these drugs were injected intraperitoneally into cats and rabbits. Increased temperature in the unrestrained cat was associated with increased locomotor activity, whilst the origin of the thermogenesis in the restrained rabbit was uncertain. That theophylline, in a dose similar to that of caffeine which produces hyperthermia, had no effect on temperature suggests that the response to caffeine may be mediated by the central nervous system (Ritchie, 1970), possibly the hypothalamus (Friedman, 1944). However, no effect on body temperature was observed when either caffeine or theophylline (as aminophylline) was injected into either the preoptic anterior hypothalamus or the cerebroventricular system of the cat. It is not possible to determine from these experiments whether or not the hyperthermic effect of caffeine is mediated by increased concentrations of cAMP.

The potentiation of a physiological response by the methylxanthine drugs has been accepted as presumptive evidence that the response is mediated by cAMP (Butcher, 1968; Goldberg and Singer, 1969). The hyperthermia produced by caffeine, which is resistant to the antipyretic effect of aspirin, augmented the febrile responses produced by bacterial and endogenous pyrogens in rabbits, but not in cats (Dascombe, 1977). No potentiation of the febrile response to pyrogens was apparent with either intraperitoneally or centrally administered caffeine or theophylline in the cat and rabbit, and hence it must be concluded that cAMP does not, as judged by this criterion, mediate fever in these species.

The results obtained by Dascombe and Milton (1975, 1976) in the cat show that the initial effect of cyclic AMP when injected into the hypothalamic area is a fall in deep body temperature. Subsequently a rise in temperature may be observed, and this is considered to be a 'non-specific fever' due to prostaglandin release following injection into the CNS. No effects on body temperature have so far been observed following the injection of cyclic GMP. Insufficient evidence is available at the present time to determine whether the rise in temperature reported in the rat and rabbit following administration of cyclic AMP into the CNS is due to a direct action of the compound or is of a non-specific nature.

Intravenous injection of bacterial pyrogen in the cat produced an increase in the cyclic AMP content of the CSF. Increases in the plasma levels were also found and, since experiments showed that cyclic AMP readily passed from the peripheral circulation to the CSF, it is uncertain whether the increased amounts of cyclic AMP found in the CSF were central or peripheral in origin.

The antipyretic drug paracetamol inhibited the increase in cyclic AMP levels in the CSF produced by intravenous bacterial pyrogen. However, although it produced antipyresis and a fall in temperature when given after fever had developed, no decrease in the elevated levels of cyclic AMP was observed.

When cats were subjected to heat and cold stress no changes were found in the cyclic AMP content of the CSF. The question, therefore, whether the cyclic nucleotides are implicated in thermoregulation, in particular, by an action on the hypothalamus, remains unanswered. The evidence so far suggests that they are not likely to be involved in pyrogen fever.

FEVER

Pathogenesis of Fever

The commonest disturbance of thermoregulation is fever, which is the increase in deep body temperature following infection and inflammation. Fever may be produced by a wide variety of organisms including both Gram-positive and Gram-negative bacteria, viruses, yeasts, fungi and protozoa, and may also occur following non-infectious pathological changes such as inflammation and tissue damage, malignancy, antigen–antibody reaction and tissue-graft rejection.

Fever almost certainly occurs as a result of changes in the central control of deep body temperature brought about by pyrogenic substances released following infection and inflammation. It therefore differs from the hyperthermia associated, for example, with exposure to a hot environment or following heavy exercise. These hyperthermic events occur as a result of the inability of the body to balance heat loss with heat gain, whereas during fever the balance is maintained but at a higher temperature than normal.

The causative agents of fever are referred to as pyrogens and many different pyrogens are known. Unfortunately, however, the nomenclature which has been used over the past 100 years is confusing and needs to be clarified. Pyrogens may be differentiated into two basic categories. First, pyrogenic substances which are external to the body such as those produced by infectious agents; these should be referred to as *exogenous* pyrogens. Secondly, there are pyrogenic substances produced by the body—*endogenous* pyrogens. The most well-known class of exogenous pyrogens comprises the endotoxins, often referred to as bacterial pyrogens. They form part of the cell wall of Gram-negative bacteria, and are thought to be present in all such bacteria. There is no evidence for the presence of endotoxins in Gram-positive bacteria (Westphal, 1957). These, together with yeasts, fungi and protozoa, must produce pyrogenic compounds of their own, which have yet to be identified.

The existence of endogenous pyrogen was first demonstrated by Beeson

(1948), who found that saline extracts of rabbit neutrophils were pyrogenic when injected intravenously into rabbits. In 1953 Bennett and Beeson showed that the pyrogenic material obtained from neutrophils was different from endotoxin, since it was heat labile, the onset of the fever response was more rapid and it was able to produce fever of equal magnitude in both normal and in endotoxin-tolerant animals. It is now recognised that endogenous pyrogen is a protein with a molecular weight in the region of 10,000 to 20,000 (see Atkins and Bodel, 1974). This pyrogenic material, first obtained from neutrophils (loosely described as polymorphonuclear leucocytes), is often referred to as leucocytic pyrogen, though since it is now recognised that not only neutrophils but also monocytes, Kupffer cells and other fixed cells of the reticular endothelial systems can produce the same or at least very similar molecules, they should all be referred to as endogenous pyrogens.

It is now generally accepted that the cells capable of producing endogenous pyrogens are activated either by exogenous pyrogens or by endogenous factors such as inflammation and tissue damage to synthesise and subsequently to release endogenous pyrogen, and it is this circulating material which is the common mediator of fever.

There is considerable evidence that both exogenous and endogenous pyrogens, injected directly into the central nervous system, produce fever in the majority of species studied (Bennett *et al.* 1957; Villablanca and Myers, 1965; King and Wood, 1958). In 1967 Cooper *et al.* prepared endogenous pyrogen from rabbit peritoneal exudate and injected it into selected areas of the brain. Fever only occurred when the pyrogen was injected into the preoptic and anterior hypothalamic areas; no effects were seen when it was injected into the posterior hypothalamus, midbrain, cerebellum or cortex. They found the dose of endogenous pyrogen injected into the hypothalamus necessary to produce fever was one-hundredth of the intravenous dose whereas the dose of endotoxin necessary to produce fever was of the same order for both routes. Jackson (1967) compared the latency of onset of the administration of endogenous pyrogen and endotoxins into the hypothalamus of the cat. He found that the latency of endogenous pyrogen was 5–6 min, compared with 30–65 min for endotoxin, and that the endogenous pyrogen only caused fever when injected into the anterior hypothalamus. There was no effect on either the preoptic area, the posterior hypothalamus or the thalamus. Endotoxin produced fever when injected into areas unresponsive to endogenous pyrogen and when injected into the cerebral ventricles, but with a considerable latency in onset. The occurrence of a fever after a long latency following the injection of sterile saline into the CNS (Jackson, 1967) is an example of the response now referred to as 'non-specific fever' (Dey *et al.,* 1974; Dascombe and Milton, 1975).

These observations all indicate that the fever resulting from the central administration of pyrogens is very similar to that resulting from peripherally induced fever and that the site of action of these pyrogens is the hypothalamus. There is still, however, one very serious gap in our knowledge of the pathogenesis of fever—whether pyrogens can cross the blood-brain barrier and pass from the periphery to the central nervous system, in particular to the hypothalamus. In 1955 Braude *et al.,* using very large (lethal) doses of endotoxin prepared from *E. coli,* labelled with ^{51}Cr, found no radioactivity in the brains of rabbits following parenteral administration; and in 1956 Rowley *et al.,* were unable to detect any radioactivity in the brains of either mice or guinea-pigs following parenteral administration of ^{32}P-labelled bacterial lipopolysaccharide.

Recently Dascombe and Milton (unpublished data) have prepared the endotoxin of *Shigella flexneri* labelled to a very high specific activity with ^{51}Cr and injected this intravenously into cats both in doses sufficient to produce a normal febrile response and in much higher doses. No radioactivity was found in any area of the brain or in the CSF following the various doses of labelled pyrogen. As well as measuring radioactivity in discrete areas of the brain, the hypothalamus was perfused following the administration of labelled pyrogen and, under these conditions, no radioactivity was found emanating from the hypothalamus.

Tritiated water injected at the same time was found to penetrate freely within the CNS. Since it is thought that endotoxins act by causing the production of endogenous pyrogen it is not necessary for endotoxins to penetrate the blood-brain barrier to produce fever. However, there is no convincing evidence that endogenous pyrogen itself can penetrate into the brain. The only experiments reported are those of Allen (1965), who prepared serum from afebrile and febrile rabbits and labelled the total serum proteins with ^{131}iodine. The labelled serum was injected into the carotid arteries of recipient rabbits, the animals were subsequently killed and autoradiographs prepared from brain slices. In rabbits receiving labelled serum from afebrile donors no radioactivity could be detected in any of the brain slices. However, in four out of five animals receiving serum from febrile animals, radioactivity was detected in the posterior hypothalamus. Radioactivity was not found in any other area of the brain, including the preoptic anterior hypothalamic area.

Allen made no attempt to purify the labelled serum and separate endogenous pyrogen from other serum proteins. Consequently she could not determine the identity of the radioactivity in the posterior hypothalamus. More recently Gander and Milton (unpublished data) have labelled endogenous pyrogen prepared from rabbit neutrophils (peritoneal exudate cells) with ^{125}Iodine. When this material was injected intravenously into conscious rabbits, no significant amounts of radioactivity could be detected in any area of the brain. Thus if endogenous pyrogen does not enter the CNS and if it is the mediator of pyrogen fever it must be initiating events outside the brain tissue, perhaps at the level of the brain capillaries. The sequence of events thought to take place in the pathogenesis of fever is summarised in figure 4.7.

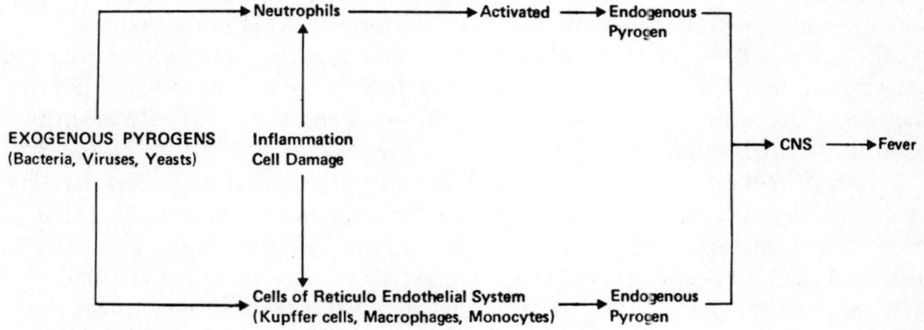

Figure 4.7 Diagrammatic representation of the sequence of events thought to occur in the development of fever

The Role of Prostaglandins in Fever and the Mode of Action of Antipyretic Drugs

In the late 1960s Milton and Wendlandt began to investigate the various transmitter substances thought to be involved in thermoregulation, with the particular aim of trying to obtain evidence for the involvement of one of these in pyrogen fever. They were seeking also to determine whether, if one such transmitter was involved, the action of antipyretic drugs in alleviating fever could be attributed to an effect on this particular transmitter. Their first experiments were concerned with the hyperthermic effect of 5-HT in the cat, and they showed that the long-lasting fevers produced by the intraventricular injection of both endotoxin and 5-HT could be prevented or abolished by paracetamol (Milton and Wendlandt, 1968).

As a result of these observations, CSF was obtained from febrile cats and assayed for 5-HT using the rat stomach strip preparation. Activity was found in some samples, but this activity was subsequently found not to be due to 5-HT. The active material was not characterised at that time, however, but it was considered to be a PG-like substance (Milton and Wendlandt, 1970). In this study on the effects of the prostaglandins on deep body temperature, it was discovered that, although the effects of PGE_1 resembled pyrogen fever in almost every way (except for the short duration of action), the PG 'fever' was not prevented or abolished by paracetamol. This observation lead to the theory that PGE_1 was a modulator in temperature regulation and that the action of antipyretic drugs was to interfere with the release of PGE_1 by pyrogen.

The subsequent observations of Milton and Wendlandt (1971*a, b*) and Feldberg and Saxena (1971*b, c*) on the effects of prostaglandins on body temperature discussed earlier indicate that a PG of the E series would be an ideal subject for modulating the increase in body temperature in fever, since PGE_1 and PGE_2 are active in very small amounts, their duration of action is short, they act on the area of the brain considered to be the centre for thermoregulation and are hyperthermic in all the placental mammals to which they have been administered.

One of the most convincing pieces of evidence connecting the prostaglandins with fever was provided by Vane (1971), who showed that the synthesis of PGE_2 and $PGF_{2\alpha}$ from arachidonic acid by guinea-pig lung homogenate was inhibited by aspirin-like drugs. Vane suggested that not only the analgesic and anti-inflammatory action of these drugs but also their antipyretic action could be explained by an inhibition of PG synthesis, a view that is now widely accepted. In 1971 Piper and Vane indicated that since there is little preformed PG in body tissues, PG synthesis could be equated with PG release. Consequently Vane's observations on the inhibition of PG synthesis by aspirin-like drugs provided the explanation to the theory previously put forward by Milton and Wendlandt (1970) that antipyretic drugs acted by preventing PG release in the CNS.

Since 1970 there have been many publications implicating the prostaglandins in fever, although there have also been those which provide evidence against their involvement. In 1973 Feldberg and Gupta obtained CSF from the third ventricle of the conscious cat and assayed it for contractile activity using the rat fundus strip preparation (Vane, 1957). They found that, whilst in afebrile animals the activity was very low or absent, during fever produced by injecting pyrogen directly into the third ventricle the activity was considerably greater. Following administration of paracetamol the fever abated and the contractile activity of the CSF was again low. From their results Feldberg and Gupta concluded that the contractile substance present in the CSF was a PG.

Feldberg *et al.* (1973) collected CSF from the cisterna magna of the conscious cat and assayed it for PGE-like activity. They found that the O-somatic antigen of *Shigella dysenteriae* produced a fever when administered into both the third ventricle and cisterna magna and also when given intravenously. In all cases during the febrile response the PGE-like activity of the CSF increased and the three antipyretic drugs acetylsalicylic acid (aspirin), paracetamol and indomethacin all abolished fever and at the same time the PGE content of the CSF fell (figure 4.8). Thin-layer chromatography of the CSF samples followed by bioassay and radio-

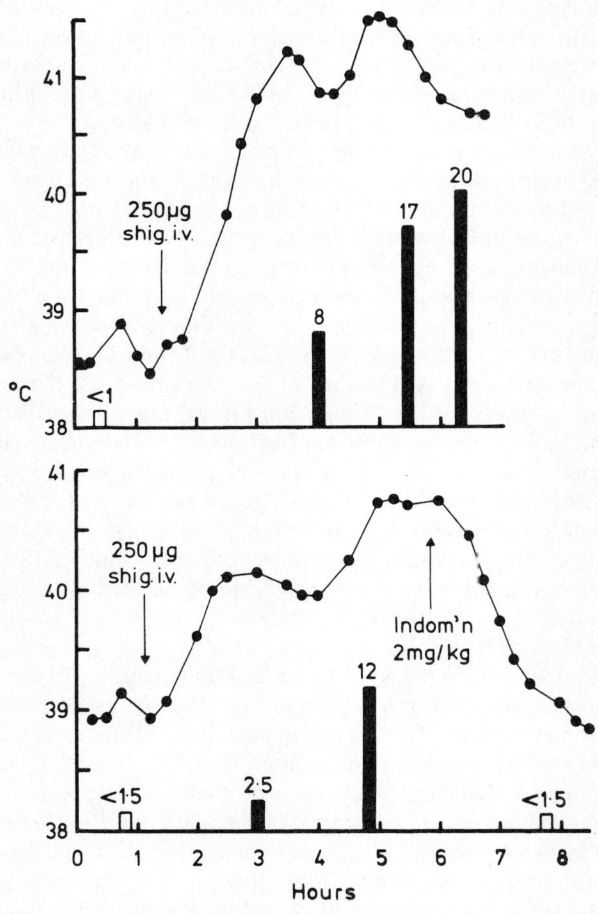

Figure 4.8 Records of rectal temperature from two unanaesthetised cats. The height of the columns and the values above refer to PGE_1-like activity in ng/ml of cisternal c.s.f.; the position of the columns refers to the time but not to the duration of the c.s.f. collection. The first arrow in both records indicates an i.v. injection of 250 μg pyrogen and the second arrow in the bottom record an i.p. injection of indomethacin, 2 mg/kg (Indom'n). (After Feldberg *et al.*, 1973, with permission)

immunoassay indicated that the prostaglandin present in the CSF of the cat during fever was PGE_2.

Similar results were obtained by Harvey *et al.* (1975) in the rabbit in which they producêd fever both with the O-somatic antigen of *Shigella dysenteriae* and with a purified pyrogen prepared from *Proteus vulgaris* ('E' Pyrogen, Organon). In addition they found that if rabbits were made tolerant to the fever-producing effect of the 'E' pyrogen by injecting it intravenously every day for ten days, then on the tenth day when the animals were refractory to the pyrogenic action no increase in the PGE content of the CSF was found. In 1974 Dey *et al.* injected purified lipid A prepared from a mutant strain of Salmonella into the ventricular system of the conscious cat and showed that this substance also produced fever accompanied by an increase in the PGE content of cisternal CSF. In addition they showed that antipyretic drugs inhibited both the fever and the release of PGE.

It is now generally accepted that bacterial pyrogens activate neutrophils and other cells to synthesise and release a low molecular weight protein known as endogenous pyrogen (leucocytic pyrogen). Harvey and Milton (1975) prepared endogenous pyrogen from cat peritoneal exudate and found that when this material was infused intravenously into a conscious cat it produced fever which was associated with the increase in the PGE content of the cisternal CSF and that this fever and the increase in PGE were inhibited by antipyretic agents. In contrast to these results, Cranston *et al.* (1975), carrying out similar experiments in the rabbit, showed that endogenous pyrogen produced fever and a rise in the PGE content of the CSF. However, when they infused aspirin at the same time as the endogenous pyrogen no increase in the PGE levels of the CSF was observed, although the dose of aspirin used had no antipyretic action and the fever still occurred. These authors maintained that, since it was possible to dissociate the effect of aspirin on PGE release from its effect on fever, the direct relationship between PGE and fever was in question. However, their experiments did not show whether, sufficient PGE was still being synthesised and released in the temperature-regulating region of the anterior hypothalamus, even though the PGE content of the CSF collected from the cisterna magna was reduced. The fact that the dose of aspirin which they infused was not antipyretic could be used as an argument to support the view that sufficient PGE was still being synthesised within the hypothalamus to produce fever.

Veale and Cooper (1975) reported that when they made extensive lesions in the preoptic area of the anterior hypothalamus of the rabbit and then applied PGE_1 locally it no longer produced a rise in deep body temperature. This observation provided further evidence that the site of action of the PGE_1 was in the preoptic anterior hypothalamus. They also observed that, in these lesioned animals, locally applied leucocytic pyrogen was also ineffective in producing a fever. In contrast to these observations, when they injected PGE_1 and leucocytic pyrogen into a lateral cerebral ventricle the hyperthermic effect of the PGE_1 was abolished by the lesion, but the fever produced by the leucocytic pyrogen was not, even though the onset of fever was more gradual. Similarly, leucocytic pyrogen given intravenously also produced fever in a lesioned animal.

The authors concluded from their results that leucocytic pyrogen also acts at a site other than the preoptic area of the anterior hypothalamus to produce fever and that this action is independent of the release of PGE_1, though in their investigations they were unable to find such a site. Unfortunately the authors did not investigate whether the fever produced by leucocytic pyrogen following lesioning of the preoptic area of the anterior hypothalamus was inhibited by antipyretic agents. If this secondary site does exist, it is possible that leucocytic pyrogen administered intravenously or intraventricularly could reach the site and release a PG. However, PGE_1, when injected into the cerebral ventricles, would be unable to reach the site.

Harvey and Milton (1975) found that if plasma obtained from a donor cat, in which fever had been produced by intravenously administered *Shigella dysenteriae*, was injected into a recipient cat which had been made refractory to the pyrogen then the recipient cat developed a fever accompanied by an increase in cisternal CSF PGE levels. These experiments showed that during bacterial pyrogen fever there was a circulating pyrogenic material in the plasma which differed from the bacterial pyrogen and which was itself capable of producing PGE release. It was concluded that this circulating pyrogen was endogenous pyrogen. In contrast, when they injected bacterial pyrogen directly into the cerebral ventricular system to produce fever no circulating pyrogenic material could be detected in the plasma. These results show, therefore, that centrally administered bacterial pyrogen does not activate the synthesis and release of endogenous pyrogen peripherally and must produce fever by acting on cells within the central nervous system.

Hypothermic Effects of Antipyretic Drugs in Afebrile States

Antipyretic drugs such as paracetamol and indomethacin (but generally not salicylates) may produce a fall in deep body temperature when administered to both man and animals in the absence of fever, particularly when given in large doses. In addition when these drugs are given as antipyretics to reduce fever the temperature may fall from the fever level to below that found in the afebrile state.

In 1973 Milton investigated this phenomenon to determine whether the fall in body temperature in the absence of fever could be attributed to an inhibition of prostaglandin synthesis and release. Indomethacin and paracetamol both produced a fall in deep body temperature when administered intraperitoneally to the conscious cat. This hypothermia was accompanied by ear skin vasodilatation and, when high doses were administered, by panting. When PGE_1 was infused into a lateral ventricle shivering and ear skin vasoconstriction occurred and deep body temperature rose. When the temperature had reached a plateau, and whilst the infusion of PGE_1 was continued, paracetamol and indomethacin were administered. Both drugs produced vasodilatation and panting but had no effect on the shivering; deep body temperature fell slightly before reaching a plateau level which was sustained until the infusion was stopped.

From these results it was considered that the effects of the two drugs to produce ear skin vasodilatation and panting were not mediated through inhibition of prostaglandin synthesis but were due to an action of the two drugs on the heat-loss mechanisms concerned. Acetylsalicylic acid did not affect deep body temperature in either the afebrile state or during PGE_1 infusion. These results are also regarded as further evidence that the prostaglandins are not involved in normal thermoregulation.

CONCLUSIONS

From this account of the actions of neurotransmitters and related drugs on the hypothalamic control of body temperature it will be obvious to the reader that we have an enormous amount of information with a minimum amount of understanding. It would be almost true to say that whatever substance we inject into the hypothalamus it will either cause the body temperature to go up or it will cause it to go down. However, every piece of information is potentially useful and it is to be hoped that in future an understanding will come from all our collective work.

This account is in no way comprehensive. The emphasis has been placed on the work and ideas of personal friends and collaborators and of the author himself. Many research workers have not been mentioned and it is hoped that they will not be offended. In particular it is hoped that this account may interest those not already working in the field, and perhaps inspire them to delve more deeply and above all to enter this fascinating research field for themselves.

REFERENCES

Allen, I. V. (1965). The cerebral effects of endogenous serum and granulocytic pyrogen. *Brit. J. Exp. Pathol.*, **46**, 25–34

Andersson, B., Jobin, M. and Olsson, K. (1966). Serotonin and temperature control. *Acta. Physiol. Scand.*, **67**, 50–56

Atkins, E. and Bodel, P. T. (1974). Fever. In L. Grand and R. T. McClusky (eds.). *The Inflammatory Process,* Vol. III, Academic Press, New York, 467–514

Banerjee, V., Burks, T. F. and Feldberg, W. (1968). Effect on temperature of 5-hydroxytryptamine injected into cerebral ventricles of cats. *J. Physiol.* (*Lond.*), **195**, 245–251

Beckman, A. L. and Carlisle, H. J. (1969). Effect of intrahypothalamic infusion of acetylcholine on behavioural and physiological thermoregulation in the rat. *Nature* (*Lond.*), **221**, 561–562

Beeson, P. B. (1948). Temperature-elevating effect of a substance obtained from polymorphonuclear leucocytes. *J. Clin. Invest.*, **27**, 524

Bennett, I. L. and Beeson, P. B. (1953). Studies in the pathogenesis of fever. Part II: Characterisation of fever-producing substances from polymorphonuclear leucocytes and from the fluid of sterile exudates. *J. exp. Med.*, **98**, 493–508

Bennett, I. L. Jr., Petersdorf, R. G. and Keene, W. R. (1957). Pathogenesis of fever: evidence for direct cerebral action of bacterial endotoxins. *Trans. Assoc. Am. Physicians,* **70**, 64–73

Bligh, J. (1966). Effects on temperature of monoamines injected into the lateral ventricles of sheep. *J. Physiol.* (*Lond.*), **185**, 46–47P

Bligh, J. (1973). *Temperature Regulation in Mammals and Other Vertebrates,* North-Holland Press, Amsterdam

Bligh, J. and Cottle, W. H. (1969). The influence of ambient temperature on thermoregulatory responses to intraventricularly injected monoamines in sheep, goats and rabbits. *Experientia,* **25**, 608–609

Bligh, J., Cottle, W. H. and Maskrey, M. (1971). Influence of ambient temperature on the thermoregulatory responses to 5-hydroxytryptamine, noradrenaline and acetylcholine injected into the lateral cerebral ventricles of sheep, goats and rabbits. *J. Physiol.* (*Lond.*), **212**, 377–392

Bligh, J. and Milton, A. S. (1973). The thermoregulatory effects of prostaglandin E_1 when infused into a lateral cerebral ventricle of the Welsh mountain sheep at different ambient temperatures. *J. Physiol. (Lond.)*, **229**, 30–31P

Braude, A. I., Carey, F. J. and Zalesky, M. (1955). Studies with radioactive endotoxin. Part II: Correlation of physiologic effects with distribution of radioactivity in rabbits injected with lethal doses of *E. coli* endotoxin labelled with radioactive sodium chromate. *J. Clin. Invest.*, **34(i)**, 858–866

Breckenridge, B. McL. and Lisk, R. D. (1969). Cyclic adenylate and hypothalamic regulatory functions. *Proc. Soc. exp. Biol. Med.*, **131**, 934–935

Brittain, R. T. and Handley, S. L. (1967). Temperature changes produced by the injection of catecholamines and 5-hydroxytryptamine into the cerebral ventricles of the conscious mouse. *J. Physiol. (Lond.)*, **192**, 805–813

Brodie, B. B. and Shore, P. A. (1957). A concept for a role of serotonin and norepinephrine as chemical mediators in the brain. *Ann. N. Y. Acad. Sci.*, **66**, 631–642

Butcher, R. W. (1968). Role of cyclic AMP in hormone actions. *New Engl. J. Med.*, **279**, 1378–1384

Butcher, R. W. and Sutherland, E. W. (1962). Adenosine 3′, 5′-phosphate in biological materials. Part I: Purification and properties of cyclic 3′, 5′-nucleotide phosphodiesterase and use of this enzyme to characterize adenosine 3′, 5′-phosphate in human urine. *J. Biol. Chem.*, **237**, 1244–1250

Cabanac, M., Stolwijk, J. A. J. and Hardy, J. D. (1968). Effect of temperature and pyrogens on single-unit activity in the rabbit's brainstem. *J. appl. Physiol.*, **24**, 645–652

Cammock, S., Dascombe, M. J. and Milton, A. S. (1976). Prostaglandins in thermoregulation. In B. Samuellson and R. Paoletti (eds.). *Advances in Prostaglandin and Thromboxane Research*, Vol. 1, Raven Press, New York, 375–380

Canal, N. and Ornesi, A. (1961). La Serotonina quale agente ipertermizzante. *Atti Accad. Med. Lomb.*, **16**, 64–69

Cheung, W. Y. (1970). Cyclic nucleotide phosphodiesterase. In P. Greengard and E. Costa (eds.). *Role of cyclic AMP in Cell Function*, Raven Press, New York, 51–65

Clark, W. G. and Cumby, H. R. (1975). The antipyretic effect of indomethacin. *J. Physiol. (Lond.)*, **248**, 625–638

Clark, W. G., Cumby, H. R. and Davis, H. E. (1974). The hyperthermic effect of intracerebroventricular cholera enterotoxin in the unanesthetized cat. *J. Physiol. (Lond.)*, **175** 493–504

Cooper, K. E. (1966). Temperature regulation and the hypothalamus. *Br. med. Bull.*, **22**, 238–242

Cooper, K. E., Cranston, W. I. and Honour, A. J. (1964). Temperature changes induced by 5-HT, noradrenaline and pyrogens injected into the rabbit brain. *J. Physiol. (Lond.)*, **240**, 68–69P

Cooper, K. E., Cranston, W. I. and Honour, A. J. (1965). Effects of intraventricular and intrahypothalamic injection of noradrenaline and 5-HT on body temperature in conscious rabbits. *J. Physiol. (Lond.)*, **181**, 852–864

Cooper, K. E., Cranston, W. I. and Honour. A. J. (1967). Observations on the site and mode of action of pyrogens in the rabbit brain. *J. Physiol. (Lond.)*, **191**, 325–337

Cox, B. and Lee, T. F. (1977*a*). Interactions between cholinergic and dopaminergic systems in thermoregulation. In K. E. Cooper, P. Lomax and E. Schönbaum (eds.). *Drugs, Biogenic Amines and Body Temperature*, Karger, Basle, 30–31

Cox, B. and Lee, T. F. (1977*b*). Do central dopamine receptors have a physiological role in thermoregulation? *Br. J. Pharmac.*, **61**, 83–86

Cranston, W. I., Hellon, R. F. and Mitchell, D. (1975). Fever and brain prostaglandin release. *J. Physiol. (Lond.)*, **248**, 27–29P

Cushing, H. (1931*a*). The similarity in the response to posterior lobe extract (pituitrin) and to pilocarpine when injected into the cerebral ventricles. *Proc. Nat. Acad. Sci.*, **17**, 171–177

Cushing, H. (1931*b*). The action of atropine in counteracting the effects of pituitrin and of pilocarpine injected into the cerebral ventricles. *Proc. Nat. Acad. Sci.*, **17**, 178–180

Cushing, H. (1931*c*). The method of action of pituitrin introduced into the ventricle. *Proc. Nat. Acad. Sci.*, **17**, 239–247

Dascombe, M. J. (1977). Effects of methylxanthine drugs on pyrogen-induced hyperthermia. *Eur. J. Pharmac.*, **45**, 389–392

Dascombe, M. J. and Milton, A. S. (1972). The effect of caffeine on the antipyretic action of aspirin administered during endotoxin induced fever. *Br. J. Pharmac.*, **46**, 548–549P

Dascombe, M. J. and Milton, A. S. (1975). The effects of cyclic adenosine 3′, 5′-monophosphate and other adenine nucleotides on body temperature. *J. Physiol. (Lond.)*, **250**, 143–160

Dascombe, M. J. and Milton, A. S. (1976). Cyclic adenosine 3′, 5′-monophosphate in cerebrospinal fluid during thermoregulation and fever. *J. Physiol. (Lond.)*, **263**, 441–463

Dey, P. K., Feldberg, W., Gupta, K. P., Milton, A. S. and Wendlandt, S. (1974). Further studies on the role of prostaglandins in fever. *J. Physiol. (Lond.)*, **241**, 629–646

Downey, J. A., Mottram, R. F. and Pickering, G. W. (1964). The location by regional cooling of central temperature receptors in the conscious rabbit. *J. Physiol. (Lond.)*, **170**, 415–441

Duffy, M. J. and Schwarz, V. (1974). The effects of adenosine 3′, 5′-cyclic monophosphate and adenosine triphosphate on calcium ion binding in erythrocyte membranes. *Biochem. Soc. Trans.*, **2**, 406–407

Eisenman, J. S. (1969). Pyrogen-induced changes in the thermosensitivity of septal and preoptic neurons. *Am. J. Physiol.*, **216**, 330–334

Eisenman, J. S. and Jackson, D. C. (1967). Thermal response patterns of septal and preoptic neurons in cats. *Exptl. Neurol.*, **19**, 33–45

Eliasson, S. and Ström, G. (1950). On the localization in the cat of hypothalamic and cortical structures influencing cutaneous blood flow. *Acta. Physiol. Scand.*, **20**, suppl. 70, 113–118

von Euler, C., Linder, E. and Myrin, S.-O. (1943). Über die Fiebererregende Wirkung des Adrenalins. *Acta. Physiol. Scand.*, **5**, 85–96

Ewen, L., Milton, A. S. and Smith, S. (1976). Effects of prostaglandin $F_{2\alpha}$ and prostaglandin D_2 on the body temperature of conscious cats. *J. Physiol. (Lond.)*, **258**, 121–122P

Feldberg, W. and Gupta, K. P. (1973). Pyrogen fever and prostaglandin-like activity in cerebrospinal fluid. *J. Physiol. (Lond.)*, **288**, 41–53

Feldberg, W., Gupta, K. P., Milton, A. S. and Wendlandt, S. (1973). Effect of pyrogen and antipyretics on prostaglandin activity in cisternal CSF of unanaesthetized cats. *J. Physiol. (Lond.)*, **234**, 279–303

Feldberg, W., Hellon, R. F. and Lotti, V. J. (1967). Temperature effects produced in dogs and monkeys by injections of monoamines and related substances into the third ventricle. *J. Physiol. (Lond.)*, **191**, 501–515

Feldberg, W., Hellon, R. F. and Myers, R. D. (1966). Effects on temperature of monoamines injected into the cerebral ventricles of anaesthetized dogs. *J. Physiol, (Lond.)*, **186**, 416–423

Feldberg, W. and Lotti, V. J. (1967). Temperature responses to monoamines and an inhibitor of monoamine oxidase injected into the cerebral ventricles of rats. *Br. J. Pharmac. Chemother.*, **31**, 152–161

Feldberg, W. and Myers, R. D. (1963). A new concept of temperature regulation by amines in the hypothalamus. *Nature (Lond.)*, **200**, 1325

Feldberg, W. and Myers, R. D. (1964). Effect on temperature of amines injected into the cerebral ventricles. A new concept of temperature regulation. *J. Physiol. (Lond.)*, **173**, 226–237

Feldberg, W. and Myers, R. D. (1965). Changes in temperature produced by microinjection of amines into the anterior hypothalamus of cats. *J. Physiol. (Lond.)*, **177**, 239–245

Feldberg, W., Myers, R. D. and Veale, W. L. (1970). Perfusion from cerebral ventricle to cisterna magna in the unanaesthetized cat. Effect of calcium on body temperature. *J. Physiol. (Lond.)*, **207**, 403–416

Feldberg, W. and Saxena, P. N. (1970). Mechanism of action of pyrogen. *J. Physiol. (Lond.)*, **211**, 245–261

Feldberg, W. and Saxena, P. N. (1971*a*). Effects of adrenoceptor blocking agents on body temperature. *Brit. J. Pharmac.*, **43**, 543–554

Feldberg, W. and Saxena, P. N. (1971*b*). Fever studies on prostaglandin E_1 fever in cats. *J.* Physiol. (*Lond.*), **219**, 739–745

Feldberg, W. and Saxena, P. N. (1971*c*). Fever produced by prostaglandin E_1. *J. Physiol. (Lond.)*, **217**, 547–556

Findlay, J. D. and Robertshaw, D. (1967). The mechanism of body temperature changes induced by intraventricular injections of adrenaline, noradrenaline and 5-hydroxytryptamine in the ox (*Bos taurus*). *J. Physiol. (Lond.)*, **189**, 329–336

Findlay, J. D. and Thompson, G. E. (1968). The effect of intraventricular injections of noradrenaline, 5-hydroxytryptamine, acetylcholine and tranylcypromine on the ox (*Bos taurus*) at different environmental temperatures. *J. Physiol. (Lond.)*, **194**, 809–816

Flower, R. J. and Vane, J. R. (1972). Inhibition of prostaglandin synthesis in brain explains the antipyretic activity of paracetamol (4-acetamidophenol). *Nature*, **240**, 410–411

Friedman, M. (1944). Etiology and pathogenesis of neurocirculatory asthenia. Part I: Hyperthermia as one of the manifestations of neurocirculatory asthenia. *War Med.*, **6**, 221–227

Gessa, G. L., Krishna, G., Forn, J., Tagliamonte, A. and Brodie, B. B. (1970). Behavioral and vegetative effects produced by dibutyryl cyclic AMP injected into different areas of the brain. In F. Greengard and E. Costa (eds.). *Role of Cyclic AMP in Cell Function,* Raven Press, New York, 371–381

Goldberg, A. L. and Singer, J. J. (1969). Evidence for a role of cyclic AMP in neuromuscular transmission. *Proc. Natn. Acad, Sci.* USA, **64**, 134–141

Hall, G. H. (1972). Changes in body temperature produced by cholinomimetic substances injected into the cerebral ventricles of unanaesthetized cats. *Br. J. Pharmac.*, **44**, 634–641

Hammel, H. T., Hardy, J. D. and Fusco, M. M. (1960). Thermoregulatory responses to hypothalamic cooling in unanaesthetised dogs. *Am. J. Physiol.*, **198**, 481–486

Harvey, C. A. and Milton, A. S. (1974). The effect of parachlorophenylalanine on the response of the conscious cat to intravenous and intraventricular bacterial pyrogen and to intraventricular prostaglandin E_1. *J. Physiol. (Lond.)*, **236**, 14–15P

Harvey, C. A. and Milton, A. S. (1975). Endogenous pyrogen fever, prostaglandin release and prostaglandin synthetase inhibitors. *J. Physiol. (Lond.)*, **250**, 18–20P

Harvey, C. A. and Milton, A. S. (1976). The effects of parachlorophenylalanine and 6-hydroxydopamine on thermoregulatory responses to heat and cold stress. *J. Physiol. (Lond.)*, **263**, 208–209P

Harvey, C. A., Milton, A. S. and Straughan, D. W. (1975). Prostaglandin E levels in cerebrospinal fluid of rabbits and the effects of bacterial pyrogen and antipyretic drugs. *J. Physiol. (Lond.)*, **248**, 26–27P

Hellon, R. F. (1967). Thermal stimulation of hypothalamic neurones in unanaesthetized rabbits. *J. Physiol. (Lond.)*, **193**, 381–395

Hellon, R. F. (1970). The stimulation of hypothalamic neurons by changes in ambient temperature. *Pflügers. Arch.*, **321**, 56–66

Hemingway, A., Rasmussen, T., Wikoff, H. and Rasmussen, A. T. (1940). Effects of heating hypothalamus of dogs by diathermy. *J. Neurophysiol.*, **3**, 329–338

Henderson, W. R. and Wilson, W. C. (1936). Intraventricular injection of acetylcholine and eserine in man. *Quart. J. exptl. Physiol.*, **26**, 83–95

Hill, H. F. and Horita, A. (1971). Inhibition of D-amphetamine hyperthermia by blockade of dopamine receptors. *J. Pharm. Pharmacol.*, **23**, 715–717

Hill, H. F. and Horita, A. (1972). A pimozide sensitive effect of apomorphine on body temperature of the rabbit. *J. Pharm. Pharmacol.*, **24**, 490–491

Horita, A. and Gogerty, J. H. (1958). The pyretogenic effect of 5-hydroxytryptophan and its comparison with that of LSD. *J. Pharmacol. exp. Ther.*, **122**, 195–200

Jackson, D. L. (1967). A hypothalamic region responsive to localized injection of pyrogens. *J. Neurophysiol.*, **30**, 586–602.

Jacobson, F. H. (1967). 'Warmth response' evoked by preoptic injections of serotonin. *Fedn. Proc.*, **26**, 555

Kahn, R. H. (1904). Über die Erwärmung des Carotidblutes. *Arch. Anat. Physiol.*, **28**, 81–134

Kennedy, M. S. and Burks, T. F. (1974). Dopamine receptors in the central thermoregulatory mechanism of the cat. *Neuropharmac.*, **13**, 119–128

King, M. K. and Wood, W. B. Jr. (1958). Studies on the pathogenesis of fever. Part IV: The site of action of leucocytic and circulating endogenous pyrogen. *J. exp. Med.*, **107**, 291–303

Komiskey, H. L. and Rudy, T. A. (1977). Serotoninergic influences on brainstem thermoregulatory mechanisms in the cat. *Brain Res.*, **134**, 297–315

Kruk, Z. L. (1972). The effect of drugs acting on dopamine receptors on the body temperature of the rat. *Life Sci.*, **11**, Part I, 845–850

Kulkarni, A. S. (1967). Effects on temperature of serotonin and epinephrine injected into the lateral cerebral ventricle of the cat. *J. Pharmacol. Exp. Ther.*, **157**, 541–545

Laburn, H., Rosendorff, C., Willies, G. and Woolf, C. (1974). A role for noradrenaline and cyclic AMP in prostaglandin E_1 fever. *J. Physiol. (Lond.)*, **240**, 49–50P

Magoun, H. W., Harrison, F., Brobeck, J. R. and Ranson, S. W. (1938). Activation of heat-loss mechanisms by local heating of the brain. *J. Neurophysiol.*, **1**, 101–114

Milton, A. S. (1973). Prostaglandin E_1 and endotoxin fever, and the effects of aspirin, indomethacin and 4-acetamidophenol. *Advs. Biosciences*, **9**, 495–500

Milton, A. S. and Harvey, C. A. (1975). Prostaglandins and monoamines in fever. In P. Lomax, E. Schönbaum and J. Jacob (eds.). *Temperature Regulation and Drug Action*, Karger, Basel, 133–142

Milton, A. S. and Wendlandt, S. (1968). The effect of 4-acetamidophenol in reducing fever produced by the intracerebral injection of 5-hydroxytryptamine and pyrogen in the conscious cat. *Brit. J. Pharmac.*, **34**, 215P

Milton, A. S. and Wendlandt, S. (1970). A possible role for prostaglandin E_1 as a modulator for temperature regulation in the central nervous system of the cat. *J. Physiol. (Lond.)*, **207**, 76–77P

Milton, A. S. and Wendlandt, S. (1971*a*). The effects of 4-acetamidophenol (paracetamol) on the temperature response of the conscious rat to the intracerebral injection of prostaglandin E_1, adrenaline and pyrogen. *J. Physiol. (Lond.)*, **217**, 33–34P

Milton, A. S. and Wendlandt, S. (1971*b*). Effects on body temperature of prostaglandins of the A, E and F series on injection into the third ventricle of unanaesthetized cats and rabbits. *J. Physiol. (Lond.)*, **218**, 325–336

Murakami, N. (1973). Effects of iontophoretic application of 5-hydroxytryptamine, noradrenaline and acetylcholine upon hypothalamic temperature-sensitive neurones in rats. *Jap. J. Physiol.*, **23**, 435–466

Myers, R. D. (1966). Release of chemical factors from the diencephalic region of the unanaesthetized monkey during changes in body temperature. *J. Physiol. (Lond.)*, **188**, 50–51P

Myers, R. D. (1971). Hypothalamic mechanisms of pyrogen action in the cat and monkey. In G. E. W. Wolstenholme and J. Birch (eds.). *Pyrogens and Fever*, Churchill Livingstone, Edinburgh and London, 131–146

Myers, R. D. and Chinn, C. (1973). Evoked release of hypothalamic norephinephrine during thermoregulation in the cat. *Am. J. Physiol.*, **224**, 230–236

Myers, R. D. and Sharpe, L. G. (1967). Intracerebral injections and perfusions in the conscious monkey. In H. Vagtborg (ed.). *The Use of Subhuman Primates in Drug Evaluation.* Univ. Texas Press, Austin

Myers, R. D. and Yaksh, T. L. (1968). Feeding and temperature responses in the unrestrained rat after injections of cholinergic and aminergic substances into the cerebral ventricles. *Physiol. Behav.*, **3**, 917–928

Myers, R. D. and Yaksh, T. L. (1969). Control of body temperature in the unanaesthetized monkey by cholinergic and aminergic systems in the hypothalamus. *J. Physiol. (Lond.)*, **202**, 483–500

Nakayama, T., Eisenman, J. S. and Hardy, J. D. (1961). Single-unit activity of anterior hypothalamus during local heating. *Science N. Y.*, **134**, 560–561

Nakayama, T., Hammel, H. T., Hardy, J. D. and Eisenman, J. S. (1963). Thermal stimulation of electrical activity of single units of the preoptic region. *Am. J. Physiol.*, **204**, 1122–1126

Pettinger, W. A., Bautz, G. T., Wiggan, G. A. and Sheppard, H. (1970). Cyclic AMP as a mediator of vasodilatation: indirect evidence. *Pharmacologist*, **12**, 291

Piper, P. and Vane, J. (1971). The release of prostaglandins from lung and other tissues. *Ann. N. Y. Acad. Sci.*, **180**, 363–385

Pittman, Q. J., Veale, W. L. and Cooper, K. E. (1976). Observations on the effect of salicylate in fever and the regulation of body temperature against cold. *Can. J. Physiol. and Pharmacol.*, **54**, 101–106

Rall, T. W. and Sattin, A. (1970). Factors influencing the accumulation of cyclic AMP in brain tissue. In F. Greengard and E. Costa (eds.). *Role of Cyclic AMP in Cell Function*, Raven Press, New York, 113–133

Reimann, H. A. (1967). Caffeinism: a cause of long-continued, low-grade fever. *J. Am. Med. Assoc.*, **202**, 1105–1106

Ritchie, J. M. (1970). Central nervous system stimulants. Part II: the xanthines. In L. S. Goodman and A. Gilman (eds.). *The Pharmacological Basis of Therapeutics*, Macmillan, London, 4th edn., 358–370

Robison, G. A., Butcher, R. W. and Sutherland, E. W. (1971). *Cyclic AMP*, Academic Press, New York

Rowley, D., Howard, J. G. and Jenkin, C. R. (1956). The fate of ^{32}P-labelled bacterial lipopolysaccharide in laboratory animals. *Lancet*, **1**, 336–367

Ruckebusch, Y., Grivel, M. L. and Laplace, J. P. (1965). Variations interspécifiques des modifications de la température centrale liées a l'injection cérébroventriculaire de catécholamines et de 5-hydroxytryptamine. *C. R. seanc. Soc. Biol.*, **159**, 8–9

Satinoff, E. (1964). Behavioural thermoregulation in response to local cooling of the rat brain. *Am. J. Physiol.*, **206**, 1389–1394

Simmonds, M. A. (1970). Effect of environmental temperature on the turnover of 5-hydroxytryptamine in various areas of rat brain. *J. Physiol. (Lond.)*, **211**, 93–108

Soifer, H. (1957). Aminophylline toxicity. *J. Pediat.*, **50**, 657–669

Stitt, J. T. (1973). Prostaglandin E_1 fever induced in rabbits. *J. Physiol. (Lond.)*, **232**, 163–179

Stuart, D. G., Maxwell, D. S., Hayward, J. N., Fairchild, M. D., Adey, W. R. and Porter, R. W. (1963). Unit activity in the hypothalamus. *Bol. Inst. Estud. méd. Biol.*, **21**, 349–370

Sutherland, E. W. and Rall, T. W. (1958). Fractionation and characterization of a cyclic adenine ribonucleotide formed by tissue particles. *J. biol. Chem.*, **232**, 1077–1091

Tagliamonte, A., Tagliamonte, P., Perez-Cruet, J., Stern, S. and Gessa, G. L. (1971). Effect of psychotropic drugs on tryptophan concentration in the rat brain. *J. Pharmacol. Exp. Ther.*, **177**, 475–480

Toivola, P. and Gale, C. C. (1970). Effect on temperature of biogenic amine infusion into hypothalamus of baboon. *Neuroendocrinology*, **6**, 210–219

Uchimura, H. and Murakami, N. (1973). Effects of adrenoceptor-blocking agent phenoxybenzamine on rectal temperature and skin temperature of rabbits. *J. Physiol. Soc. Japan*, **35**, 435–436

Vane, J. R. (1957). A sensitive method for the assay of 5-hydroxytryptamine. *Brit. J. Pharmac. Chemother.*, **12**, 344–349

Vane, J. R. (1971). Inhibition of prostaglandin synthesis as a mechanism of action of aspirin-like drugs. *Nature, New Biology*, **231**, 232–235

Varagić, V. M. and Beleslin, D. B. (1973). The effect of cyclic *N*-2-*o*-dibutyryl-adenosine-3′, 5′-monophosphate, adenosine triphosphate and butyrate on the body temperature of conscious cats. *Brain Res.*, **57**, 252–254

Veale, W. F. and Cooper, K. E. (1975). Comparison of sites of action of prostaglandin and leucocyte pyrogen in brain. In P. Lomax, E. Schönbaum and J. Jacob (eds.). *Temperature Regulation and Drug Action*, Karger, Basel, 218–226

Villablanca, J. and Myers, R. D. (1965). Fever produced by microinjection of typhoid vaccine into hypothalamus of cats. *Am. J. Physiol.*, **208**, 703–707

Westphal, O. (1957). Pyrogen. In G. F. Springer (ed.). *Polysaccharides in Biology*, Macy, New York, 115

Wit, W. and Wang, S. C. (1968). Temperature-sensitive neurons in preoptic/anterior hypothalamic region: effects on increasing ambient temperature. *Am. J. Physiol.*, **215**, 1151–1169

Discussion

Morley (Alderley Park)
Have you investigated the action of endorphins?

Milton
Yes, but we haven't been able to involve them yet. We wondered whether there would be any release of enkephalins during pyrogen fever, but we have not seen any consistent release. There may however be increases during brain damage.

Feldberg (Mill Hill)
May I offer a partial answer to your question. Morphine itself is pyrogenic in the cat; enkephalins haven't been tried but they would be expected to have the same effect as morphine.

Milton
If I could add to that. Morphine hyperthermia in the cat is not blocked by antipyretic drugs but it is blocked by nalaxone. The hyperthermia is unusual in that it may be due to increased motor activity. Also the hypothalamus is unlikely to be the site of action because morphine injected directly into the hypothalamus is ineffective and there is no increased motor activity. After intraventricular injection, when morphine reaches other brain structures, there is both an increased motor activity and hyperthermia.

Barratt (Leeds)
In the central part of your review you talked of cholinomimetics, catecholamines, 5-HT and other postulated neurotransmitters. I fail to see that this kind of evidence supports a physiologic involvement. It only shows that you are able to manipulate temperature pharmacologically. I think you are right. As I said at the end of my presentation, the pharmacologist using these techniques alone doesn't accomplish very much.

Milton
However, one should consider the use of blocking drugs where the evidence is perhaps more impressive.

Matthews (Matthew)
Can Professor Milton throw any light on progesterone thermogenesis and its mechanism?

Milton
I'm afraid I know nothing about that subject. Some work is appearing on releasing substances such as TRF, but that is all the work I know about.

Bisset (London)
I wonder whether Professor Milton would like to comment of the controversy about the involvement of prostaglandins in pyrogen fever.

Milton
It has been reported that prostaglandin blocking drugs prevent prostaglandin hyperthermia but have no effect on pyrogen fever, which argues against a prostaglandin E being involved in the pyrogen response. The problem is that the blocking drug needs to be given locally in concentrations 10,000 times that of the agonist and this concentration may cause damage at the injection site. Also the pyrogens may be releasing prostaglandins from other sites. It has been shown that if the hypothalamus is destroyed, then locally applied prostaglandin and pyrogen are ineffective, yet intravenous pyrogen causes a fever. Therefore it looks as if pyrogen can produce fever by acting at sites other than the hypothalamus. However, as prostaglandins injected intraventricularly are ineffective it is argued this fever does not involve prostaglandins. My view is that two other experiments are required. First, prostaglandin should be injected locally into other brain sites which it may not reach after intraventricular injection. Secondly, the fever after intravenous pyrogen should be challenged with an antipyretic drug, because if the antipyretic were effective the chances would be that it was a prostaglandin fever. I find it inconceivable not to believe that prostaglandins are involved in fever. They are the most active substances we know which increase deep body temperature. During fever prostaglandin levels in the cerebrospinal fluid increase. The antipyretic drugs are prostaglandin synthetase inhibitors: during fever they reduce both body temperature and the increases in the prostaglandin levels of the cerebrospinal fluid.

Webster (London)
Have you investigated whether you can modify prostaglandin-induced elevation of temperature by modifying other putative neurotransmitters?

Milton
Yes, we have shown that pyrogen fever can be potentiated by 6-hydroxydopamine pretreatment as can prostaglandin hyperthermia. Also if you deplete 5-HT with *p*-chlorophenylalanine the response to pyrogen is reduced and so is the response to prostaglandins.

5
Ergot alkaloids and the modulation of hypothalamic function

E. Flückiger*

INTRODUCTION

Central nervous system actions of ergot alkaloids in connection with ergot poisoning have been known for centuries, but well into the present century these effects were only considered to be signs of toxicity. Barger's classic treatise on *The Alkaloids of Ergot* (1938) reviews the early descriptions of central effects, while a review edited by Berde and Schild (1978) covers the more recent literature on ergot pharmacology.

It was recognised by Githens (1917) that in animals the alkaloids may modulate autonomic functions such as thermal homeostasis by a direct central action. An excitatory syndrome of 'sympathicotrope stimulation' after exposure to LSD 25 was described by Rothlin *et al.* (1956), which, in the rabbit, consisted of a rise in body temperature, an increase in blood-sugar concentration, mydriasis and piloerection. After wide experience with various ergot alkaloids and derivatives, Cerletti (1959) saw this excitatory syndrome as part of the general 'spectrum of activity' of ergot compounds. Individual compounds only differ in the relative contribution of the various elements to this spectrum.

Also at that time the 'spectrum of activity' of ergot drugs was expanded in a then unexpected way, when a completely new type of autonomic effect was detected. During the search for a drug to interfere with the autonomic process of ovum implantation (nidation) in the uterus of rats, ergotoxine was found to be active (Shelesnyak, 1954).Ergotoxine is a natural mixture of ergocristine, α- and β-ergokryptine and ergocornine (Stoll and Hofmann, 1943; Schlientz *et al.*, 1968). Shelesnyak concluded from his experimental observations and a critical appraisal of the scanty literature that ergotoxine probably acts via the hypothalamus and the pituitary to inhibit luteotropine (prolactin) secretion. Ten years later Zeilmaker (1964) reported that ergocornine inhibits ovulation when administered to rats in the pro-oestrus phase of the ovarian cycle. Thus there is

*Biological and Medical Research Division, Sandoz Ltd, CH-4002 Basle, Switzerland

ample evidence that ergot alkaloids modulate diverse functions controlled by the hypothalamus.

The following review will concentrate on the endocrinological aspects in the 'spectrum of activities' of ergot alkaloids and their derivatives.

STRUCTURAL CHARACTERISTICS

Endocrine activities have been described for a great variety of ergot derivatives and they have recently been discussed in great detail (Flückiger and Del Pozo, 1978). The four main classes of ergot alkaloids, the clavine alkaloids, the lysergic acids, the simple lysergic acid amides, and the peptide alkaloids or ergopeptines,

Table 5.1 List of official and non-official names to identify compounds discussed in this chapter

Class 1: Clavines	
lergotrile Lilly 83636	2-Chloro-6-methyl-8β-cyanomethyl- ergoline
methergoline MCE	1-Methyl-*N*-carbobenzoxy-dihydroly- sergylamine
Class 2: Simple lysergic acid amides	
LSD 25 *d*-LSD lysergidum Delyside^{R}	*d*-Lysergic acid diethylamide
methysergide UML-491 Deseril^{R}, Sansert^{R}	1-Methyl-lysergic acid-L-2-butanolamide
methylergometrine (-basine, -novine) Methergine^{R}	*d*-Lysergic acid-L-2-butanolamide hydrogen maleinate
Class 3: Ergopeptines	
ergotamine tartrate Gynergene^{R}	2′β-Methyl-5′α-benzyl-ergopeptine tartrate
α-ergokryptine	2′β-Isopropyl-5′α-sec.butyl-ergopeptine
ergocornine	2′β-5′*d*-Diisopropyl-ergopeptine
2-Br-α-ergokryptine mesylate CB154 bromocriptine Parlodel^{R}, Pravidel^{R}	2′β-Isopropyl-5′α-sec. butyl-2-bromo- -ergopeptine mesylate
dihydro-ergotamine mesylate DHE, DHE-45 Dihydergot^{R}	2′β-Methyl-5′α-benzyl-9, 10-dihydro- ergopeptine mesylate
Class 4: 8α-Amino-ergolines	
lysuride, lisuride mesorgydine Lysenyl^{R}	*N*-[6-Methyl-8α-(9-ergolenyl)]-*N*′, *N*′- diethylurea hydrogen maleinate

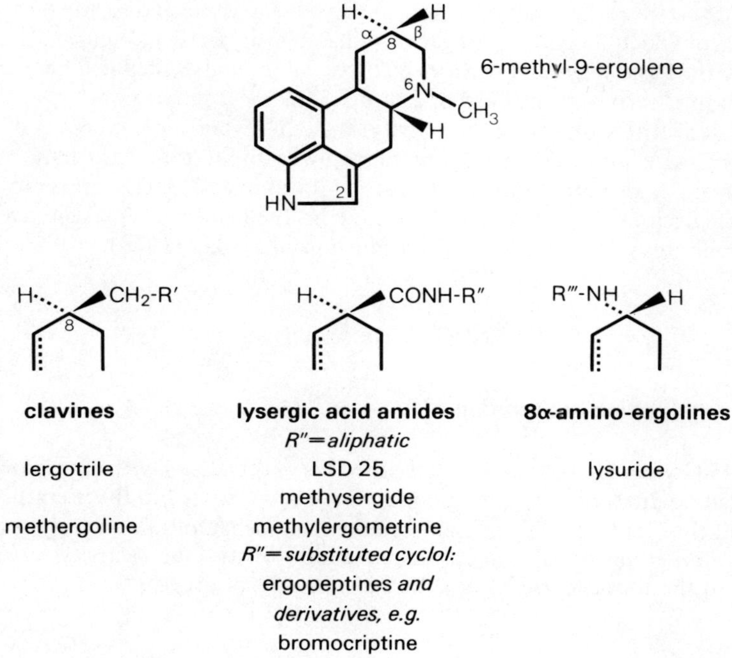

Figure 5.1 Basic chemical structures of the various groups of ergot compounds

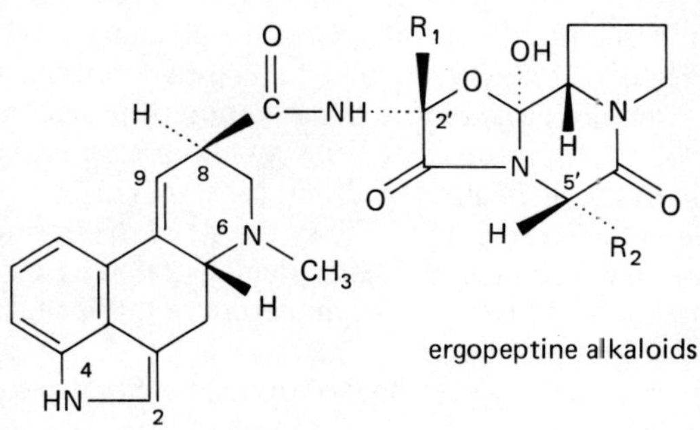

R_1	R_2			
	$-CH_2 \cdot C_6H_5$	$-CH_2 \cdot CH(CH_3)_2$	$-CH(CH_3) \cdot CH_2CH_3$	$-CH(CH_3)_2$
$-CH_3$	ergotamine	α-ergosine	β-ergosine	ergovaline
$-CH(CH_3)_2$	ergocristine	α-ergokryptine	β-ergokryptine	ergocornine
$-CH_2 \cdot CH_3$	ergostine	α-ergoptine	β-ergoptine	ergonine

Figure 5.2 Chemical structures of the ergopeptine alkaloids

all provide active compounds. To these a fifth class must be added, the 8-amino-ergolines, of which no natural congeners (alkaloids) are known. Figures 5.1 and 5.2 show the basic structural features of those compounds which will be mentioned in this review. At the head of figure 5.1 the most simple 6-methyl-ergoline derivative with endocrine activity, 6-methyl-9-ergolene, is depicted. This compound, which is of non-natural origin, is a weak inhibitor of prolactin secretion and is also uterotonic, lowers blood pressure and blocks serotonin receptors (Bach *et al.*, 1974). Table 5.1 contains the names of representative compounds grouped according to Rutschmann and Stadler (1978).

ENDOCRINE ACTIONS

Prolactin–Inhibition of Secretion

Since 1954, when Shelesnyak first described his observations with ergotoxine, more than 60 different ergot compounds have been reported in the literature to inhibit the secretion of prolactin from the mammalian pituitary. Only very few contribute substantially to our understanding of structure–activity relationships or of the sites and modes of action of this class of drugs.

Sites of action

Shelesnyak (1954) considered two sites of action at which the ergot alkaloids could suppress prolactin secretion–the hypothalamus and the pituitary. Zeilmaker and Carlsen (1962) presented the first evidence, based on *in vivo* experiments, that ergocornine acts directly at the pituitary level. Later *in vitro* experiments by Pasteels *et al.* (1971), using a bioassay method, showed that ergocornine and bromocriptine reduce the secretion of prolactin-like material from rat and foetal human pituitaries. Using a radioimmunoassay system to measure prolactin in the incubation medium, Lu *et al.* (1971) found that ergocornine inhibits the secretion and the synthesis of the hormone by rat pituitaries *in vitro*, and Ectors *et al.* (1972) demonstrated by electron micrography that ergocornine induces the accumulation of intracellular secretory granules in incubated rat pituitaries.

Macleod and Lehmeyer (1972) found with incubated rat pituitaries that ergocornine, α-ergokryptine and ergotamine inhibit the release of freshly labelled prolactin. Using clonal cell cultures of the rat pituitary tumour, type GH_3, Tashjian and Hoyt (1972) demonstrated with 1–10μg bromocriptine a dose-dependent inhibition of prolactin secretion into the medium, while the secretion of GH remained unaltered. Also Vale *et al.* (1976) reported a dose-dependent reduction of prolactin release with low concentrations of bromocriptine using cultured normal rat pituitary cells. This action is not specific to lysergic acid amide derivatives, since Clemens *et al.* (1975), using the clavine derivative lergotrile, also observed an inhibitory action on prolactin secretion from incubated rat pituitaries. Of the two sites of action considered by Shelesnyak (1954), the direct action at the pituitary level is thus well documented.

Other observations suggest that these drugs may also act at the hypothalamic level of prolactin control: Fuxe and Hökfelt (1970) first reported that in lactating rats treatment with either ergocornine or bromocriptine reduces the dopamine turnover of the tubero-infundibular dopaminergic (TIDA) neurones of the arcuate nucleus. Wuttke *et al.* (1971) found increased PIF activity in the hypothalamus of rats treated with ergocornine. These observations were interpreted as showing a hypothalamic site of action for these compounds (Hökfelt and Fuxe, 1972*b*).

These workers also made the important observation that in male and female rats hypophysectomy decreases transmitter turnover rate of the TIDA neurones and that treatment of such animals with prolactin stimulates the transmitter turnover rate of the TIDA neurones. It is thus obvious that the activity of these TIDA neurones is responsive to alterations in circulating prolactin levels, exhibiting the same change irrespective of whether the serum prolactin concentration is reduced by pharmacological means or by surgical intervention. Thus Dang and Voogt (1977) reported interruption of pseudopregnancy in the rat after hypothalamic implantation of prolactin. These observations do not disprove the assumption that the TIDA neurones take part in the control of prolactin secretion, but they suggest that they are in turn strongly influenced by (short loop) feedback effects of the circulating prolactin concentration.

Prolactin has been known for some time to inhibit its own secretion (Meites, 1973; Neill, 1974) and prolactin-sensitive neurones in the hypothalamus of rabbits have been detected by Clemens *et al.* (1971) using electrophysiological methods. Poulain and Carette (1976), using guinea-pigs, have iontophoretically applied prolactin to individual neurones of the hypothalamus and found neurones responding by a change in firing rate both in the septum and preoptic area (POA). Thus observations of augmented or attenuated prolactin secretion after electrical stimulation of the POA, or the medial basal prechiasmatic area, have now been given a neuronal basis.

It must be concluded that changes in transmitter turnover of the TIDA neurones induced by systemic administration of certain prolactin-inhibitory drugs do not necessarily reflect a direct action of the drugs on these neurones.

Mechanism of action

Ergot alkaloids and related compounds have traditionally been known to act on α-adrenoceptors and 5-HT receptors. A new dimension was added by Corrodi *et al.* (1973), who provided biochemical evidence that systemically administered ergocornine or bromocriptine reduced the turnover-rate of dopamine (DA) and increased the DA content in the brain of rats. These findings are interpreted as being due to a presynaptic stimulation of dopaminergic neurones by such drugs. Kehr (1977) also observed increased DA content and decreased DA turnover after treating rats with lisuride. In man the concentration of homovanillic acid (HVA), a metabolite of DA, has been found to be lowered in the cerebrospinal fluid after treatment with bromocriptine, adding further support to this concept (Curzon, 1975). On the other hand, Fuxe *et al.* (1975*a*) could not find a reduction of DA-turnover in rat brains after administration of metergoline, which is also a potent inhibitor of prolactin secretion.

In mammals prolactin secretion is considered to be controlled by the hypothalamus, mainly by a tonic inhibitory neurosecretory efferent, the prolactin inhibitory factor, or PIF. This factor has not yet been identified, but DA has been isolated from hypothalamic extracts with PIF activity. DA injected into the portal system connecting the hypothalamus and the anterior pituitary has been shown to attenuate prolactin secretion (Takahara *et al.*, 1974). In particular the *in vitro* work of Macleod (1976) has provided positive evidence that prolactin cells are equipped with DA-sensitive receptors, stimulation of which reduces the release of prolactin, an action which is inhibited by DA-receptor blocking agents. Thus haloperidol and perphenazine antagonise, in a dose-dependent manner, the inhibitory action of α-ergokryptine (Macleod, 1976) and bromocriptine (Macleod, personal communication) on prolactin secretion. Similar interactions have been observed by Caron *et al.* (1976) using the technique of specific binding measurements. Clemens (1976) demonstrated that the action of lergotrile *in vitro* could be antagonised by pimozide, but not by phentolamine or propranolol.

These findings demonstrate that control of prolactin cell function by the hypothalamus may well be based on a DA-like neuroendocrine transmitter and that certain ergot alkaloids and related drugs may act directly at the pituitary level by mimicking the effects of hypothalamically released DA at the prolactin cell membrane. On the other hand, as the existence of a hypothalamic PIF which differs in chemical nature from DA has not been ruled out under *in vivo* conditions, DA-receptor stimulants might release PIF by acting at the dopaminergic synapses thought to exist between the TIDA neurones and PIF-releasing neurones (Greibrokk *et al.*, 1974; Schally *et al.*, 1977).

As the inhibition of prolactin secretion by bromocriptine is not only demonstrable *in vivo* and with fresh *in vitro* preparations of anterior pituitaries but also in clonal cell cultures (Tashjian and Hoyt, 1972), it seems evident that bromocriptine acts directly to inhibit prolactin release and not indirectly by the release of stored catecholamines. On the other hand, it has been observed in rats that the inhibitory system is attenuated if the biogenic amine stores are depleted and their synthesis is reduced by pretreatment with reserpine (RES) and α-methyl-*p*-tyrosine (AMPT) (Loew *et al.*, 1976). It was therefore of interest to see how such pretreatment influences the *in vivo* prolactin-secretion-inhibitory property of this drug. As shown in figure 5.3, RES plus AMPT treatment of rats elevates their serum prolactin levels greatly, but bromocriptine reduces these levels by about 90 per cent to near-normal levels. Similar observations have been reported for lisuride (Horowski and Gräf, 1975; Gräf *et al.*, 1976), thus demonstrating that the actions of the two drugs do not depend on intact stores of biogenic amines.

Neurotransmitters other than DA are also implicated in the physiological control of prolactin secretion. There is no evidence that ergot alkaloids and related compounds stimulate or block acetylcholine, H_1- or H_2- receptors or β-adrenoceptors (Müller-Schweinitzer and Weidmann, 1978), and we can neglect these otherwise important aspects of prolactin control mechanisms (Lawson and Gala, 1975; Meites *et al.*, 1976; Arakelian and Libertun, 1977). However, the role of noradrenaline and serotonin must be considered.

Noradrenaline (NA) inhibits prolactin secretion if incubated together with rat pituitaries, but the concentrations needed for such an effect are much higher than for DA (Clemens, 1976; Caron *et al.*, 1976). In intact rats Donoso

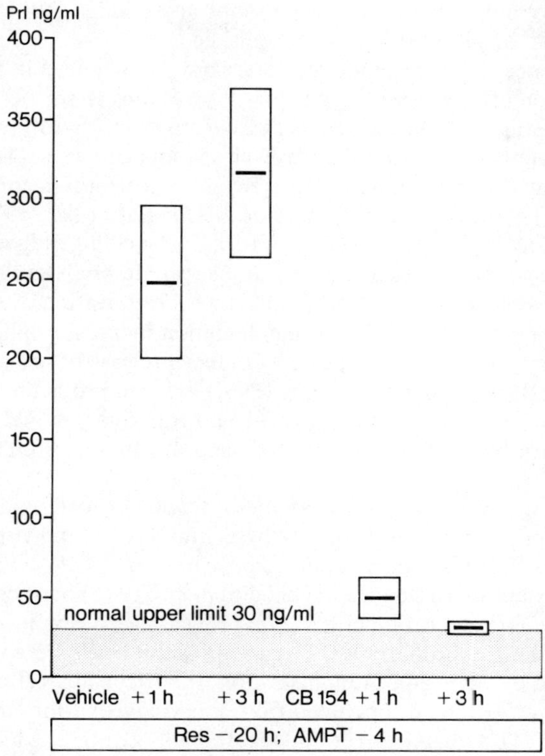

Figure 5.3 The inhibition of prolactin (Prl) secretion by bromocriptine (CB 154) in rats pretreated with reserpine (Res) and α-methyl-*p*-tyrosine (AMPT). All animals were treated with reserpine for 20h and then with AMPT 4h before the start of the experiment. At t_0, animals were treated with either vehicle or CB 154 and after 1h or 3h rats were decapitated for blood collection. Vertical rectangles indicate s.e.m. ($n \geqslant 6$)

et al. (1971) observed that dihydroxyphenyl serine (DOPS), a precursor of NA, increases prolactin serum levels. Clonidine, an α-adrenoceptor agonist was found either to decrease prolactin secretion (Meites *et al.*, 1976) or to increase it (Lawson and Gala, 1975), but phenoxybenzamine and phentolamine, two α-adrenoceptor-blocking agents were also found to augment the plasma prolactin concentration in rats (Lawson and Gala, 1975). Bromocriptine has negligible α-adrenoceptor agonist activity but blocks α-adrenoceptors in spiral strips of canine femoral veins (pA_2 9.1), being equipotent with dihydroergotamine (DHE) (pA_2 9.2) (Müller-Schweinitzer and Weidmann, 1978). In contrast to this finding, DHE is more than 20 times less potent an inhibitor of prolactin secretion than bromocriptine (Flückiger *et al.*, 1976). Ergocornine and bromocriptine were observed by Corrodi *et al.* (1973) and lisuride by Kehr (1977) to enhance NA-depletion in rat brains after NA-synthesis inhibition, suggesting some NA-receptor blocking action. In view of the collected evidence, participation of the α-adrenoceptor blocking property in the intact organism cannot be excluded, but the importance of this mechanism of action is not yet established.

There is evidence that serotoninergic nerves are also involved in the control of prolactin secretion (Tindal, 1974; Clemens, 1976; Meites *et al.*, 1976), but the physiological relevance of this system is difficult to assess. Serotonin (5-HT), 5-hydroxytryptophan and tryptophan, as well as quipazine, a 5-HT-receptor agonist, have been found to stimulate the release of prolactin in animals and man (Rose and Ganong, 1976; Clemens *et al.*, 1977; Kato *et al.*, 1974; Macindoe and Turkington, 1973; Del Pozo *et al.*, 1977*a*). The suckling-induced prolactin release in lactating rats is accompanied by an increase in brain levels of 5-hydroxy-indolacetic acid, a metabolite of 5-HT, indicating the activation of serotoninergic neurones (Kordon *et al.*, 1976). Serotonin depletion by *p*-chlorophenylalanine (PCPA) was found to diminish the suckling-induced release of prolactin (Kordon *et al.*, 1973/74). Parachloroamphetamine (PCA) reduced prolactin levels in oestrogen-primed rats (Chen and Meites, 1975). From this it would be expected that 5-HT-receptor blockade would also reduce prolactin secretion under certain conditions.

Treatment with 5-HT antagonists like methysergide or metergoline produced lowering of serum prolactin levels in healthy volunteers and in hyperprolactinaemic patients (Del Pozo *et al.* 1977*b*; Peracchi *et al.* 1977), while tricyclic 5-HT antagonists produced no effect. Chiodini *et al.* (1976) have compared the action of metergoline and bromocriptine on prolactin secretion in acromegalic patients. The actions of both drugs on serum prolactin levels were similar and could be blocked by pimozide. Administration of a tricyclic 5-HT antagonist was without effect. It can be concluded from this study that the 'spectrum of activity' of the 5-HT antagonist metergoline probably contains a hitherto unrecognised DA-receptor agonist component. It may be assumed that this will also be found true for other ergot derivatives which were originally developed because of their 5-HT antagonistic properties. Lisuride is one such example which has recently been shown to have an important DA-receptor agonist activity coupled with a high prolactin secretion inhibitory potency (Horowski *et al.*, 1975; Horowski and Wachtel, 1976). A recent biochemical study by Kehr (1977) confirmed the conclusions based on behavioural and endocrine experiments and showed lisuride to be a stimulator of central DA-receptors. In contrast to this are observations by Fuxe *et al.* (1975*a*) with metergoline, which did not produce signs of central DA-receptor stimulation but which clearly acted as a 5-HT-receptor blocker using biochemical criteria. In view of the findings that pimozide blocks the prolactin lowering effect of metergoline in man (Chiodini *et al.*, 1976), one must conclude that the mode of action of metergoline is not yet clear.

Another aspect of 5-HT antagonism should be mentioned in this context. Biochemical studies on the changes of biogenic amine metabolism in the brain by Corrodi *et al.* (1975) after ergocornine, by Snider *et al.* (1975) using bromocriptine and by Kehr (1977) after lisuride have shown that these drugs, which are potent 5-HT antagonists in the periphery (Cerletti and Doepfner, 1958; Flückiger, *et al.*, 1976; Horowski *et al.*, 1975), do not seem to block but rather to stimulate 5-HT-receptors in the CNS. The lack of evidence for 5-HT blockade is in agreement with the results of an electrophysiological analysis of the central actions of methysergide by Haigler and Aghajanian (1974). The negative results of these authors with metergoline are in conflict with the biochemical data produced by Fuxe *et al.* (1975*a*) mentioned above.

In summary, many ergot alkaloids and related compounds inhibit prolactin secretion. The mechanism of this action is best explained on the basis of DA-receptor stimulant activity, but the possibility of a participation *in vivo* of the 5-HT blocking component of certain derivatives cannot be excluded.

Gonadotrophin–Inhibition of Secretion

As first described by Zeilmaker (1964) and later by Kraicer and Strauss (1970), ovulation may be inhibited in rats after ergocornine administration. This effect has not raised the same interest as the inhibitory effect on ovum implantation and therefore our understanding of this aspect of ergot pharmacology is very unsatisfactory. It is clear that ergocornine (Wuttke *et al.*, 1971) and bromocriptine (Marko and Flückiger, 1974) may reduce dose-dependently the preovulatory LH surge in the afternoon of pro-oestrus. This is the immediate cause of the suppression of ovulation in the night from pro-oestrus to oestrus. Table 5.2 provides a summary of the reports on ovulation inhibitory ergot compounds. No information is available on the effect of ergot compounds on the tonic gonadotrophin secretion in female laboratory animals. It should also be mentioned that, apart from a few contradictory publications, nothing definite is known about actions of ergot drugs on gonadotrophin secretion in the male.

Sites of action

It is not clear from the literature where the inhibitory action on the ovulatory process takes place. Three sites of action should be considered–the ovarian, the pituitary and the hypothalamic sites. Nooter and Zeilmaker (1970) reported that the inhibitory action of ergocornine can be overcome by electrical stimulation of the preoptic area of the hypothalamus. Wuttke *et al.* (1971) found a reduced LH-RF-like activity in the hypothalamus of rats treated in pro-oestrus with ergocornine. These observations suggest a hypothalamic effect of ergocornine rather than an action on the pituitary or the ovarian level. In contrast, Seki *et al.* (1974) have observed a slightly reduced sensitivity of bromocriptine-pretreated rats towards LH-RF or towards rat hypothalamic extracts with LH-RF-like activity. This suggests that the pituitary should also be considered in future studies of altered gonadotrophin secretion after the administration of ergot derivatives.

Mechanism of action

None of the ergot compounds has been systematically studied for the pharmacological mechanism which causes ovulation inhibition. Because of this the following can only be regarded as a search for trends which might give a basis for further studies.

Table 5.2 Ovulation-inhibiting ergot alkaloids and derivatives

Name	*Animal and dose*	*Reference*
Alkaloids:		
ergotamine	Adult rat, fully effective after 10 mg/kg i.p. 4 mg/kg i.p.: partially effective	Raj and Greep (1973)
	Adult rat, ED_{50} = 0.3 mg/kg s.c.	Marko (unpublished)
ergosine (α-)	Adult rat, ED_{50} = 0.34 (0.18–0.63) mg/kg s.c.	Marko (unpublished)
ergovaline	Adult rat, ED_{50} = 0.7 (0.4–1.2) mg/kg s.c.	Marko (unpublished)
ergocristine	Adult rat, ED_{50} = 1.3 (0.9–1.8) mg/kg s.c.	Flückiger *et al.* (1976)
α-ergokryptine	Adult rat, ED_{50} = 1.7 (1.4–2.1) mg/kg s.c.	Flückiger *et al.* (1976)
β-ergokryptine	Adult rat, ED_{50} = 1.7 (1.2–2.3) mg/kg s.c.	Flückiger *et al.* (1976)
ergocornine	Adult rat, 1 mg/rat s.c.: effective	Zeilmaker (1964)
	Adult rat, 0.5 mg/rat s.c.: effective	Kraicer and Strauss (1970)
	Juvenile PMS rat	Hökfelt and Fuxe (1972*b*)
	Adult rat, ED_{50} = 1.7 (1.2–2.4) mg/kg s.c.	Marko and Flückiger (1974)
	Juvenile PMS rat, ED_{50} = 3.2 (2.2–7.0) mg/kg s.c.	
ergostine	Adult rat, ED_{50} = 1.8 (1.1–2.8) mg/kg s.c.	Marko (unpublished)
ergonine	Adult rat, ED_{50} = 1.8 (1.1–2.8) mg/kg s.c.	Marko (unpublished)
Derivatives:		
LSD 25	Adult mouse, 2 × 100 μg no effect	Brown (1967)
	Juvenile PMS mouse, 30 μg partially effective	
methysergide	Adult mouse, 100 μg no effect	Brown (1967)
	Juvenile PMS mouse, 5–20 μg acting dose-dependently	
bromocriptine	Juvenile PMS rat	Hökfelt and Fuxe (1972*b*)
	Adult rat, ED_{50} = 20 (13.7–29.0) mg/kg s.c.	Marko and Flückiger (1974)
	Juvenile PMS rat, ED_{50} = 1.5 (0.5–5.9) mg/kg s.c.	
2-Br-ergocornine	Adult rat, ED_{50} > 20 mg/kg s.c.	Flückiger *et al.* (1976)

Table 5.3 Comparison of implantation and ovulation inhibitory activity with α-adrenoceptor and 5-HT antagonistic activity at vascular receptors

	ED_{50} mg/kg s.c.		pA_2*	pD_2'*
	Implantation inhibition	Ovulation inhibition	α-Adrenoceptor blockade	5-HT-Receptor blockade
Ergotamine	14	0.3	8.8	9.6
Ergocristine	4.2	1.3	–	–
α-Ergokryptine	1.1	1.7	8.9	8.6
β-Ergokryptine	2.8	1.7	–	–
Ergocornine	2.7	1.7	8.8	–
Methysergide	>10	1	5.5	9.1
2-Br-Ergocornine	>5	>20	–	–
Bromocriptine	0.75	20	8.9	7.6

*These data are taken from Müller-Schweinitzer and Weidmann (1978).

(1) A comparison of the ovulation inhibitory potency of some ergot compounds with their prolactin secretion inhibitory activity (assessed in the implantation inhibition test) in the adult rat (table 5.3) shows that the two activities vary independently. This suggests that the two actions represent two different basic properties of the drugs. It has been shown in the previous section that prolactin secretion inhibition is best explained by a DA-receptor stimulatory action. Therefore, DA-receptor stimulation cannot easily be used to explain ovulation inhibition.

(2) Table 5.3 also shows the α-adrenoceptor-blocking activity of some of the ergot compounds. It can be seen that whereas the individual values for ovulation inhibitory potency cover more than two orders of magnitude, α-adrenoceptor blockade is a rather constant quality of the compounds listed in table 5.3, except for methysergide. This ergot derivative is of medium potency as an ovulation inhibitor but is about three orders of magnitude less active as an α-adrenoceptor blocker than the rest of the series. Thus these data do not lend support to the idea that the ovulation inhibition induced by these drugs is correlated to α-adrenoceptor blockade.

(3) Table 5.3 also contains data on the potency of the drugs as 5-HT receptor antagonists. It is evident that ergotamine and methysergide, the two most potent ovulation inhibitors, are also the most active inhibitors of 5-HT, while bromocriptine is least active by both criteria and α-ergokryptine is intermediate to both. Thus 5-HT antagonism and ovulation inhibition vary in parallel in this small series of compounds, suggesting that ovulation inhibition may be somehow related to serotonin antagonism.

How do the conclusions in respect of the importance of these three basic pharmacological properties (i.e. DA-agonist, α-antagonist and 5-HT antagonist) relate to the results of other workers? The lack of correlation between DA-agonist activity and inhibition of ovulation (table 5.3) seems in contrast to the conclusion drawn by Fuxe *et al.* (1975*b*) on the basis of their experiments using pregnant mare serum gonadotrophin (PMSG)-induced ovulation in juvenile rats. With this model these workers observed an important inhibitory dopaminergic input on the pre-ovulatory PMSG-induced LH surge.

In our experience this model produces qualitatively different pharmacological results in comparison to the spontaneously ovulating adult rat (Marko and Flückiger, 1974). Whereas in the model of PMSG-induced ovulation bromocriptine is similarly active in inhibiting LH and prolactin secretion, the adult cycling rat shows a much reduced sensitivity towards the LH inhibitory activity of bromocriptine, while the sensitivity of the prolactin system is as high as in the juvenile rat. A similar difference of sensitivity between ovulation in cycling adult mice and PMSG-induced ovulation in juvenile mice has been observed by Brown (1967) using methysergide and LSD 25. For ergocornine the reverse has been observed, with adult rats proving somewhat more sensitive than the juvenile PMSG-treated animals (Marko and Flückiger, 1974). PMSG-induced ovulation may therefore not reflect the same hypothalamic gonadotrophin regulatory situation as is found in the spontaneously cycling adult rat.

Using ovariectomised adult female rats, Drouva and Gallo (1977) produced further evidence of an inhibitory dopaminergic influence on LH secretion, but these authors seem to be of the opinion that this effect may be restricted to the special situation of the ovariectomised rat. Recently the discussion about the dopaminergic involvement in gonadotrophin secretion has been revived by two clinical studies. In one the infusion of DA to normal volunteers and to hyperprolactinaemic patients not only reduced serum prolactin levels but also levels of LH (Leblanc *et al.*, 1976). Similar results were obtained after L-dopa ingestion (Lachelin *et al.*, 1977). In hyperprolactinaemic patients, bromocriptine (2.5mg orally) not only lowered serum levels of prolactin and LH but also those of FSH, these effects lasting for more than 10 h (Lachelin *et al.*, 1977). These findings seem to be in accord with the observations in PMSG-treated juvenile rats (Fuxe *et al.*, 1976), but in view of other clinical experience the observations of Yen's group do not seem to be functionally relevant (e.g. Del Pozo *et al.*, 1975).

The importance of α-adrenoceptor blockade is difficult to assess. Many experimental data point to a stimulatory role of NA in the release of gonadotrophins (Rose and Ganong, 1976; Fuxe *et al.* 1975*b*) and therefore α-adrenoceptor blockade should inhibit ovulation (Everett, 1964; Piva *et al.*, 1969). Ergocornine and to a lesser extent bromocriptine have been found to cause a partial depletion of NA stores in the whole brain of rats and to enhance the H44/68-induced NA release in hypothalamic tissue (Corrodi *et al.*, 1973). Lisuride acts similarly (Kehr, 1977). Therefore, using biochemical criteria, these drugs seem to block NA-receptors in the CNS but the data are unfortunately insufficient to allow a comparison of these effects with their ovulation-inhibitory potency. Thus the participation of NA-receptor blockade cannot be discussed in more depth.

Serotonin is also implicated as a transmitter in the hypothalamic control of phasic gonadotrophin secretion (Wilson, 1974; Hery *et al.*, 1975). Depletion of 5-HT stores by PCPA or the administration of the 5-HT receptor antagonist methiothepin inhibits the preovulatory LH surge and ovulation in rats (Hery *et al.*, 1976). The finding that ovulation inhibition by ergot compounds seems to be related to their 5-HT antagonistic potency would be consistent with this, except for the fact that ergot derivatives with the exception of metergoline (Fuxe *et al.*,1975*a*) are characterised as stimulators rather than as inhibitors of 5-HT-receptors at the CNS level (Corrodi *et al.*, 1975; Kehr, 1977).

Gonadotrophin–Stimulation of Secretion

If an ergot derivative interrupts pseudopregnancy or prevents ovum implantation in the rat this is followed by recurrence of a new ovarian cycle with ovulation within about three days (Shelesnyak, 1955; Kraicer and Shelesnyak, 1968). Thus interruption of a prolactin-dominated endocrine situation with suppressed phasic release of gonadotrophins is followed by resumption of the cyclic phenomena of reproduction. The question is, are there two separate phenomena or is the latter the consequence of the former? No ergot alkaloid has been described as inducing a sexual cycle in rats without at the same time suppressing a state of prolactin dominance. Thus ergot alkaloids which inhibit prolactin secretion do not induce precocious puberty in juvenile rodents. In post-partum sheep, treatment with bromocriptine shortens the anoestrous period (Kann and Martinet, 1975) while suppressing concomitantly the otherwise high levels of circulating prolactin. In normal cycling sheep the attenuation of normal prolactin does not lead to a change in gonadotrophin secretion or cycle length (Niswender, 1974; Kann and Denamur, 1974).

As there is no evidence to the contrary, one could assume that inhibition of prolactin secretion and disinhibition of phasic gonadotrophin secretion are due to one and the same property of these compounds–stimulation of DA-receptors. DA has been implicated as a transmitter in the stimulatory control of gonadotrophin secretion (Kordon, 1971; Lichtensteiger and Keller, 1974; McCann, 1974), but as this concept has been disputed on the ground that the tubero-infundibular dopaminergic (TIDA) neurones were found to inhibit gonadotrophin secretion (Hökfelt and Fuxe, 1972*a*), a different explanation may have to be sought. The facilitating effect on gonadotrophin secretion could be obtained as an event secondary to inhibition of prolactin secretion alone. Hypophysectomy or treatment with ergocornine or with bromocriptine were found to decrease the activity of the TIDA neurones (Hokfelt and Fuxe, 1972*a*, *b*), an effect which could be antagonised by the injection of prolactin. Thus inhibition of TIDA neurone activity may provide disinhibition of gonadotrophin releasing hormone secretion. Borrell *et al.* (1976) have published evidence that the basomedial portion of the amygdala may also play a role in the effect of DA-receptor stimulant drugs on gonadotrophin secretion, but for the present there is not enough information to discuss this aspect.

The analysis of these interactions is complicated by the fact that, in the periphery, prolactin may influence the sensitivity of the ovary to gonadotrophin stimulation or alter steroidogenesis and thereby interfere with the steroid-gonadotrophin feedback system which forms the basis of cyclic reproductive phenomena.

Actions on the Control of Other Pituitary Hormones

Very little is known from animal studies about actions of ergot compounds on the control of pituitary hormones other than those discussed in previous sections. Interestingly, more information seems to be available from clinical studies.

Growth hormone

Dopaminergic aspects In female mice bromocriptine does not alter the pituitary growth hormone (GH) content (Yanai and Nagasawa, 1970) or serum and urine GH concentrations (Sinha *et al.*, 1974); nor does it affect the growth hormone response to oestrogens in rats (Davies *et al.*, 1974). In lactating Holstein cows, GH secretion was not altered by bromocriptine in doses which maximally suppressed prolactin secretion (Smith *et al.*, 1974). The same result was obtained in lactating goats (Hart, 1973).

In normal subjects an acute dose of bromocriptine produces an increased secretion of GH, as is also found with other DA-receptor agonists such as L-dopa or apomorphine (Camanni *et al.*, 1975; Tolis *et al.*, 1975). This effect can be blocked by pimozide (Dammacco *et al.*, 1976; Chiodini *et al.*, 1976). However, the stimulatory effect is not observed when bromocriptine is given chronically (Del Pozo *et al.*, 1977*b*) and the acute stimulatory effect probably reflects a short-lived hypothalamic action of bromocriptine and similar compounds. Acromegalic patients may react differently: in some patients bromocriptine and other DA-receptor agonists lower GH serum levels acutely and chronically (Liuzzi *et al.*, 1974; Thorner *et al.*, 1975). Massara *et al.* (1976*a, b*) observed that the infusion of 280μg/min DA, which does not modify basal GH secretion in normal subjects, lowers GH in acromegalic patients responsive to bromocriptine. It is assumed that DA does not cross the blood-brain barrier to act via the hypothalamus but acts directly on the somatotrophs of the pituitary (Verde *et al.*, 1976). Therefore it is concluded that the somatotrophs in acromegaly may acquire inhibitory DA receptors and then respond in a similar manner to prolactin cells. *In vitro* studies to test this assumption are lacking. Also, it must be pointed out that a rat pituitary tumour cell line (GH_3) producing GH and prolactin *in vitro* responded to bromocriptine by a reduced prolactin but not GH output (Tashjian and Hoyt, 1972). It is interesting to note that in anorexia nervosa the exaggerated GH response to an oral glucose tolerance test is also attenuated by bromocriptine (Harrower *et al.*, 1977).

Serotoninergic aspects Methysergide has been used as a 5-HT antagonist together with cyproheptadine to investigate the serotoninergic component in the GH response to insulin-induced hypoglycaemia (Bivens *et al.*, 1973). Both drugs reduced the GH response, 8 × 4mg cyproheptadine being more effective than 8 × 2mg methysergide. The authors concluded that serotonin is one of the factors controlling GH secretion. In contrast to these findings, bromocriptine (7 × 5mg) did not modify the insulin-induced elevation of GH and cortisol (Del Pozo *et al.*, 1977*b*). In another study the effect of methysergide on sleep-related GH secretion was investigated (Mendelson *et al.*, 1975). In addition to a lowering of prolactin the investigators found an unexpected increase of GH-levels during sleep. In view of the fact that methysergide inhibits insulin-induced GH secretion, it was concluded that GH control mechanisms differ under sleep conditions and under insulin stimulation.

In acromegalic patients, Feldman *et al.* (1976) tested methysergide and cyproheptadine acutely and observed that some patients responded to both drugs by a lowering of GH serum levels. These results demonstrate that, also in the condition of acromegaly, serotonin seems to play a role in GH control. The results should caution us against explaining the mechanism of action of ergot compounds in respect of GH secretion simply by assuming a DA-receptor stimulant action. Smythe *et al.* (1976) have stressed the serotoninergic aspects of GH control, and a detailed discussion may be found in a review by Del Pozo and Lancranjan (1977).

In summary, ergot derivatives may stimulate or inhibit GH secretion depending on the situation. GH secretion may be influenced by dopaminergic mechanisms, but serotonin antagonism may also contribute to the activity of some of the drugs.

Corticotrophin

Prolonged studies in mice with ergocornine (Sinha *et al.*, 1974) and in rats with bromocriptine (Cameron and Scarisbrick, 1973) did not show evidence of an influence of these drugs on the pituitary-adrenal axis. Burden *et al.* (1974), studying the neurotransmitter control of corticotrophin releasing factor (CRF), observed that methylergometrine antagonised the stimulatory action of serotonin on CRF release from hypothalamic fragments incubated *in vitro*. In man, 4 × 10mg methysergide reduced the ACTH response to insulin hypoglycaemia (normal subjects) but did not influence ACTH release after lysine-vasopressin (Cavagnini *et al.*, 1976). It was concluded that serotonin is involved in the hypothalamic control of ACTH secretion. Bromocriptine, which did not alter ACTH secretion in healthy subjects, was found to lower plasma ACTH levels acutely in some patients with Cushing's disease (Lamberts and Birkenhäger, 1976), and, additionally, in Nelson's and in Cushing's syndrome (Benker *et al.*, 1976). The mechanism of these actions is still unknown. Due to a lack of experimental data it is not possible to speculate on the possible relevance of dopaminergic versus serotoninergic mechanisms, although serotonin has been implicated in ACTH-secretion of Cushing's disease and Nelson's syndrome by the work of Krieger *et al.* (1975) and Krieger and Luria (1976).

Melanotrophin

Apomorphine and α-ergokryptine were found to produce pallor of the skin of frogs (*R. pipiens*), and it was concluded that MSH secretion in frogs is under dopaminergic control (Smith, 1975). Bromocriptine treatment of Syrian hamsters under the chronic influence of diethylstilboestrol was found to prevent the hyperplastic and neoplastic changes of the intermediate lobe (Hamilton *et al.*, 1975). Using a radioimmunoassay for measuring α-MSH in rats, Penny and Thody (1976) found that bromocriptine caused a diminution of circulating α-MSH, whilst haloperidol and other neuroleptics increased melanotrophin blood levels. These and *in vitro* studies (Tilders *et al.*, 1975; Baker, 1976) suggest that MSH-release from the pituitary, like the release of prolactin, is under dopaminergic inhibitory control similar to prolactin. This conclusion seems to contrast with that of Taleisnik *et al.* (1973/74), who found that methysergide inhibits depletion of melanotrophin-like activity from the pituitary of rats stimulated by intraventricular injection of hypertonic saline or by mechanical stimulation of the vagina in oestrus. They concluded that the hypothalamic control of MSH in rats also comprises serotoninergic elements. However, methysergide may have acted in these experiments as a DA-receptor agonist, as in the experiments concerning prolactin secretion (see above). The question of a serotoninergic component in the control of MSH-secretion is not yet settled.

Thyrotrophin

Studies in rats showed that 5–10mg/kg of bromocriptine given acutely reduced the cold-induced TSH response, while a smaller dose (2.5mg/kg) acutely or after repeated administration had no effect (Ranta *et al.*, 1977). Pimozide antagonised the inhibitory effect of the large dose of bromocriptine. TRH-induced TSH release was not altered by either drug and it was concluded that DA might be an inhibitory transmitter in the hypothalamic control of TRH release.

In hypothyroid patients, bromocriptine was found to attenuate the secretion of TSH (Kiyoshi *et al.*, 1974), but in normal volunteers no effect on basal TSH level or on TRH-induced TSH release was observed (Del Pozo *et al.*, 1977*b*).

Oxytocin

Ergotamine was found by Grosvenor (1956) to inhibit milk ejection acutely in lactating rats on day 9, but a second injection on day 10 produced a diminished effect. The effect of ergotamine could be overcome by the injection of oxytocin (Grosvenor and Turner, 1956). Methylergometrine, dihydroergotamine and dihydro-ergotoxine were also found to prevent milk ejection in the rat. Oxytocin counteracted the inhibitory effect of these drugs (Grosvenor and Turner, 1957). These experiments, together with experiments with further drugs, led to the conclusion that an adrenergic link exists in the reflex-arc responsible for oxytocin release. Bisset (1968) concludes that this adrenergic link must be an excitatory one. A more recent abstract (Clarke and Lincoln, 1977) reports that the α-adrenoceptor blocker phenolamine as well as the DA-receptor blockers haloperidol, pimozide and metoclopramide suppress milk ejection. The latter observation makes it unlikely that the above-mentioned ergot compounds inhibited milk ejection through a dopaminergic action.

CONCLUDING REMARKS

About seven years ago the ‘spectrum of activities’ of ergot compounds was expanded by the discovery of their DA-receptor stimulant properties and the prototype compound, bromocriptine, became available for clinical testing. At the same time the physiology and pathology of man was enriched by the discovery of the pituitary hormone prolactin, whose release is controlled by the hypothalamus via an inhibitory dopaminergic efferent. Thus the DA-receptor stimulant property occupies a prominent position today in the 'spectrum of activities' of ergot alkaloids and their derivatives. However, these drugs are also capable of an interaction with serotonin receptors and this should not be neglected when trying to elucidate the physiology of hypothalamic functions and the pharmacology of ergot drugs.

REFERENCES

Arakelian, M. C. and Libertun, C. (1977). H1 and H2 histamine receptor participation in the brain control of prolactin secretion in lactating rats. *Endocrinology*, **100**, 890–895

Bach, N. J., Hall, D. A. and Kornfeld, E. D. (1974). Descarboxylysergic acid (9, 10-didehydro-6-methylergoline). *J. Med. Chem.*, 17, 312–313

Baker, B. I. (1976). Ability of various factors to oppose the stimulatory effect of dibutyryl cyclic AMP on the release of melanocyte-stimulating hormone by the rat pituitary *in vitro. J. Endocr.*, **68**, 283–287

Barger, G. (1938). The alkaloid of ergot. In *Handb. Exp. Pharma. Ergänzungswerk*, **6**, Springer, Berlin, 84–822

Benker, G., Hackenberg, K., Hamburger, B. and Reinwein, D. (1976). Effects of growth hormone release-inhibiting hormone and bromocryptine (CB 154) in a state of abnormal pituitary-adrenal function. *Clin. Endocr.*, **5**, 187–190

Berde, B. and Schild, H. O. (1978). *Ergot Alkaloids and Related Compounds. Handbook of Experimental Pharmacology*, Springer Verlag, Heidelberg

Bisset, G. W. (1968). The milk-ejection reflex and the actions of oxytocin, vasopressin and synthetic analogues on the mammary gland. In B. Berde (ed.). *Neurohypophyseal Hormones and Similar Polypeptides, Handbook of Experimental Pharmacology* (New Series), Vol. 23, Springer Verlag, Berlin, Heidelberg and New York

Bivens, C. H., Lebovitz, H. E. and Feldman, J. M. (1973). Inhibition of hypoglycaemia-induced growth hormone secretion by the serotonin antagonist cyproheptadine and methysergide. *N. Engl. J. Med.*, **289**, 236–239

Borrell, J., Piva, F. and Martini, L. (1976). Adrenergic and cholinergic inputs to the amygdala. In F. Labrie, J. Meites and G. Pelletier (eds.). *Hypothalamus and Endocrine Function*, Plenum Press, New York, London, 37

Brown, P. S. (1967). The effect of 5-hydroxytryptamine and two of its antagonists on ovulation in the mouse. *J. Endocr.*, **37**, 327–333

Burden, J. L., Hillhouse, E. W. and Jones, M. T. (1974). A proposed model of the neurotransmitters involved in the control of corticotrophin releasing hormone. *J. Endocr.*, **63**, 20P–21P

Camanni, F., Massara, F., Fassio, V. and Molinatti, G. M. (1975). Changes in plasma growth hormone levels in normal and acromegalic subjects following administration of 2-bromo-α-ergo-cryptine. *J. Clin. Endocr. Metab.*, **40**, 363–366

Cameron E. H. D. and Scarisbrick, J. J. (1973). Determination of corticosterone in rat plasma by competitive protein-binding assay and its use in assessing the effects of CB154 and perphenazine on adrenal function. *J. Endocr.*, **58**, xxvii-xxviii

Caron, M. G., Drouin, J., Raymond, V., Kelly, P. A. and Labrie, F. (1976). Specificity of the catecholaminergic effect on prolactin secretion and (^{3}H) dihydroergocryptine binding. *Clin. Res.*, 24 , 656A

Cavagnini, F., Raggi, U., Micossi, P., Di Landro, A. and Invitti, C. (1976). Effect of an antiserotoninergic drug, metergoline, on the ACTH and cortisol response to insulin hypoglycaemia and lysine-vasopressine in man. *J. Clin. Endocr. Metab.*, **43**, 306–312

Cerletti, A. (1959). Discussion contribution. *Neuropsychopharmacology*, **1**, 117–123

Cerletti, A. and Doepfner, W. (1958). Comparative study on the serotonin antagonism of amide derivatives of lysergic acid and of ergot alkaloids. *J. Pharmacol. Exp. Ther.*, **122**, 124–136

Chen, H. J. and Meites, J. V. (1975). Effects of biogenic amines and TRH on release of prolactin and TSH in the rat. *Endocrinology*, **96**, 10–14

Chiodini, P. G., Liuzzi, A., Müller, E. E., Botalla, L., Cremascoli, G., Oppizzi, G., Verde, G. and Silvestrini, F. (1976). Inhibitory effect of an ergoline derivative, methergoline, on growth hormone and prolactin levels in acromegalic patients. *J. Clin. Endocr. Metab.*, **43**, 356–363

Clarke, G. and Lincoln, D. W. (1977). Effects of catecholamine antagonists on the milk-ejection reflex of the anaesthetized rat. *Brit. J. Pharmacol.*, **59**, 458P–459P

Clemens, J. A. (1976). Neuropharmacological aspects of the neural control of prolactin secretion. In F. Labrie, J. Meites and G. Pelletier (eds.). *Hypothalamus and Endocrine Function*, Plenum Press, New York and London, 283

Clemens, J. A., Sawyer, B. D. and Cerimele, B. (1977). Further evidence that serotonin is a neurotransmitter involved in the control of prolactin secretion. *Endocrinology*, **100**, 692–698

Clemens, J. A., Shaar, C. J., Tandy, W. A. and Roush, M. E. (1971). Effects of hypothalamic stimulation on prolactin secretion in steroid-treated rats. *Endocrinology*, **89**, 1317–1320

Clemens, J. A., Smalstig, E. B. and Shaar, C. J. (1975). Inhibition of prolactin secretion by lergotrile mesylate: mechanism of action. *Acta endocr. (Kbh)*, **79**, 230–237

Corrodi, H., Farnebo, L. O., Fuxe, K. and Hamberger, B. (1975). Effect of ergot drugs on central 5-hydroxytryptamine neurones: evidence for 5-hydroxytryptamine release or 5-hydroxytryptamine receptor stimulation. *Eur. J. Pharmacol.*, **30**, 172–181

Corrodi, H., Fuxe, K., Hökfelt, T., Lidbrink, P. and Ungerstedt, U. (1973). Effect of ergot drugs on central catecholamine neurones: evidence for a stimulation of central dopamine neurones. *J. Pharm. Pharmac.*, **25**, 409–411

Curzon, G. (1975). CSF homovanillic acid: an index of dopaminergic activity. *Advan. Neurol.*, **9**, 349–357

Dammacco, F., Rigillo, N., Tafaro, E., Gagliardi, F., Chetri, G. and Dammacco, A. (1976). Effects of 2-bromo-α-ergocryptine and pimozide on growth hormone secretion in man. *Hormone. Metab. Res.*, **8**, 247–248

Dang, B. T. and Voogt, J. L. (1977). Termination of pseudopregnancy following hypothalamic implantation of prolactin. *Endocrinology*, **100**, 873

Davies, C., Jacobi, J., Lloyd, H. M. and Meares, J. D. (1974). DNA synthesis and the secretion of prolactin and growth hormone by the pituitary gland of the male rat: effects of diethylstilboestrol and 2-bromo-α-ergocryptine methanesulphonate. *J. Endocr.*, **61**, 411–417

Del Pozo, E., Darragh, A., Lancranjan, I., Ebeling, D., Burmeister, P., Bühler, F., Marbach, P. and Braun, P. (1977*b*). Effect of bromocriptine on the endocrine system and foetal development. *Clin. Endocr.*, **6**, Suppl. 47s–55s

Del Pozo, E., Goldstein, M., Friesen, H., Brun Del Re, E. and Eppenberger, U. (1975). Lack of action of prolactin suppression on the regulation of the human menstrual cycle. *Am. J. Obstet. Gynec.*, **123**, 719–723

Del Pozo, E. and Lancranjan, I. (1978). Clinical use of drugs modifying the release of anterior pituitary hormones. In W. F. Ganong and L. Martini (eds.). *Frontiers in Neuroendocrinology*, Vol. 5, Raven Press, New York, 207–247

Del Pozo, E., Lancranjan, I., Clarenbach, P. and Wirz, A. (1977*a*). The role of serotonin in the release of prolactin: a methodological appraisal. *Acta endocr., (Kbh)*, **85**, 54 (Suppl. 212)

Donoso, A. O., Bishop, W., Fawcett, C. P., Krulich, L. and McCann, S. M. (1971). Effects of drugs that modify brain monoamine concentrations on plasma gonadotrophin and prolactin levels in the rat. *Endocrinology*, **89**, 774–784

Drouva, S. V. and Gallo, R. V. (1977). Further evidence for inhibition of episodic luteinizing hormone release in ovariectomized rats by stimulation of dopamine receptors. *Endocrinology*, **100**, 702–798

Ectors, F., Danguy, A. and Pasteels, J. L. (1972). Ultrastructure of organ cultures of rat hypophyses exposed to ergocornine. *J. Endocr.*, **52**, 211–212

Everett, J. W. (1964). Central neural control of reproductive functions of the adenohypophysis. *Physiol. Rev.*, **44**, 373–431

Feldman, J. M., Plonk, J. W. and Bivens, C. H. (1976). Inhibitory effect of serotonin antagonists on growth hormone release in acromegalic patients. *Clin. Endocr.*, **5**, 71–78

Flückiger, E. and Del Pozo, E. (1978). Influence on the endocrine system. In B. Berde and H. O. Schild (eds.). *Ergot Alkaloids and Related Compounds. Handbook of Experimental Pharmacology*, Springer Verlag, Heidelberg, 615–690

Flückiger, E., Marko, M., Doepfner, W. and Niederer, W. (1976). Effects of ergot alkaloids on the hypothalamo-pituitary axis. *Postgrad. Med. J.*, (Suppl. 1) **52**, 57–61

Fuxe, K., Agnati, L. and Everitt, B. (1975*a*). Effect of methergoline on central monoamine neurons. Evidence for a selective blockade of central 5-HT receptors. *Neuroscience Letts.*, **1**, 283–290

Fuxe, K. and Hökfelt, T. (1970). Central monoaminergic systems and hypothalamic function. In L. Martini, M. Motta and F. Fraschini (eds.). *The Hypothalamus*, Academic Press, New York, 123–138

Fuxe, K., Löfström, A., Agnati, L. F., Everitt, B. J., Hökfelt, T., Jonsson, G. and Wiesel, F. A. (1975*b*). On the role of central catecholamine and 5-hydroxytryptamine neurons in neuroendocrine regulation. In W. E. Stumpf and L. D. Grand (eds.), Karger, Basel, **420**

Fuxe, K., Löfström, A., Agnati, L. F., Everitt, B. J., Johansson, O., Jonsson, G., Wuttke, W. and Goldstein, M. (1976). Role of monoamines in the control of gonadotrophin secretion. In T. C. A. Kumar (ed.). *Neuroendocrine Regulation of Fertility, International Symposium, Simla, 1974*, Karger, Basel, 124–140

Githens, T. S. (1917). The influence of ergotoxine on body temperature. *J. Pharmac. Exp. Ther.*, **10**, 327–340

Gräf, K. J., Neumann, F. and Horowski, R. (1976). Effect of the ergot derivative lisuride hydrogen maleate on serum prolactin concentrations in female rats. *Endocrinology*, **98**, 598–605

Greibrokk, T., Currie, B. L., Johansson, K. N. G., Hansen, J. J., Folkers, K. and Bowers, C. Y. (1974). Purification of a prolactin inhibiting hormone and the revealing of hormone *d*-GHIH which inhibits the release of growth hormone. *Biochem. Biophys. Res. Commun.*, **59**, 704–709

Grosvenor, C. E. (1956). Effect of ergotamine on milk-ejection in lactating rats. *Proc. Soc. Exp. Biol. Med.*, **91**, 294–296

Grosvenor, C. E. and Turner, C. W. (1956). Ergotamine oxytocin and milk let-down in lactating rats. *Proc. Soc. Exp. Biol. Med.*, **93**, 466–468

Grosvenor, C. E. and Turner, C. W. (1957). Evidence for adrenergic and cholinergic components in milk let-down reflex in lactating rats. *Proc. Soc. Exp. Biol. Med.*, **95**, 719–722

Haigler, H. J. and Aghajanian, G. K. (1974). Peripheral serotonin antagonists: failure to antagonize serotonin in brain areas receiving a prominent serotoninergic input. *J. Neural Transm.*, **35**, 257–273

Hamilton, J. M., Flaks, A., Saluja, P. G. and Maguire, S. (1975). Hormonally induced renal neoplasia in the male Syrian hamster and the inhibitory effect of 2-bromo-α-ergocryptine methane-sulfonate. *J. Natn. Cancer Inst.*, **54**, 1385–1400

Harrower, A. D. B., Yap, P. L., Nairn, J. M., Walton, H. Y. and Craig, A. (1977). Growth hormone, insulin and prolactin secretion in anorexia nervosa and obesity during bromocriptine treatment. *Br. Med. J.*, **II**, 156–159

Hart, I. C. (1973). Effect of 2-Br-α-ergocryptine on milk yield and the level of prolactin and growth hormone in the blood of the goat at milking. *J. Endocr.*, 57, 179–180

Hery, M., Laplante, E. and Kordon, C. (1975). Role of pituitary sensitivity and adrenal secretion in the effect of serotonin depletion on luteinising hormone regulation. *J. Endocr.*, **67**, 463–464

Hery, M., Laplante, E. and Kordon, C. (1976). Participation of serotonin in the phasic release of LH. Part I: Evidence from pharmacological experiments. *Endocrinology*, **99**, 496–503

Hökfelt, T. and Fuxe, K. (1972*a*). On the morphology and the neuroendocrine role of the hypothalamic catecholamine neurons. In K. M. Knigge, D. E. Scott and A. Weindl (eds.). *Brain-endocrine Interaction, Median Eminence: Structure and Function,* Karger, Basel, 181–223

Hökfelt, T. and Fuxe, K. (1972*b*). Effects of prolactin and ergot alkaloids on the tubero-infundibular dopamine (DA) neurons. *Neuroendocrinology*, 9, 100–122

Horowski, R. and Gräf, K.-J. (1975). Prolactin secretion in rats under the influence of different agents acting on the dopaminergic system. *Acta endocr. (Kbh.),* Suppl. **199**, 203

Horowski, R., Neumann, F. and Gräf, K.-J. (1975). Influence of apomorphine hydrochloride dibutyryl-apomorphine and lysenyl on plasma prolactin concentrations in the rat. *J. Pharm. Pharmac.*, 27, 532–534

Horowski, R. and Wachtel, H. (1976). Direct dopaminergic action of lisuride hydrogen maleate, an ergot derivative, in mice. *Eur. J. Pharmacol.*, **36**, 373–376

Kann, G. and Denamur, R. (1974). Possible role of prolactin during the oestrous cycle and gestation in the ewe. *J. Reprod. Fert.*, **39**, 473–483

Kann, G. and Martinet, J. (1975). Prolactin levels and duration of post-partum anoestrus in lactating ewes. *Nature*, **257**, 63–64

Kato, Y., Nakai, Y., Imura, H., Chihara, K. and Ohgo, S. (1974). Effect of 5-hydroxytryptophan (5-HTP) on plasma prolactin levels in man. *J. Clin. Endocr. Metab.*, **38**, 695–697

Kehr, W. (1977). Effect of lisuride and other ergot derivatives on monoaminergic mechanisms in rat brain. *Eur. J. Pharmacol.*, **41**, 261–273

Kiyoshi, M., Zoshio, O., Mitsuko, H., Kaichiro, I. and Yuichi, K., (1974). Inhibition of thyrotropin and prolactin secretion in primary hypothyroidism by 2-Br-α-ergocryptine. *J. Clin. Endocr. Metab.*, **39**, 391–394

Kordon, C. (1971). Blockade of ovulation in the immature rat by local microinjection of α-methyl-dopa into the arcuate region of the hypothalamus. *Neuroendocrinology*, 7, 202–209

Kordon, C., Blake, C. A., Terkel, J. and Sawyer, C. H. (1973/74). Participation of serotonin containing neurons in the suckling-induced rise in plasma prolactin levels in lactating rats. *Neuroendocrinology,* **13**, 213–223

Kordon, C., Hery, M. and Enjalbert, A. (1976). Neurotransmitters and control of pituitary function. In F. Labrie, J. Meites and G. Pelletier (eds.). *Hypothalamus and Endocrine Function,* Plenum Press, New York and London, 51

Kraicer, P. F. and Shelesnyak, M. C. (1968). Interruption of pregnancy, induction of ovulation and delayed pseudopregnancy following suppression of luteal function. *Acta endocr. (Kbh.),* **58**, 251–260

Kraicer, P. F. and Strauss, J. F. (1970). Ovulation block produced by an inhibitor of luteotrophin, ergocornine. *Acta endocr. (Kbh.),* **65**, 698–706

Krieger, D. T., Amorosa, L. and Linick, F. (1975). Cyproheptadine induced remission of Cushing's disease. *New Engl. J. Med.*, **293**, 893–896

Krieger, D. T. and Luria, M. (1976). Effectiveness of cyproheptadine in decreasing plasma ACTH concentration in Nelson's syndrome. *J. Clin. Endocr. Metab.*, **43**, 1179–1182

Lachelin, G. C. L., Leblanc, H. and Yen, S. S. C. (1977). The inhibitory effect of dopamine agonists on LH release in women. *J. Clin. Endocr. Metab.*, **44**, 728–732

Lamberts, S. W. J. and Birkenhäger, J. C. (1976). Effect of bromocriptine in pituitary-dependent Cushing's syndrome. *J. Endocr.*, 70, 315–316

Lawson, D. M. and Gala, R. R. (1975). The influence of adrenergic, dopaminergic, cholinergic and serotoninergic drugs on plasma protein levels in ovariectomized, oestrogen-treated rats. *Endocrinology*, **96**, 313–318

Leblanc, H., Lachelin, G. C. L., Abu-Fadil, S. and Yen, S. S. C. (1976). Effects of dopamine infusion on pituitary hormone secretion in humans. *J. Clin. Endocr. Metab.*, **43**, 668–674

Lichtensteiger, W. and Keller, P. J. (1974). Tubero-infundibular dopamine neurones and the secretion of luteinizing hormone and prolactin: extrahypothalamic influences, interaction with cholinergic systems and the effect of urethane anesthesia. *Brain Res.*, 74, 279–303

Liuzzi, A., Chiodini, P. G., Botalla, L., Cremascoli, G., Müller, E. E. and Silvestrini, F. (1974). Decreased plasma growth hormone (GH) levels in acromegalics following CB 154 (2-Br-α-ergocryptine) administration. *J. Clin. Endocr. Metab.*, **38**, 910–912

Loew, D. M., Vigouret, J. M. and Jaton, A. L. (1976). Neuropharmacological investigations with two ergot alkaloids, hydergine and bromocriptine. *Postgrad. Med. J.*, **52**, Suppl. 1, 40–46

Lu, K. H., Koch, Y. and Meites, J. (1971). Direct inhibition by ergocornine of pituitary prolactin release. *Endocrinology*, 89, 229–233

Macindoe, J. H. and Turkington, R. W. (1973). Stimulation of human prolactin secretion by intravenous infusion of l-tryptophan. *J. Clin. Invest.*, **52**, 1972–1978

MacLeod, R. M. (1976). Regulation of prolactin secretion. In L. Martini and W. F. Ganong (eds.). *Frontiers in Neuroendocrinology*, Raven Press, New York, 169

MacLeod, R. M. and Lehmeyer, J. E. (1972). Regulation of the synthesis and release of prolactin. In G. E. W. Wolstenholme and J. Knight (eds.). *Lactogenic Hormones*, Churchill Livingstone, Edinburgh and London, 53–82

McCann, S. M. (1974). Regulation of secretion of follicle-stimulating hormone and luteinizing hormone. In E. Knobil and W. H. Sawyer (eds.). *Handbook of Physiology, Section 7: Endocrinology, Vol. 4, The Pituitary Gland and its Neuroendocrine Control Part 2*, American Physiological Society, Washington, DC, 489–518

Marko, M. and Flückiger, E. (1974). Inhibition of spontaneous and induced ovulation in rats by nonsteroidal agents. *Experientia*, **30**, 1174–1176

Massara, F., Camanni, F., Belforte, L. and Molinatti, G. M. (1976*a*). Dopamine-induced inhibition of prolactin and growth hormone secretion in acromegaly. *Lancet*, **I**, 485

Massara, F., Camanni, F., Belforte, L. and Molinatti, G. M. (1976*b*). Dopamine and inhibition of prolactin and growth-hormone secretion. *Lancet*, **I**, 913

Meites, J. (1973). Control of prolactin secretion in animals. In J. L Pasteels and C. Robin (eds.). *Human Prolactin*, Excerpta Medica, Amsterdam, 105–118

Meites, J., Simpkins, J., Bruni, J. and Advis, J. (1976). Role of biogenic amines in control of anterior pituitary hormones. *Internat. Res. Commun. Syst. Med. Sci.*, **5**, 1–7

Mendelson, W. B., Jacobs, L. S., Reichman, J. D., Othmer, E., Cryer, P. E., Trivedi, B. and Daughaday, W. H. (1975). Methysergide. Suppression of sleep-related prolactin secretion and enhancement of sleep-related growth hormone secretion. *J. Clin. Invest.*, **56**, 690–697

Müller-Schweinitzer, E. and Weidmann, H. (1978). Basic pharmacological properties. In B. Berde and H. O. Schild (eds.). *Ergot Alkaloids and Related Compounds. Handbook of Experimental Pharmacology*, Springer Verlag, Heidelberg, 87–232

Neill, J. D. (1974). Prolactin: its secretion and control. In E. Knobil and W. H. Sawyer (eds.). *Handbook of Physiology Section 7: Endocrinology, Vol. 4. The Pituitary Gland and its Neuroendocrine Control Part 2*, American Physiological Society, Washington DC, 469–488

Niswender, G. D. (1974). Influence of 2-Br-α-ergocryptine on serum levels of prolactin and the oestrous cycle in sheep. *Endocrinology*, **94**, 612–615 (1974)

Nooter, K. and Zeilmaker, G. H. (1970). Effects of ergocornine and hypothalamic stimulation on ovulation in the rat. *J. Endocr.*, **48**, lxiv

Pasteels, J. L., Danguy, A., Frerotte, M. and Ectors, F. (1971). Inhibition de la sécrétion de prolactine par l'ergocornine et la 2-Br-α-ergocryptine: action directe sur l'hypophyse en culture. *Annls. Endocr. (Paris)*, **32**, 188–192

Penny, R. J. and Thody, A. J. (1976). Preliminary studies on the control of α-melanocyte-stimulating hormone secretion in the rat. *J. Endocr.*, **69**, 2P–3P

Peracchi, M., Lombroso, G. C., D'Alberton, A., Cammareri, G., Caccamo, A. and Crosignani, P. G. (1977). Antiserotonin drugs in the treatment of the amenorrhea-galactorrhea syndrome. *Acta endocr. (Kbh)*, **85**, 54 (Suppl. 212)

Piva, F., Sterescu, N., Zanisi, M. and Martini, L. (1969). Non-steroidal antifertility agents affecting brain mechanisms. *Bull. Wld. Hlth. Org.*, **41**, 275–288

Poulain, P. and Carette, B. (1976). Actions of iontophoretically applied prolactin on septal and preoptic neurons in the guinea-pig. *Brain Res.*, **116**, 172–176

Raj, H. G. and Greep, R. O. (1973). Inhibition of ovulation and luteinizing hormone secretion in the cyclic rat by ergotamine tartrate. *Proc. Soc. exp. Biol. Med.*, **144**, 960–962

Ranta, T., Männistö, P. and Tuomisto, J. (1977). Evidence for dopaminergic control of thyrotrophin secretion in the rat. *J. Endocr.*, **72**, 329–335

Rose, J. C. and Ganong, W. F. (1976). Neurotransmitter regulation of pituitary secretion. In W. B. Essman and L. Valzelli (eds.). *Current Developments in Psychopharmacology*, Vol. 3, Spectrum Publications Inc., New York, 86

Rothlin, E., Cerletti, A., Konzett, H., Schalch, W. R. and Taeschler, M. (1956). Zentrale vegetative LSD-Effekte. *Experientia*, **12**, 154–155

Rutschmann, J. and Stadler, P. (1978). Chemical background. In B. Berde and H. O. Schild (eds.). *Ergot Alkaloids and Related Compounds. Handbook of Experimental Pharmacology*, Springer Verlag, Heidelberg, 29–85

Schally, A. V., Redding, T. W., Arimura, A., Dupont, A. and Linthcum, G. L. (1977). Isolation of gamma-amino butyric acid from pig hypothalami and demonstration of its prolactin release inhibiting (PIF) activity *in vivo* and *in vitro*. *Endocrinology*, **100**, 681–691

Schlientz, W., Brunner, R., Rüegger, A., Berde, B., Stürmer, E. and Hofmann, A. (1968). β-Ergokryptin, ein neues Alkaloid der Ergotoxin-Gruppe. *Pharm. Acta Helv.*, **43**, 497–509

Seki, M., Seki, K., Yoshihara, T., Watanabe, N., Okumura, T., Tajima, C., Huan, S.-Y. and Kuo, C.-C. (1974). Direct inhibition of pituitary LH secretion in rats by CB 154 (2-Br-α-ergocryptine). *Endocrinology*, **94**, 911–914

Shelesnyak, M. C. (1954). Ergotoxine inhibition of deciduoma formation and its reversal by progesterone. *Am. J. Physiol.*, **179**, 301–304

Shelesnyak, M. C. (1955). Disturbance of hormone balance in the female rat by a single injection of ergotoxine ethanesulfonate. *Am. J. Physiol.*, **180**, 47–49 (1955)

Sinha, Y. N., Selby, F. W. and Vanderlaan, W. P. (1974). Effects of ergot drugs on prolactin and growth hormone secretion and on mammary nucleic acid content in C3H/Bi mice, *J. Natn. Cancer Inst.*, **52**, 189–191

Smith, A. F. (1975). The effect of apomorphine and ergocryptine on the release of MSH by the Pars intermedia of *Rana pipiens. Neuroendocrinology*, **19**, 363–376.

Smith, V. G., Beck, T. W., Convey, E. M. and Tucker, H. A. (1974). Bovine serum prolactin, growth hormone, cortisol and milk yield after ergocryptine. *Neuroendocrinology*, **15**, 172–181

Smythe, G. A., Compton, P. J. and Lazarus, L. (1976). Serotoninergic control of human growth hormone secretion: the actions of L-dopa and 2-bromo-α-ergocryptine. In A. Pecile and E. E. Müller (eds.). *Growth Hormone and Related Peptides*, Excerpta Medica, Amsterdam, 222–235

Snider, S. R., Hutt, CH., Stein, B. and Fahn, S. (1975). Increase in brain serotonin produced by bromocriptine. *Neuroscience Letts.*, **1**, 237–241

Stoll, W. A. and Hofmann, A. (1943). Die Alkaloide der Ergotoxingruppe: Ergocristin, Ergokryptin, Ergocornin (7. Mitteilung über Ergot-Alkaloide). *Helv. Chim. Acta*, **26**, 1570–1601

Takahara, J., Arimura, A. and Schally, A. V. (1974). Suppression of prolactin release by a purified porcine PIF preparation and catecholamines infused into a rat hypophyseal portal vessel. *Endocrinology,* **95**, 462–465

Taleisnik, S., Celis, M. E. and Tomatis, M. E. (1973/74). Release of melanocyte-stimulating hormone by several stimuli through the activation of a 5-hydroxytryptamine-mediated inhibitory neuronal mechanism. *Neuroendocrinology,* **13**, 327–388

Tashjian, A. H. and Hoyt, R. F. (1972). Transient control of organ-specific functions in pituitary cells in culture. In M. Sussman (ed.). *Molecular Genetics and Developmental Biology*, Prentice Hall Inc., Englewood Cliffs, N. J., 353–387

Thorner, M. O., Chait, A., Aitken, M., Benker, G., Bloom, S. R., Mortimer, C. H., Sanders, P., Stuart Mason, A. and Besser, G. M. (1975). Bromocriptine treatment of acromegaly. *Br Med. J.,* **1**, 299–303

Tilders, F. J. H., Mulder, A. H. and Smelik, P. G. (1975). On the presence of a MSH-release inhibiting system in the rat neurointermediate lobe. *Neuroendocrinology,* **18**, 125–130

Tindal, J. S. (1974). Hypothalamic control of secretion and release of prolactin. *J. Reprod. Fert.,* **39**, 437–461

Tolis, G., Pinter, E. J. and Friesen, H. (1975). The acute effect of 2-bromo-α-ergocryptine (CB 154) on anterior pituitary hormones and free fatty acids in man. *Int. J. Clin. Pharmacol.,* **12**, 281–283

Vale, W., Rivier, C., Brown, M., Chan, L., Ling, N. and Rivier, J. (1976). Application of adenohypophyseal cell cultures to neuroendocrine studies. In F. Labrie, J. Meites and G. Pelletier (eds.). *Hypothalamus and Endocrine Function*, Plenum Press, New York and London, 397

Verde, G., Oppizzi, G., Colussi, G., Cremascoli, G., Botalla, L., Müller, E. E., Silvestrini, F., Chiodini, P. G. and Liuzzi, A. (1976). Effect of dopamine infusion on plasma levels of growth hormone in normal subjects and in acromegalic patients. *Clin. Endocr.,* **5**, 419–423

Wilson, C. A. (1974). Hypothalamic amines and the release of gonadotrophins and other anterior pituitary hormones. In A. B. Simmonds (ed.). *Advances in Drug Research*, Vol. 8, Academic Press, London and New York, 120–204

Wuttke, W., Cassell, E. and Meites, J. (1971). Effects of ergocornine on serum prolactin and LH, and on hypothalamic content of PIF and LRF. *Endocrinology*, **88**, 737–741

Yanai, R. and Nagasawa, H. (1970). Effect of ergocornine and 2-Br-α-ergocryptine (CB 154) on the formation of mammary hyperplastic alveolar nodules and the pituitary prolactin levels in mice. *Experientia,* **26**, 649–650

Zeilmaker, G. H. (1964). Experimentele Onderzoekingen over let eerste Stadium van de zwangerschap bij de Rat. Academisch Proefshrift, Amsterdam

Zeilmaker, G. H. and Carlsen, R. A. (1962). Experimental studies on the effect of ergocornine methanesulfonate on the luteotrophic function of the rat pituitary gland. *Acta. endocr. (Kbh),* **41**, 321–335

Discussion

Martini (Milan)
When considering the structures of the ergot derivatives one wonders how these could fit into a dopamine receptor. The molecular weight is different, the configuration of the molecule is totally different. Have you any comment about this?

Flückiger
This is a very tricky question because little work has been done. Using peripheral tissues as models the idea has formed that these drugs have affinity for both pre- and post-synaptic α-adrenoceptors and the 5-HT receptor. Most likely the affinity is not just for the receptor but also for the neighbourhood of the receptor. Thus the agonist–antagonist reactions are not simple as with lower molecular weight compounds. The dopamine receptor has only received attention in the last few years and here we lack good peripheral models. Perhaps receptor ligand binding studies will provide more information in the future.

6
Hypothalamic peptide hormones and their analogues

A. V. Schally*, D. H. Coy*, A. Arimura*, T. W. Redding*, A. J. Kastin*, C. Meyers*, J. Seprodi*, R. Chang*, W.-Y. Huang*, K. Chihara*, E. Pedroza*, J. Vilchez* and R. Millar†

INTRODUCTION

Although various anatomical and physiological data discussed elsewhere in this book supported the concept of hypothalamic control of the pituitary gland (Green and Harris, 1947), direct evidence for the existence of hypothalamic hormones involved in the release of pituitary hormones was lacking until Saffran and Schally (1955) demonstrated the presence of a corticotrophin releasing factor (CRF) in hypothalamic and neurohypophyseal extracts. This opened the way for subsequent discoveries of other hypothalamic regulatory substances. The existence of at least nine hypothalamic regulatory hormones is now established with some certainty (see table 6.1), and four of these substances, thyrotrophin releasing hormone (TRH), luteinising hormone-releasing hormone/follicle stimulating hormone-releasing hormone (LH-RH/FSH-RH), somatostatin, and MSH-release inhibiting factor (MIF), have been isolated in pure state, structurally identified, and synthesised. These substances were also previously called "factors".

In this chapter we use the word *hormone* for substances which have had their structures determined and which have been demonstrated to be the physiological regulators of the secretion of the appropriate anterior pituitary hormones. Other hypothalamic substances whose structures have not been clearly determined will be referred to as *factors*, since their physiological activity cannot be ascribed to a specific structure. For at least four pituitary hormones, growth hormone (GH), prolactin, thyrotrophin (TSH) and MSH, there is apparently a dual system of hypothalamic control, one system being stimulatory and one inhibitory. Much work is now under way on these hypothalamic hormones and their synthetic analogues, including their clinical evaluation as diagnostic and therapeutic

*Endocrine and Polypeptide Laboratory, Veterans' Administration Hospital, and Department of Medicine, Tulane University School of Medicine, New Orleans, Louisiana, USA
†Department of Chemical Pathology, Medical School, Observatory, Cape Town, South Africa

Table 6.1 Hypothalamic hormones or factors controlling the release of pituitary hormones

	Abbreviation
Corticotrophin (ACTH)-releasing factor	CRF
Thyrotrophin (TSH)-releasing hormone	TRH
Luteinising hormone (LH)-releasing hormone/ follicle-stimulating hormone (FSH)-releasing hormone	LH-RH/FSH-RH
Growth hormone (GH)-release inhibiting hormone	GH-RIH; somatostatin
Growth hormone (GH)-releasing factor	GH-RF
Prolactin release-inhibiting factor	PIF
Prolactin releasing factor	PRF
Melanocyte-stimulating hormone (MSH)-release-inhibiting factor	MIF
Melanocyte-stimulating hormone (MSH)-releasing factor	MRF

agents. Indeed the attention of many scientists and clinicians is at present focused on the hypothalamus.

We report here the most recent biochemical, physiological and pharmacological findings relating to each of the known hypothalamic regulatory peptides, including their direct effects on the CNS. Other details can be found elsewhere (Schally *et al.*, 1973*a*, 1976*d*; Schally and Arimura, 1978; Coy *et al.*, 1975*a*,*c*, 1976*a*; Kastin *et al.*, 1976, 1977; Vale *et al.*, 1977).

CORTICOTROPHIN RELEASING FACTOR (CRF)

CRF appears to be the neurochemical mediator of the classical response to 'stress'. Thus external environmental factors, various stimuli, emotions, or indeed any interference with the body's capability to maintain homeostasis (heat, cold, infections, injury, toxins, lack of oxygen), acting via the CNS, may cause a liberation of CRF, which in turn will stimulate the release of ACTH from the pituitary. Consequently with *in vivo* assays it is not always possible to ensure that the anterior pituitary tissue is the primary target organ for the substance tested and that the injected substance does not induce a non-specific stress response. To circumvent the non-specificity of *in vivo* assays, Saffran and Schally (1955) developed *in vitro* methods to guide attempts at isolation of the factor(s) responsible for the release of ACTH.

Utilising isolated rat anterior pituitary tissue as a test system for the detection of substances with CRF activity, Saffran and Schally (1955) showed that hypothalamic or posterior pituitary extracts significantly increased the rate of liberation of ACTH. Subsequently they purified small amounts of CRF from porcine neurohypophyseal powders (Schally *et al.*, 1958). Guillemin *et al.* (1957) reached essentially the same conclusions as Saffran and Schally, namely that hypothalamic and posterior pituitary tissue contained a peptide different from oxytocin and vasopressin which stimulated the release of ACTH *in vitro.*

Recent Studies on CRF

Although CRF was the first hypothalamic hypophysiotrophic hormone to be detected, attempts to isolate it in pure form were severely hampered by the difficulty of assays and by the loss of activity during the purification. However, the work on its purification has been resumed recently by several laboratories. The success now being achieved can be explained by rapid and sensitive methods for the detection of ACTH—for example, radioimmunoassays (RIA) and the ability to use modern techniques for purification of peptides, such as counter-current distribution (CCD), chromatography on Sephadex and its derivatives, high-pressure liquid chromatography and other techniques. Thus several studies on purification of CRF have been reported recently.

Cooper *et al.* (1976) described the purification of an acid extract of porcine hypothalami by ultrafiltration and chromatography on Sephadex G-50. Two materials with CRF activity were obtained with molecular weights (MW) of 30,000 and 1500 daltons, and with the characteristics of peptides, the former possibly representing the aggregated form or a precursor peptide. Jones *et al.* (1977) also found two forms of CRF in media in which rat hypothalami were incubated with serotonin. The purification of CRF from incubation medium consisted of gel filtration, chromatography on carboxymethylcellulose (CMC) and high-voltage electrophoresis.

In our most recent studies (Schally *et al.*, 1977*a*), hypothalamic extracts from nearly half a million pig hypothalami were first separated into 14 fractions by gel filtration on a preparative scale column of Sephadex G-25 in 1.0 M acetic acid. CRF activity was detected *in vitro* by stimulation of the release of immunoreactive ACTH (Rees *et al.*, 1971) from rat pituitary quarters (Saffran and Schally, 1955), or monolayer cultures of pituitaries from adrenalectomised rats (Vale *et al.*, 1972*b*). CRF was also measured *in vivo* after i/v or i/c injection in rats or mice under pharmacological blockade by pretreatment with chlorpromazine, morphine and Nembutal (Arimura *et al.*, 1967), followed by fluorometric measurement, competitive binding assay, or RIA of plasma corticosterone (Vecsei, 1974). Significant CRF activity was found in three fractions: fraction 2, $R_f = 0.76$, emerging just after the void volume; fraction 8 with an $R_f = 0.4$ and MW about 1000, and in fractions 11 and 12 which also contained catecholamines and had a low R_f of 0.3.

Fractions from Sephadex with $R_f = 0.76$ containing high MW CRF were purified by column chromatography on CMC. CRF activity was eluted in fractions with a conductivity of 7000–10,000 μS. The CRF-active area, still contaminated with ACTH-like activity, was subjected to countercurrent distribution (CCD) in a system of 0.1 per cent acetic acid: *l*-butanol: pyridine (11:5:3). The CRF-active fractions, $K = 1.1$, released ACTH *in vivo* and *in vitro* in doses of 1μg. These fractions were repurified by chromatography on SE-

Sephadex using a linear gradient of pyridine acetate buffers. This step yielded two areas with CRF activity (conductivities = 4000–5000μS), detectable in doses of 0.1μg/ml *in vitro* and devoid of any inherent ACTH activity. These fractions were also active *in vivo* in doses of 1μg. After rechromatography on SE-Sephadex, some CRF fractions were active *in vitro* in doses of 0.03 μg. Measurements of molecular weight and amino acid analysis indicated an MW for this CRF of 3200 daltons. CRF activity was completely lost after 16 hours' digestion with trypsin and partially destroyed by thermolysin. The results indicate that CRF activity in this fraction is due to a basic polypeptide with a MW of about 3200.

CRF contained in fraction 8 with an $R_f = 0.4$ from Sephadex was also repurified by chromatography on CMC. CRF activity was found in well-separated acidic fractions, neutral fractions and basic fractions. The basic fractions had the highest CRF activity, increasing ACTH release *in vitro* more than ten-fold. However, these fractions had the highest contamination with an ACTH-like peptide. Most of this contamination was eliminated after repurification by CCD. Some of these fractions increased ACTH release ten-fold.

The same methods (CMC and CCD) were used for the purification of fractions 11 and 12 from Sephadex with $R_f = 0.3$, containing small MW or retarded CRF. After rechromatography on Sephadex G-25 or ion-exchange chromatography on SE-Sephadex, a tetra-decapeptide with moderate CRF activity was isolated and its amino acid composition can be seen in table 6.2. However, its physiological importance remains to be determined and it is difficult, if not impossible to interpret at present the finding of multiple CRF activities. It is possible that: (1) CRF purified from high MW Sephadex fractions is pro-CRF (a precursor of CRF); (2) CRF of the intermediate MW is the physiological CRF (CRH); (3) in addition to the tetra-decapeptide CRF discussed above, CRF activity of low MW or retarded fractions is due to molecules such as noradrenaline. Similarly, it cannot be excluded that the body's defences against stress have evolved in such

Table 6.2 Amino acid composition of a CRF (AVS-49-151 Nos. 204–225)*

Amino acid	nmol	*Ratio*	*Integer*
Thr	37.8	3.0	3
Pro	23.4	1.9	2
Gly	14.0	1.1	1
Leu	16.2	1.2	1
Tyr	12.5	1.0	1
Phe	48.8	3.9	4
His	12.0	1.0	1
Lys	11.62	0.9	1
NH_3	29.2	–	–

*40 ng in 4N MSA for 24 h at 110°C; % peptide = 59%
MW ≈ 1701; total residues = 14

an exquisite way that different stresses may result in the liberation of different type of CRF, not necessarily exclusively from the hypothalamus, all being able to stimulate the release of ACTH. It may be pertinent to recall here the hypothesis, formulated by us more than 14 years ago, that several molecules with CRF activity exist (Schally *et al.*, 1960). The results of Cooper *et al.* (1976), Jones *et al.* (1977) and our studies confirm this hypothesis. Further work is needed for the isolation and characterisation of other CRFs and determination of their physiological role. Some preliminary results also indicate that CRF preparations can release β-endorphin in addition to ACTH.'

THYROTROPHIN RELEASING HORMONE (TRH)

The release of thyrotrophin (TSH) by the anterior pituitary gland is regulated by an interaction between hypothalamic TRH, which stimulates TSH release, and the thyroid hormones thyroxine (T_4) and triiodothyronine (T_3) which inhibit it. Both actions are exerted directly on the pituitary tissue. Recent work indicates that somatostatin, because of its suppressing action on the liberation of TSH,

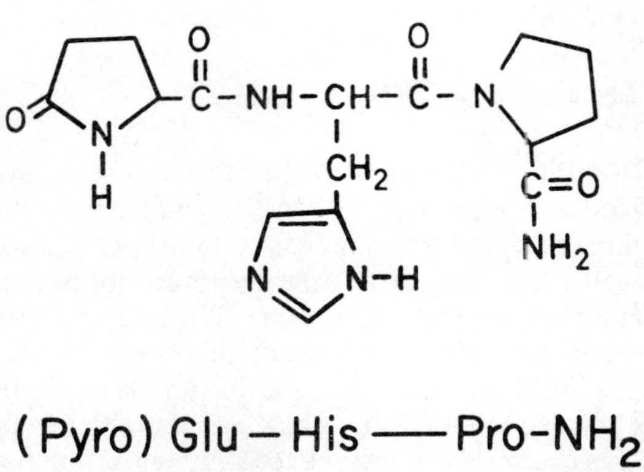

Figure 6.1 Structure of thyrotrophin releasing hormone (TRH)

may also be another regulator of TSH release (Arimura and Schally, 1976). The existence of TRH was first demonstrated in the early 1960s, but the first hint as to its chemical structure came in 1966 when it was shown that TRH isolated from porcine hypothalami contained three amino acids–histidine, proline and glutamic acid–in equimolar ratios (Schally *et al.*, 1966). After the painstaking processing of additional tons of pig and sheep brain tissue to obtain more material the amino acid sequence and the structure of porcine and ovine TRH was determined (Folkers *et al.*, 1969; Bøler *et al.*, 1969; Nair *et al.*, 1970; Burgus *et al.*, 1970) and the structure (pyro)Glu-His-Pro-NH_2 was confirmed by synthesis (figure 6.1).

Biological Effects of TRH

A variety of basic and clinical studies on TRH have been carried out during the past eight years (Schally *et al.*, 1973*a*; Vale *et al.*, 1977; Hall and Gomez-Pan, 1976; Schally and Arimura, 1977). Synthetic (pyro)Glu-His-Pro-NH_2 has been shown to stimulate TSH release in mammals such as mice, rats, nutria, sheep, goats, cows, humans, as well as in birds. However, it was inactive in the tadpole and lungfish (Gorbman and Hyder, 1973). TRH may also stimulate the synthesis of TSH. Among the physiological stimuli which appear to release TRH is exposure to mild cold. TRH can be administered intravenously, subcutaneously, intraperitoneally or orally. It will also liberate prolactin in rats, sheep and humans (Jacobs *et al.*, 1971; Tashjian *et al.*, 1971; Bowers *et al.*, 1971; Debeljuk *et al.*, 1973), but it remains to be established whether this effect is physiological or pharmacological.

In addition, TRH will stimulate GH secretion in animals under certain conditions, and in patients with acromegaly or renal failure (Schalch *et al.*, 1972). Recently it has been reported that TRH will increase colonic activity in rabbits (Smith *et al.*, 1977). This response, which may explain side-effects such as the nausea and gastric cramp seen occasionally after administration of TRH, appears to be mediated by the CNS and to involve cholinergic receptors.

CNS Effects and Localisation of TRH

The effects of TRH on the CNS have been the subject of many investigations (Kastin *et al.*, 1976, 1977; Vale *et al.*, 1977). The antidepressant activity of TRH in several animal models, such as reduction in sleeping time induced by ethanol and barbiturates, and behavioural events cannot be attributed to its action on the pituitary–thyroid axis since some effects can be obtained in hypophysectomised animals. Various neuropharmacological studies suggest that TRH may act on the brain and on the spinal cord (Kastin *et al.*, 1977; Martin *et al.*, 1975). This possible role of TRH as a neurotransmitter in the CNS is supported by the presence of significant concentrations of immunoreactive TRH in the extrahypothalamic brain areas of various vertebrates examined, including rat, chicken, snake, frog, tadpole, salmon and lamprey (Jackson and Reichlin, 1974). For instance in the rat the concentration of TRH is highest in the hypothalamus, but the hypothalamic TRH constitutes only about 30 per cent of the total brain content of TRH. Networks of TRH-positive nerve terminals have also been localised by immunofluorescence techniques around the motor neurones of the rat (Hökfelt *et al.*, 1975*b*).

These studies on the regional and phylogenetic distribution of TRH have been made possible by the generation of antibodies to TRH and the development of sensitive and specific radioimmunoassays (RIA) for this hormone (Jackson and Reichlin, 1974; Bassiri and Utiger, 1972; Oliver *et al.*, 1974). Immunoreactive TRH has also been detected in blood, cerebrospinal fluid and urine from the rat and man. However, the measurement of TRH in body fluids is complicated and the interpretation of the results difficult because of its rapid inactivation by proteolytic enzymes.

Metabolism

TRH is rapidly inactivated by rat and human plasma, its half-life in the blood of the rat being about 4 min (Redding and Schally, 1971). Among the products of digestion of TRH with plasma or hypothalamic fragments are (pyro)Glu-His-Pro, proline and proline amide (Schally *et al.*, 1973*a*; Bauer and Lipmann, 1976; Nair *et al.*, 1971*b*).

Mechanism of Action of TRH

^{3}H-TRH is specifically bound by plasma membrane receptors of bovine and murine anterior pituitary glands (Barden and Labrie, 1973; Labrie *et al.*, 1976). Derivatives of cyclic AMP and theophylline stimulate TSH and prolactin release *in vitro* and the intracellular mechanisms which are initiated by the interaction between TRH and the membrane receptor appear to be similar to those which mediate secretory responses of other endocrine cells; that is, they may be mediated by cyclic AMP (Labrie *et al.*, 1976).

Table 6.3 TSH and prolactin releasing activity and CNS activity of TRH analogues

	% activity			
Analogue	TRH	PRH	CNS	*Investigators*
(pyro) Glu-His-Pro-NH_2	100	100	100	
(pyro) Glu-His-Pro	0.5	0.5		Nair *et al.*, 1971*b*
(pyro) Glu-His-Pro-$NHCH_3$	3–10	4		Chang *et al.*, 1971
(pyro) Glu-His-OCH_3	2–15			Chang *et al.*, 1971
d-(pyro) Glu-*d*-His-*d*-Pro-NH_2	<0.5	<0.5		Wilber and Flouret, 1971; Hinkle *et al.*, 1974
(pyro) Glu-Phe-Pro-NH_2	10			Sievertsson *et al.*, 1972
(pyro) Glu-His-Pro-Gly-NH_2	30			Sievertsson *et al.*, 1974
(pyro) Glu-β-(2-thienyl)-Ala-Pro-NH_2	30			Sievertsson *et al.*, 1974
β-(pyrazolyl-3)-Ala-2-TRH	5			Hofmann and Bowers, 1970; Gillessen *et al.*, 1971
β-(pyrazolyl-1)-Ala-2-TRH	150			Coy *et al.*, 1975*b*
3-N-Imidazole-Methyl-His-2-TRH	800	700	100*	Rivier *et al.*, 1972; Hinkle *et al.*, 1974
Homo (pyro) Glu-His-Pro-NH_2	100		400†	Hirschmann, 1977
Homo (pyro) Glu-His-Thio-Pro-NH_2	100		1500‡	Hirschmann, 1977

*Reduction in alcohol-induced sleeping time
†Restoration of anticonvulsant activity of methazolamide in reserpinised mice
‡Restoration of flexor reflex in spinalised rats

Analogues of TRH

In comparison with LH-RH and somatostatin, relatively few analogues of TRH have been synthesised because of its small size and also because early work indicated that the structural requirements are very rigorous (Chang *et al.*, 1971; Bowers *et al.*, 1970; Rivier *et al.*, 1972). Thus most analogues have little TRH activity except for (pyro)Glu-Phe-Pro-NH_2 and (pyro)Glu-His-Pro-Gly-NH_2 which possess 10 per cent and 30 per cent TRH activity, respectively (table 6.3) (Sievertsson *et al.*, 1972, 1974). [N^{3im}-Me-His]^2-TRH has 3–10 times greater activity than the natural product (Rivier *et al.*, 1972) and β-(pyrazolyl-1)-alanine-2-TRH is also more active than TRH in rats and mice (Coy *et al.*, 1975*b*). This is in contrast to [N^{1im}-Me-His]^2-TRH and β-(pyrazolyl-3)-alanine-2-TRH, which have only 0.1 per cent and 5 per cent activity respectively (Rivier *et al.*, 1972; Gillesen *et al.*, 1971; Hofmann and Bowers, 1970). The activities of TRH analogues for stimulating the release of thyrotrophin and prolactin are similar (table 6.3) and a good correlation exists between binding affinity and TRH and PRF activity (Hinkle *et al.*, 1974). However, CNS and pituitary receptors for TRH may be different, as indicated by significant dissociation between TRH and the activity in the central nervous sytem of analogues such as homo-*p*Glu-His-Pro-NH_2 and homo-*p*Glu-His-Thiopro-NH_2, which are equipotent with TRH in stimulating TSH but have 4–15 times as much activity on the CNS (Hirschmann, 1977).

PROLACTIN RELEASING FACTOR (PRF)

It is well established that the hypothalamus can both stimulate and inhibit prolactin secretion. The inhibitory influence may predominate in man, rat, rabbit and other mammals, and probably in reptiles and amphibians (Meites and Clemens, 1972), but not in birds. The galactorrhoea seen in women after section of the pituitary stalk or administration of certain tranquillisers may result from a suppression of the inhibitory influence of the hypothalamus on prolactin secretion. That the control of prolactin secretion is mediated by hypothalamic neurohumoral substances was shown by various studies *in vitro* and *in vivo* which revealed the presence of both a prolactin releasing factor (PRF) and prolactin release inhibiting factor (PIF) in hypothalamic extracts of rats and domestic animals (Meites and Clemens, 1972; Schally *et al.*, 1973*a*, 1977*b*). Studies on the localisation of these neurohumors indicate that PIF and PRF activities appear to be situated in different areas of the hypothalamus. A part of the PRF activity in the hypothalamic extracts is unquestionably due to TRH, which under certain conditions can stimulate prolactin secretion *in vivo* and *in vitro* in rats, sheep and humans (Koch *et al.*, 1977*b*) (see also the section on TRH). The PRF activities of many analogues of TRH are related to their thyrotrophin releasing potencies (Hinkle *et al.*, 1974). It has also been reported that administration of an antiserum to TRH will greatly suppress serum prolactin levels in male and female rats (Koch *et al.*, 1977*b*). This finding, if confirmed, would indicate that TRH might play a physiological role in the control of prolactin secretion.

However, the experimental and clinical evidence that a PRF different from

TRH exists may be summarised as follows: (1) prolactin and TSH can be released independently under various physiological conditions; (2) there is a dissociation between TSH and prolactin levels in patients with various thyroid diseases (l'Hermite *et al.*, 1974) (moreover, pretreatment with 100μg T_3 inhibits the TSH, but not the prolactin release, in response to TRH in normal men); (3) our own work and that of several other groups of investigators indicates that partially purified hypothalamic fractions, apparently different from TRH, can still stimulate the release of prolactin from rat pituitaries *in vivo* and *in vitro* (Dular *et al.*, 1974; Szabo and Frohman, 1976; Boyd *et al.*, 1976).

In our own laboratory the purification of PRF has been seriously impeded by the presence in hypothalamic fractions of substances such as TRH which increase the release of prolactin and of those which inhibit it such as catecholamines and gamma-amino butyric acid (Schally *et al.*, 1976*c*, 1977*b*). Thus on the one hand TRH can obscure the PRF activity of some fractions under most assay conditions, especially if the two activities are related chemically and overlap, and it can also at least in part neutralise the PIF activity. On the other hand, compounds with PIF activity such as catecholamines and gamma-aminobutyric acid (GABA) can also cancel in part the stimulatory effect of PRF on prolactin. We have shown that when one of these PRF fractions from Sephadex is chromatographed on CMC or subjected to CCD, the PRF activity is found in the areas corresponding to those of TRH. This could indicate that this PRF is chemically similar to TRH or that all the activity is due to TRH. However, estimates of TRH activity by RIA indicate that this PRF activity cannot be accounted for by TRH contamination. Further purification of this fraction is in progress. We have also established the presence of several fractions with different physicochemical properties from TRH (as shown by different partition coefficients on CCD, basicity on CMC, and SE-Sephadex and behaviour on molecular sieving on Sephadex G-25), which stimulate the release of prolactin from rat pituitary fragments or from monolayer cultures of pituitary cells (Schally, Arimura, Redding *et al.*, unpublished). These fractions are also being repurified.

A variety of natural substances of central and peripheral origin can augment prolactin release *in vivo* and even *in vitro*. Among these substances are β-endorphin, enkephalins, vasopressin, oestrogens, monoiodotyrosine, histamine, cyclic AMP, prostaglandins E_1 and E_2, and 5-hydroxytryptophan. The effects of the 5-hydroxytryptophan can be explained by stimulation of the biosynthesis of serotonin which may be the neurotransmitter responsible for the release of PRF and those of vasopressin and histamine by stress or an action mediated via the CNS. Similarly, the opiate peptides β-endorphin, Met-enkephalin and some of their analogues will also release prolactin if given into the third ventricle, but their effects are exerted through CNS centres (Dupont *et al.*, 1977*b*), and they also do not represent the physiological PRF, the chemistry of which remains to be elucidated.

A variety of agents produce an increase in prolactin levels. They include inhibitors of the action, synthesis or storage of catecholamines such as reserpine, methyl-dopa, α-methyl-*p*-tyrosine, and phenothiazines such as chlorpormazine and perphenazine, haloperidol together with other drugs like tricyclic antidepressants, some benzamides such as sulpiride and nicotine. Some of these drugs act in part directly on the pituitary, but most of them appear to affect the turnover of putative synaptic neurotransmitters controlling PRF and PIF secretion.

PROLACTIN RELEASE-INHIBITING FACTOR (PIF)

The presence of PIF activity in hypothalamic extracts was demonstrated many years ago (Meites and Clemens, 1972), but the nature of physiological PIF is still not clear. Many substances present in the hypothalamic or brain extracts can inhibit prolactin secretion *in vitro* and *in vivo* (Schally *et al.*, 1976*c*, 1977*b*; Dular *et al.*, 1974). The presence of materials with PRF activity such as TRH which could nullify the PIF activity under certain conditions has seriously hampered the work on PIF. Somatostatin has some inhibitory effect on prolactin secretion under some conditions such as *in vitro* monolayer cultures of pituitary cells (Vale *et al.*, 1977). However, its presence can be readily detected by the RIA for somatostatin.

Among the most powerful inhibitors of prolactin release are catecholamines (MacLeod, 1969; Birge *et al.*, 1970). It has been suggested that catecholamines might exert this effect by stimulating the release of PIF from the hypothalamus (Kamberi *et al.*, 1971) but our recent work has show that at least a part of the PIF activity of catecholamines is due to a direct action on the pituitary gland. Our conclusions are based on tests with purified hypothalamic catecholamines and with synthetic dopamine and noradrenaline. When PIF activity present in acetic acid extracts of pig hypothalami was purified by gel filtration on Sephadex G-25, chromatography on CMC, CCD, rechromatography on Sephadex G-25 and partition chromatography, some of the highly purified fractions which greatly inhibited the release of prolactin *in vitro* and *in vivo*, were found to contain a significant percentage of noradrenaline and dopamine (Schally *et al.*, 1976*c*). The magnitude of inhibition of release of prolactin was related to the noradrenaline content of these fractions.

Synthetic noradrenaline and dopamine in doses of 10–100ng also strongly inhibited the release of prolactin *in vitro* (Schally *et al.*, 1976*c*). When dopamine or noradrenaline were dissolved in 5 per cent glucose solution and infused into a hypophyseal portal vessel of the rat, prolactin secretion was significantly suppressed. The suppressive effect of catecholamines *in vivo* was also dose-dependent (Takahara *et al.*, 1974*a*). These results showed that either synthetic catecholamines, or those purified from hypothalamic tissue, inhibit the release of prolactin by an action exerted directly on the pituitary gland. However, whether dopamine or noradrenaline represents a physiological prolactin release inhibiting hormone remains still to be determined. Moreover, the predominant component of the catecholamine fraction purified by us from pig hypothalami was noradrenaline, whilst in the hypophyseal portal blood only dopamine is found. Recent pharmacological evidence (Enjalbert *et al.*, 1977) suggests that hypothalamic PIF is different from dopamine. Various other drugs can suppress prolactin release, among them apomorphine (which mimics the action of dopamine), L-dopa (which elevates cerebral catecholamines) and monoamine oxidase inhibitors such as iproniazid and pargyline. Catecholamine-mediated inhibition of prolactin release can be nullified by dopamine receptor blocking agents such as perphenazine, pimozide, or haloperidol.

Another hypothalamic substance with PIF activity, the effect of which, in contrast to catecholamines, cannot be blocked by perphenazine is GABA

(Schally *et al.*, 1977*b*). GABA was isolated by us from a fraction with PIF activity obtained by chromatography on CMC of a concentrate of catecholamines from pig hypothalamic extracts. This fraction, chromatographically distinct from catecholamines, was further purified by six steps involving chromatography on Sephadex G-25, CCD in two different solvent systems, free-flow electrophoresis, chromatography on triethylaminoethyl cellulose and rechromatography on Sephadex G-25 to yield more than 700mg of material which was identified as GABA. Natural and synthetic GABA were found to inhibit prolactin release *in vitro* from isolated rat pituitary halves in doses as low as 0.1μg/ml. The extent of inhibition was proportional to the dose, and natural and synthetic GABA possessed identical PIF activity.

Synthetic GABA also suppressed prolactin release in monolayer cultures of rat pituitary cells and nullified the TRH-stimulated prolactin liberation (Schally *et al.*, 1977*b*). GABA also had PIF activity *in vivo*, although large doses were needed for an effect. After serum prolactin in male or female rats was elevated by administration of drugs such as monoiodotyrosine, perphenazine, chlorpromazine, haloperidol, or sulpiride, intravenous injection or infusion of GABA in doses of 1–100mg significantly decreased serum prolactin levels. Oral administration of 300mg GABA similarly completely suppressed the monoiodotyrosine-induced elevation in prolactin levels. It is of interest that β-hydroxy-GABA also significantly depressed prolactin release, but β-(*p*-chlorophenyl)-GABA (Lioresal^{R}) and four other analogues of GABA were not effective. These results show that GABA inhibits prolactin release by a direct action on the pituitary gland, but it is not known whether this effect is physiological or pharmacological, as the doses needed to obtain an effect are rather large (Schally *et al.*, 1977*b*).

Other compounds with PIF activity different from catecholamines and GABA, but still chemically unidentified, are present in pig hypothalamic extracts. One of them appears to be a polyamine and two others are polypeptides.

CB-154 (bromocriptine, an ergot alkaloid) is now used to inhibit the release of prolactin and to suppress undesired lactation, for instance in cases of galactorrhoea. A naturally occurring compound with PIF activity could be useful clinically in the situation described above, and speculatively in other situations such as for inhibiting the growth of prolactin-dependent mammary or prostate tumours, since this hormone has been linked with tumourigenesis of these organs. However, the chemistry of physiological prolactin release inhibiting hormone still remains to be elucidated.

MELANOCYTE STIMULATING HORMONE (MSH) –RELEASING FACTOR (MRF), AND –RELEASE INHIBITING FACTOR (MIF)

The release of MSH from the pars intermedia of the pituitary gland appears to be controlled by a hypothalamic stimulating factor (MRF) and inhibitory factor (MIF), the latter having a predominant role (Kastin and Schally, 1967; Kastin *et al.*, 1976, 1977). However, the physiological MIF and MRF have still not been identified with certainty. The earliest of these prospective MIFs was isolated from bovine hypothalami and identified as H-Pro-Leu-Gly-NH_2 in our laboratory (Schally and Kastin, 1966; Nair *et al.*, 1971*a*). Walter and associates (Celis *et al.*, 1971*b*) originally observed that Pro-Leu-Gly-NH_2 is formed by incubating oxytocin with an enzyme present in hypothalamic tissue and that this tripeptide inhibits MSH release in the rat. Various studies in animal models of Parkinsonism and mental depression (Kastin *et al.*, 1976, 1977) suggest a direct CNS action of Pro-Leu-Gly-NH_2 independent of its effect on MSH secretion, since the effects can be obtained in hypophysectomised animals. H-Pro-Leu-Gly-NH_2 has been shown to be effective alone or in conjunction with L-dopa in Parkinson's disease and at low doses seems to improve mental depression (Kastin *et al.*, 1977). Thus this MIF served to introduce the concept of a direct CNS effect of hypothalamic peptides.

Several analogues of this MIF have been synthesised recently by the group at Ayerst Research Laboratories, including H-Pro-*N*-isobutyl-Gly-Gly-NH_2 and stereoisomers of both Pro-Me-Leu-Gly-NH_2 and Pro-Me-Leu-Ala-NH_2 (Failli *et al.*, 1977). All analogues antagonised fluphenazine-induced catalepsy in rats and most potentiated the behavioural effects of L-dopa in mice. L-Pro-*N*-methyl-D-Leu-Gly-NH_2 was the most active of all these analogues after parenteral or oral administration.

However, not everybody agrees that Pro-Leu-Gly-NH_2 is in fact MIF. Other substances such as tocinoic acid, the cyclic pentapeptide ring of oxytocin, and tocinamide have been proposed as MIFs (Bower *et al.*, 1971), but these peptides are said to be active in mammals such as the rat and hamster and less active or inactive in amphibia such as the frog. Moreover, neither tocinoic acid nor tocinamide has ever been identified in the hypothalamus or the pituitary. The report on the MIF activity of tocinoic acid has not been confirmed by others and the proposal has now been withdrawn. It has also been suggested that catecholamines directly inhibit pituitary MSH secretion, the inhibition being mediated by dopaminergic innervation of pars intermedia cells (Tilders *et al.*, 1975). MSH itself has considerable effect in the CNS, improving the attention process in both rats and humans, even in those retarded mentally (Kastin *et al.*, 1977). In conclusion, considerable confusion still exists as to the identity of physiological MIF, although not in the evidence for the inhibitory control of MSH release.

There is also evidence for substances with MRF activity. Celis *et al.* (1971*a*) proposed that the opened *N*-terminal ring portion of oxytocin H-Cys-Tyr-Ile-Gln-Asn-OH constitutes MRF, but again there is little confirmation of this.

LH- AND FSH-RELEASING HORMONE (LH-RH/FSH-RH)

It is well established now that the hypothalamus controls the secretion of LH and FSH from the pituitary gland and through LH and FSH regulates the gonadal function. The release of LH and FSH from the anterior pituitary gland is regulated by a complex interaction between LH-RH/FSH-RH and sex steroids (Schally *et al.*, 1971*d*, 1973*a*, 1976*d*). The feedback effects of sex steroids are principally inhibitory (negative feedback), but can also be stimulatory (positive feedback), especially in the case of oestrogen. Thus the peak in the oestrogen concentration in plasma which precedes the ovulatory surge of LH in rats, monkeys and women appears to augment pituitary responsiveness to LH-RH (Vilchez-Martinez *et al.*, 1974*a*; Nillius and Wide, 1972) and to stimulate LH-RH release from the hypothalamus (Neill *et al.*, 1977). These effects of sex steroids are exerted in part on the hypothalamus and other CNS centres and in part on the pituitary (Hilliard *et al.*, 1971) where their direct action, which is now unequivocally established, appears to involve the receptor binding sites for LH-RH.

After the demonstration of the presence of LH-RH and FSH-RH activity in hypothalamic extracts of rats, domestic animals and humans, extensive physiological and biochemical work on both activities was then vigorously pursued by several laboratories in the 1960s. We used porcine hypothalami for purification of LH-RH and FSH-RH (Schally *et al.*, 1968). The efforts of our group led to the isolation of LH-RH, the determination of its composition (Schally *et al.*, 1971*a*, *b, d, e*), the elucidation of its amino acid sequence (Matsuo *et al.*, 1971*b*; Baba *et al.*, 1971) and finally its synthesis (Matsuo *et al.*, 1971*a*). The structure of LH-RH/FSH-RH is shown in figure 6.2. The structure of ovine LH-RH was later found to be identical to that of the porcine hormone (Burgus *et al.*, 1972). Subsequent studies have suggested that bovine, human and rat LH-RH are identical with the porcine and ovine hormone (Schally *et al.*, 1973*a*, 1976*d*; Mortimer *et al.*, 1976).

Our isolation, determination of structure and synthesis of LH-RH/FSH-RH opened up the vast field of reproduction to a novel approach in physiological, immunological, biochemical, behavioural, immunohistological, veterinary and clinical investigations. The synthesis of LH-RH by our group (Matsuo *et al.*, 1971*a*) and others (Geiger *et al.*, 1971; Yanaihara *et al.*, 1973; Immer *et al.*,

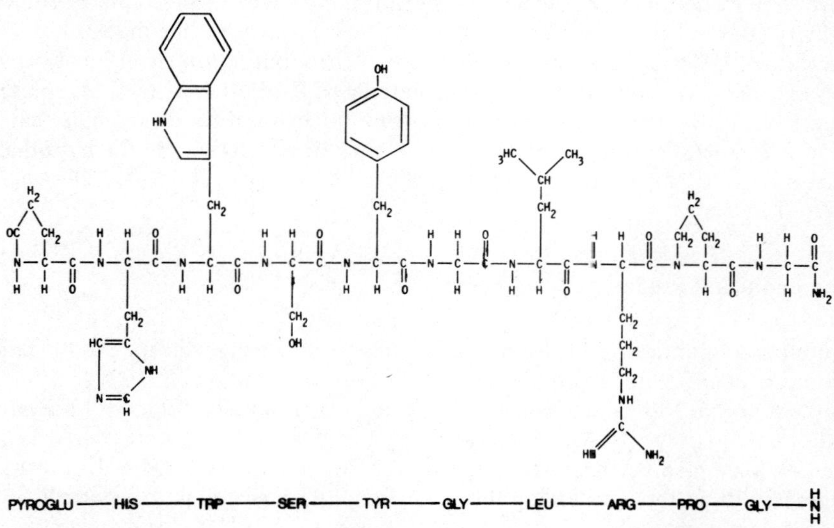

Figure 6.2 Structure of luteinising hormone-releasing hormone/follicle stimulating hormone-releasing hormone (LH-RH/FSH-RH)

1974*b*) made this hormone readily available for a variety of studies. LH-RH was found greatly to enhance the release of both LH and FSH and to induce ovulation in a variety of mammals, including rats, mice, rabbits, golden hamsters, mink, spotted skunk, impala, rock hyrax, sheep, cattle, pigs, horses, monkeys and humans. In rats or humans LH-RH raises plasma LH and FSH not only after intravenous, subcutaneous, or intracarotid injection, but also after intravaginal, intrarectal, intranasal, cutaneous or oral administration. However, the intranasal, rectal and oral doses are 100 to 10,000 times larger than those required parentally.

LH-RH was also found to be active in non-mammalian species. Thus chickens, pigeons and some species of fish such as brown trout and carp respond to LH-RH. These studies in mammals, birds and fish indicate that species-specificity does not occur with LH-RH. However, dogfish and goldfish LH-RH appears to be immunologically different from the porcine and ovine hormone (Schally *et al.*, 1976*d*), and it has also been reported recently that the bonnet monkey is insensitive to D-Ala^{6}, desGly^{10}-LH-RH ethylamide (Levitan *et al.*, 1977), an analogue which is a highly effective stimulant of LH and FSH release in rats, mice, humans, sheep and other species (see below).

In 1971, because both natural LH-RH and the synthetic decapeptide corresponding to its structure possessed major FSH-RH as well as LH-RH activity, we took the bold step of proposing that one hypothalamic hormone, designated LH-RH/FSH-RH, could be responsible for this dual effect (Schally *et al.*, 1971*b*). This concept is now supported by much physiological as well as immunological data and the evidence that LH-RH releases FSH is now indisputable. The fact that ovulation has been induced by LH-RH in most of these species demonstrates that this decapeptide can release enough FSH to induce follicular maturation. Claims have been made for the existence of other substances with LH-RH and FSH-RH activity (Bowers *et al.*, 1973; Fawcett *et al.*, 1975). However, the material of Bowers *et al.* (1973) stimulates FSH release *in vitro*, but not *in vivo* (Schally *et al.*, 1976*a*), and it is also probable that the material of Fawcett *et al.* (1975) represents an artifact of the LH-RH decapeptide. Recent biochemical results indicate that the LH-RH decapeptide represents the bulk of FSH-RH activity in the hypothalamus and it does appear to be the principal FSH releasing hormone (Schally *et al.*, 1976*a*).

Immunological Studies

Production of antisera to LH-RH in rats, rabbits, guinea-pigs, sheep, and humans (Arimura *et al.*, 1973; Mortimer *et al.*, 1976; Brown, G. M. *et al.*, 1977; Kerdelhué *et al.*, 1973; Koch *et al.*, 1973) permitted the establishment of several radioimmunoassays (RIA) and the performance of a variety of immunological studies. Male rabbits which have been actively immunised with LH-RH and have generated its antibodies develop testicular atrophy associated with aspermatogenesis (Arimura *et al.*, 1973). The pituitary content of LH is also reduced in these rabbits (Arimura *et al.*, 1973).

Castrated rats actively immunised with LH-RH show parallel decreases in serum LH and FSH levels associated with a rise in serum antibody titre to LH-RH.

Administration of anti-LH-RH gamma-globulin to castrated rats prevents the rise in serum LH and FSH levels normally seen after such an operation, and also the development of castration cells in the pituitary (Arimura *et al.*, 1976*b*).

Passive immunisation of normal cycling rats with LH-RH prevents the preovulatory surge of LH and FSH, blocks ovulation (Arimura *et al.*, 1974*a*; Koch *et al.*, 1973) and induces hyperprolactinaemia (Kerdelhué *et al.*, 1976). In cycling hamsters, injection of antiserum to LH-RH also blocks the pre-ovulatory surge of LH and ovulation (de la Cruz *et al.*, 1976*a*), arrests follicular maturation and reduces serum oestradiol levels (de la Cruz *et al.*, 1976*a*). Thus antiserum to LH-RH can also suppress follicular development. This reduction in serum FSH in addition to LH after passive or active immunisation with LH-RH supports the physiological role of the LH-RH decapeptide in the regulation of FSH secretion. Hypothalamic LH-RH is also necessary for normal implantation and the maintenance of pregnancy (Arimura *et al.*, 1976*a*; Nishi *et al.*, 1976), since passive immunisation with LH-RH in early pregnancy causes a delay in implantation of fertilised ova or termination of pregnancy in rats, depending on the time the antibody to LH-RH is injected.

A peak of LH-RH levels in the peripheral plasma and in the hypophyseal portal blood before and during the pre-ovulatory surge of LH release in rats, sheep, rabbits, monkeys and women can be detected by RIA (Tsou *et al.*, 1977; Arimura *et al.*, 1974*b*; Kerdelhué *et al.*, 1973; Neill *et al.*, 1977), indicating that this decapeptide is the mediator responsible for the release of the ovulatory quota of LH. LH-RH has also been found in the hypophyseal portal blood of castrated monkeys and rats, and in the peripheral circulation of post-menopausal women (Mortimer *et al.*, 1976).

Localisation of LH-RH

The availability of antisera to LH-RH made possible various studies on the localisation of LH-RH by RIA or immunohistochemical methods (Sétáló *et al.*, 1975, 1978; Zimmerman, 1977). The bulk of LH-RH appears to be localised in the median eminence and in the arcuate nucleus. The pathway of LH-RH-containing nerve fibres in the median eminence of rats coincides with the course of the nerve fibres of the tubero-infundibular tract. The production of LH-RH in neuronal cell bodies, especially in the medial preoptic and the suprachiasmatic area, seems to be well documented (Sétáló *et al.*, 1978). However, immunohistological studies indicate that extrahypothalamic brain areas are also involved in the synthesis of LH-RH. This could suggest that, in addition to being the regulator of the release of LH and FSH, LH-RH might act as a central neurotransmitter (Kastin *et al.*, 1976, 1977). This view is supported by: (1) various neuropharmacological studies (Kastin *et al.*, 1976, 1977); (2) the significant LH-RH content of extrahypothalamic brain areas, particularly the mesencephalon and organum vasculosum; (3) the fact that some axons which appear to be carrying LH-RH terminate outside the median eminence; and (4) the location of LH-RH in synaptosomes (Sétáló *et al.*, 1975, 1978; Zimmerman, 1977). LH-RH has indeed been shown to excite sexual behaviour (Moss and McCann, 1973), and to modulate the electrical activity of neurones in the CNS (Moss, 1977).

Mechanism of Action

Although the mechanism of action of LH-RH is not completely clear, there is much evidence indicating that cyclic AMP may mediate its effects (Labrie *et al.*, 1976), since: (1) cyclic AMP, its derivatives or agents (such as prostaglandins) which increase intracellular cyclic AMP stimulate LH and FSH secretion *in vitro* (Borgeat *et al.*, 1972); (2) LH-RH stimulates the accumulation of cyclic AMP in rat anterior pituitary tissue *in vitro*; (3) there is a close parallelism between the LH- and FSH-releasing activity of various agonistic analogues of LH-RH and their abilities to induce accumulation of cyclic AMP; and (4) antagonistic analogues of LH-RH inhibit this accumulation of cyclic AMP (Labrie *et al.*, 1976). LH-RH appears to exert its effect by activating adenylcyclase which may lead to phosphorylation of physiologically important protein substrates. However, not all investigators agree that cyclic AMP participates in the acute release of LH (U. Zor, personal communication).

Since the initial step in the mechanism of action of LH-RH should involve an interaction with pituitary plasma membrane receptors, a receptor assay using highly purified pituitary cell membrane preparations has been developed in our laboratory. We have determined that the number of binding sites for LH-RH is approximately 2.4 pM/mg of protein with a high affinity constant of 7.1×10^9 M^{-1}. We have also observed that the superactive agonist D-Trp^6-LH-RH and the antagonist D-Phe^2, D-Trp^3, D-Phe^6-LH-RH compete with LH-RH for its pituitary plasma membrane receptors, displacing the ^{125}I-LH-RH more strongly than its parent hormone (Pedroza *et al.*, 1977). Therefore, both stimulatory and inhibitory analogues of LH-RH may exert their action on the same pituitary plasma membrane receptors as those for LH-RH. The powerful and protracted effect of such superactive analogues as D-Trp^6-LH-RH could be due to their greater ability to bind to the pituitary LH-RH receptors. Similarly, the inhibition of LH release induced by several inhibitory analogues of LH-RH such as D-Phe^2, D-Trp^3, D-Phe^6-LH-RH, could be mediated by the same mechanism of binding to LH-RH pituitary receptors.

Analogues of LH-RH

The interest in possible veterinary and medical applications of LH-RH analogues stimulated the search for new derivatives (Coy *et al.*, 1975*a*, *c*, 1976*b*; Schally *et al.*, 1976*d*). The principal targets were: (1) the development of analogues with prolonged biological activity which would be more useful therapeutically than LH-RH itself (half-life $\approx$ 2 min); and (2) development of inhibitory (antagonistic) analogues which would form the basis of new birth-control methods. The synthesis of many hundreds of LH-RH analogues was made possible by the use of rapid solid-phase techniques. The studies on these peptides have shed much light on the relationship between biological activity and structure and provided information into the role played by individual amino acids in preserving overall conformation and binding affinity to pituitary receptor sites and in triggering gonadotrophin release.

Early results showed that the amino-terminal tripeptide and tetrapeptide fragments of LH-RH as well as the carboxyl-terminal nonapeptide and octapeptide of LH-RH have very little or no LH-RH activity (Schally *et al.*, 1972*a*). Thus very active small fragments cannot be obtained from LH-RH. However, certain amino acids such as tyrosine in position 5 can be replaced by closely related amino acids in the LH-RH molecule without a major loss of activity. Most substitutions for the pyroglutamic acid residue result in almost complete loss of activity. However, acylated glycine^{1}-analogues lacking the (pyro)Glu- ring also have a small but definite activity. Of these, formylsarcosine^{1}-LH-RH and acetyl-sarcosine^{1}-LH-RH have 64 per cent and 72 per cent respectively, of the activity of the parent molecule. Incorporation of 2-pyrrolidone-4-carboxylic acid, an isomer of (pyro)Glu, or its *N*-methyl analogue into LH-RH in place of pyroglutamic acid gives rise to compounds with 19 per cent and 58 per cent LH-RH activity, respectively. In general, amino acids in position 1 and from positions 4 to 10 appear to be involved only in binding to the receptors and in exerting conformational effects. However, histidine and tryptophan probably exert a functional effect in addition to providing receptor-binding capacity, since simple substitutions or deletions in positions 2 or 3 greatly decrease or abolish LH-RH activity. Nevertheless, substitution in position 2 with analogues related to L-histidine and possessing acid–base properties and hydrogen donor and acceptor capability, such as (β-pyrazolyl-3)-Ala^{2}-LH-RH, results in considerable LH-RH activity (Coy *et al.*, 1974*a*). Deletion or replacement of tryptophan in position 3 results in nearly complete loss of activity, but pentamethylphenylalanine^{3}-LH-RH possesses high LH-RH activity, probably because of its electron transfer capability in position 3 (Coy *et al.*, 1974*b*).

Superactive analogues of LH-RH

The C-terminal residue of LH-RH can be altered in various ways without loss of activity and Fujino *et al.* (1972) were the first to report that replacement of glycine-amide alkylamide groups produces analogues more active than LH-RH. Thus des-Gly^{10}-LH-RH ethylamide (EA) is 3–5 times more potent than LH-RH and its activity is also prolonged, and des-Gly^{10}-LH-RH-2, 2, 2-trifluoroethylamide is 5–9 times as active as LH-RH (Coy *et al.*, 1975*d*).

Replacement of glycine in the 6 position of LH-RH can also lead to an increase in biological activity (Monahan *et al.*, 1973), D-Ala^{6}-LH-RH being 6–7 times more potent than LH-RH, a phenomenon attributed to a better binding conformation than with LH-RH. We have synthesised D-Leu^{6}-LH-RH, which is 5–9 times more active than LH-RH (Vilchez-Martinez *et al.*, 1974*b*) and D-Phe^{6}-LH-RH and D-Trp^{6}-LH-RH, which are 10 and 13 times, respectively, more potent *in vivo* than LH-RH, and which show prolonged activity (Coy *et al.*, 1976*e*).

The incorporation of both changes, ethylamide in the 10 position and a D-amino acid in the 6 position, produces analogues 30–60 times more potent than LH-RH and which cause prolonged release of LH and FSH (Sandow *et al.*, 1977; König *et al.*, 1975; Fujino *et al.*, 1974; Coy *et al.*, 1974*c*; Vilchez-Martinez *et al.*, 1974*b*). D-Ala^{6}, des-Gly^{10}-LH-RH ethylamide, D-Leu^{6}, des-Gly^{10}-LH-RH ethylamide, and D-Ser(Bu^{t})6-des-Gly^{10}-LH-RH ethylamide, release 50–60 times as much LH and 15 times as much FSH as do similar doses of LH-RH, and are active in rats when administered orally and intravaginally. However, the doses needed for oral, rectal, vaginal and nasal administration are about 1000, 400, 50 and 10 times larger, respectively than those causing comparable elevation by the subcutaneous route. In sheep D-Leu^6- and D-Ala^6-ethylamide analogues are also 18–50 times more potent than LH-RH.

D-Ala^6-, D-Leu^6-, and D-Trp^6- analogues are more resistant to the action of brain and hypothalamic enzymes than LH-RH (Koch *et al.*, 1977*a*). However, the magnitude and duration of binding of D-Ala^6- and D-Leu^6, desGly^{10}-LH-RH ethylamide and related analogues to the pituitary receptors appear to be much greater than for LH-RH and may explain the prolonged activity of these analogues better than decreased inactivation (Reeves *et al.*, 1977).

Precoital and postcoital contraceptive activity of LH-RH and its highly active agonistic analogues

Recent studies have demonstrated that prolonged (chronic) treatment with large doses of LH-RH and some long-acting, superactive stimulatory analogues of LH-RH causes impairment of reproductive functions. Thus large doses of LH-RH caused gonadal inhibition in golden hamsters (Sandow *et al.*, 1977). In male rats and hamsters with gonadal atrophy induced by blinding, spermatogenesis is resumed after administration of 50–100ng of LH-RH daily for four weeks, whereas doses of 500–1000ng severely impair spermatogenesis, possibly by a negative feedback (Sandow *et al.*, 1977).

Banik and Givner (1975) have reported that induction of premature ovulation in dioestrus rats by administration of 10ng D-Ala^6, des-Gly^{10}-LH-RH ethylamide is associated with interference with the mating behaviour and pregnancy during the ensuing pro-oestrus and oestrus. These antifertility effects of the analogue disappear in the subsequent cycle. When this compound is given in 20–120ng doses every third day starting on the afternoon of dioestrus, it causes a rhythmical antifertility effect in 4-day cyclic rats which have been allowed cohabitation with fertile males except for the first 24 hours after treatment (Banik and Givner, 1976). However, the use of this compound every fourth day does not interfere with their fertility. This antifertility effect is probably achieved by inducing ovulation at a physiologically 'wrong time' (Banik and Givner, 1976).

D-Leu^6, des-Gly^{10}-LH-RH ethylamide, given in doses of 0.2–5 μg has been reported to inhibit the human chorionic gonadotrophin (HCG)-induced ovarian and uterine weight augmentation in rats (Rippel and Johnson, 1976). Chronic administration of D-Leu^6, des-Gly^{10}-LH-RH ethylamide in doses of 0.5–3 μg to immature rats inhibited normal ovarian growth and maturation as evidenced by arrest of uterine growth, delay of vaginal opening and absence of normal cycling. In mature rats, cessation of cycling, atrophy of the ovaries to a 'prepubertal' size and uterine weight regression occurred in response to 1–10 μg of this analogue (Johnson *et al.*, 1976*a*). In both groups termination of treatment was followed by prompt restoration of normal ovarian weight and function. Thus chronic administration of large doses of this analogue may result in the inhibition of reproductive processes in the rat. Long-term administration of large doses of D-Leu^6, desGly^{10}-LH-RH ethylamide has also been reported to cause regression of endocrine-dependent mammary tumours, probably through suppression of ovarian steroid function (Johnson *et al.*, 1976*b*) There is evidence that ^{125}I-labelled D-Leu^6, desGly-NH$_2^{10}$

-LH-RH ethylamide is specifically bound to ovarian receptors. It is, therefore, possible that these superactive analogues and, presumably LH-RH itself, could have a direct inhibitory influence on ovarian growth (Mayar *et al.*, 1977). Recent studies (Auclair *et al.*, 1977) have also shown that daily administration of as little as 8–40 ng of D-Leu^6, desGly-NH_2^{10}-LH-RH ethylamide three times a day for one week to male rats results in a 30–80 per cent reduction of testicular LH/HCG and prolactin receptors. Testosterone levels are also reduced. These results are consistent with the marked loss of LH receptors and steroidogenic response to gonadotrophins observed in rats after daily injections of LH or HCG (Hsueh *et al.*, 1976). Equally dramatic is the ability of large doses of LH-RH (100–1000 μg/day) and smaller but still pharmacological doses (1–6 μg/day) of the superactive analogues D-Ala^6-LH-RH ethylamide D-Leu^6-LH-RH ethylamide and D-Trp^6-LH-RH to block implantation and terminate gestation when given daily postcoitally to rats (Corbin *et al.*, 1976, 1977; Humphrey *et al.*, 1977; Arimura *et al.*, 1978). This effect is also dose dependent and appears to be directly related to hypersecretion of LH, functional luteolysis and/or inhibition of oestrogen and progesterone secretion.

Clinical uses of LH-RH and its superactive agonistic analogues

LH-RH has been used diagnostically to determine the pituitary LH and FSH reserve. Used in combination with the clomiphene test, it may be helpful in differentiating pituitary and hypothalamic causes of hypogonadism. LH-RH has also been used therapeutically to induce ovulation in amenorrhoeic women and to treat oligospermia in men. The use of LH-RH and its analogues can avoid superovulation and the resultant multiple births, which are not uncommon after administration of preparations of human menopausal gonadotrophins (HMG-Pergonal^R) followed by HCG.

Since one injection of the superactive analogues D-Ala^6, desGly-NH_2^{10}-LH-RH ethylamide, D-Leu^6, desGly-NH_2^{10}-LH-RH ethylamide, D-Ser(Bu^t)^6, desGly-NH_2^{10}-LH-RH ethylamide, or D-Trp^6-LH-RH can induce protracted stimulation of the release of LH and FSH lasting as long as 24 hours, these analogues should be more convenient and practical to use than LH-RH which has occasionally been given three times daily for therapeutic purposes (Nillius *et al.*, 1975). Moreover, the analogues are active not only after parenteral but also after intranasal, intravaginal, intrarectal and oral administration if suitable doses are given. However, in view of the paradoxical antifertility effects of large doses of LH-RH and its long-acting superactive analogues, caution must be exercised in devising clinical protocols. Nevertheless, preliminary results with D-Trp^6-LH-RH indicate its great potential for the induction of ovulation in amenorrhoeic women and the treatment of men with hypogonadotrophic hypogonadism (Jaramillo-Jaramillo *et al.*, 1978*a*, *b*).

Inhibitory analogues of LH-RH

The concept of antagonists of LH-RH was based on the assumption that replacement or deletion of some amino acids in LH-RH, especially those in positions 2 and 3, might result in analogues possessing features requisite for effective binding, but lacking those which are necessary for a functional effect. Such analogues would be competitive inhibitors of LH-RH; that is, they would be devoid of LH-RH activity but, by competing for attachment to the receptor site with endogenous LH-RH, they would reduce the secretion of LH and FSH. The information gained from studies on various stimulatory analogues of LH-RH has been used to guide attempts to create synthetic inhibitors of LH-RH.

The first inhibitory peptide claimed to be active was desHis^2-LH-RH, (Vale *et al.*, 1972*c*), which was found to be at best a weak inhibitor of LH-RH in the monolayer cell system. It was also found that when tryptophan in position 3 was replaced by Leu, the resulting peptide Leu^3-LH-RH (Vilchez-Martinez *et al.*, 1975) was mildly inhibitory. Attempts were made to increase the inhibitory activities of some of these weak inhibitors by introducing some of the modifications used for superactive analogues, that is by incorporating either a C-terminal ethylamide modification or the D-amino acid in the 6 position. Thus, desHis^2, desGly^{10}-LH-RH ethylamide containing the C-terminus modified with ethylamide and desHis^2, D-Ala^6-LH-RH (Monahan *et al.*, 1973; Coy *et al.*, 1973*b*) inhibited more effectively LH-RH-induced LH release than desHis^2-LH-RH. However, if both these modifications were incorporated within the same analogue, increased inhibition was often masked by an increase in inherent gonadotrophin-releasing activity.

Another important discovery in the antagonist field is that D-Phe^2-LH-RH is a far more effective inhibitor than desHis^2-LH-RH (Rees *et al.*, 1974). In general, the D-Phe^2 analogues are about three times more active than the desHis^2-analogues *in vitro* and more potent as well as longer-acting *in vivo* (Coy *et al.*, 1976*d*). Analogues such as D-Phe^2, D-Ala^6- and D-Phe^2, D-Leu^6-LH-RH inhibit the LH-RH-induced LH and FSH release in immature male rats and are capable of blocking ovulation and the pre-ovulatory gonadotrophin surge in cyclic rats when multiple doses of 12 mg/kg or greater are given during the afternoon of pro-oestrus (Beattie *et al.*, 1976). Analogues such as D-Phe^2, D-Phe^6- and D-Phe^2, D-Trp^6-LH-RH are even more potent, blocking ovulation in single doses of about 6 mg/kg and producing inhibition of LH for six hours and FSH for eight hours after injection (de la Cruz *et al.*, 1976*b*; Coy *et al.*, 1976*b*).

Attempts have also been made to synthesise peptides with lower LH-RH activity but which retain the desirable properties of D-Phe^2, D-Phe^6-LH-RH and D-Phe^2, D-Trp^6-LH-RH. Since Phe^3-LH-RH has only 0.1 per cent of the LH-releasing activity of LH-RH itself, this change was incorporated into D-Phe^2, D-Phe^6-LH-RH in an attempt to decrease its gonadotrophin-releasing activity. The resulting derivative, D-Phe^2, Phe^3, D-Phe^6-LH-RH, suppresses LH and FSH release in male rats in response to LH-RH for at least four hours, and, given at noon on the pro-oestrus day, results in a 95 per cent reduction of the pre-ovulatory LH surge and an 84 per cent reduction of the FSH surge (de la Cruz *et al.*, 1976*b*). Later it was repeatedly found that a single injection of 1.5 mg of this peptide at noon is sufficient to block spontaneous ovulation completely.

Since Phe^5-LH-RH has only 50 per cent LH-RH activity, attempts were also made to lower the inherent LH-RH activity of D-Phe^2, D-Phe^6-LH-RH by incorporating Phe at position 5. D-Phe^2, Phe^5, Phe^6-LH-RH behaved quite similarly to D-Phe^2, Phe^3, D-Phe^6-LH-RH (Ferland *et al.*, 1976*b*).

However, the replacement of Trp by D-Trp or Pro in position 3 appeared to increase significantly the inhibitory potency of the peptides (Coy *et al.*, 1977; Humphries *et al.*, 1976). D-Phe^2, D-Trp^3, D-Phe^6-LH-RH is both longer-acting (nearly 10 hours in the rat) and more potent than D-Phe^2, Phe^3, D-Phe^6-LH-RH. In tests for inhibition of ovulation, D-Phe^2, D-Trp^3, D-Phe^6-LH-RH was approximately twice as effective as D-Phe^2, Phe^3, D-Phe^6-LH-RH (Coy *et al.*, 1977). D-Phe^2, D-Trp^3, D-Phe^6-LH-RH (Coy *et al.*, 1977*b*) and D-Phe^2, Pro^3, D-Phe^6-LH-RH (Humphries *et al.*, 1976) completely block ovulation at single doses in the range of 1 mg per rat. Work is presently in progress in this laboratory on certain dimeric forms of some of these inhibitory peptides which are up to four times more active than the best ones reported here. Thus the progress being made in this area could possibly lead eventually to the development of new birth-control methods.

GROWTH HORMONE-RELEASING FACTOR (GH-RF)

The secretion of growth hormone (GH) from the anterior pituitary is regulated by a dual system of hypothalamic control, one system being inhibitory and one being stimulatory (Schally *et al.*, 1973*a*; Reichlin *et al.*, 1976; Vale *et al.*, 1977). The inhibition of GH secretion is mediated by somatostatin (GH-RIH), which is described in the next section. The stimulatory effect on GH release of some hypothalamic fractions (Takahara *et al.*, 1975) appears to be due to a GH-RF, which, under some conditions, might predominate over the inhibitory action of somatostatin. It would be very desirable clinically to obtain GH-RH because of a world shortage of human GH, and because of the assumed capability of GH-RH to stimulate growth in young pituitary dwarfs and to serve as an anabolic agent free of androgenic effects. GH-RH activity was first demonstrated over ten years ago, but, despite intense efforts by several groups to purify it and determine its structure, the nature of the physiological GH-RF is still unknown.

Several years ago we isolated from porcine hypothalami a decapeptide which stimulated, in certain *in vitro* and *in vivo* rat systems, the release of bioassayable GH, but did not have an effect on the levels of immunoreactive GH in rats, sheep, pigs, monkeys and humans (Schally *et al.*, 1969, 1971*c*, 1972*b*; Kastin *et al.*, 1972). Consequently we withdrew the proposal that this decapeptide might represent GH-RH (Kastin *et al.*, 1972). Similarly, the claims of Youdaev *et al.* (1973) that *p*Glu-Ser-Gly-NH_2 might be GH-RF were dispelled when we (Schally *et al.*, 1973*b*) and others reported that this tripeptide neither stimulated GH release *in vitro* or *in vivo* in rats, nor increased serum GH levels in sheep or humans.

Several substances found in the hypothalamus and/or higher brain centres such as catecholamines, serotonin, TRH, vasopressin, substance P, endorphins and enkephalins, arginine, prostaglandins, cyclic guanosine monophosphate, and other natural or synthetic substances such as insulin, 2-deoxyglucose, apomorphine, analogues of enkephalins, L-dopa and cyclic AMP derivatives can, under certain conditions, stimulate the release of GH *in vivo* and *in vitro* (Cehovic *et al.*, 1972; Takahara *et al.*, 1974*b*). However, the effect of these substances on GH release could be explained in various ways: (1) vasopressin could liberate GH in primates, but not in rats or mice, by non-specific stress; (2) hypoglycaemia induced by insulin or 2-deoxyglucose might stimulate glucose-sensitive hypothalamic receptors; (3) cyclic AMP, cyclic guanosine monophosphate and prostaglandin could be messengers for the action of GH-RH; (4) in acromegaly the stimulatory effects of TRH on GH release might be explained by loss of receptor specificity; (5) catecholamines and serotonin may be neurotransmitters involved in the stimulation of GH-RF; (6) apomorphine and L-dopa stimulate dopamine receptors; and (7) enkephalins and endorphins act via higher brain centres.

Attempts to purify GH-RF have been hindered by the presence of large amounts of somatostatin and somatostatin-like substances in hypothalamic extracts. The availability of antiserum to somatostatin (Arimura *et al.*, 1975*a*) has now greatly facilitated the search for GH-RF by permitting the neutralisation of somatostatin activity in fractions being assayed for GH-RF. Thus, using antisera to somatostatin or columns of cyanogen bromide-activated Sepharose coupled with anti-somatostatin-gamma-globulin to eliminate or reduce somatostatin, we have succeeded in purifying several fractions with GH-RF activity (Redding and Schally, 1977). These fractions stimulate the release of growth hormone from monolayer cultures of rat anterior pituitaries and increase plasma GH in mice pretreated with chlorpromazine-morphine-Nembutal. However, further work is needed for the isolation and structural elucidation of GH-RF.

GROWTH HORMONE-RELEASE INHIBITING HORMONE (GH-RIH ; SOMATOSTATIN)

Isolation, Structure and Synthesis

The presence in ovine hypothalamic extracts of a substance which inhibits the release of GH from the anterior pituitary gland was observed by Krulich *et al.* (1968). This observation was confirmed by the subsequent isolation from sheep and pig hypothalami of a peptide named GH-RIH, or somatostatin, which was capable of inhibiting the release of immunoreactive GH *in vitro* and *in vivo* (Brazeau *et al.*, 1973; Schally *et al.*, 1976*b*). The primary structure of this tetradecapeptide (figure 6.3) is identical in both species (Brazeau *et al.*, 1973;

H-Ala-Gly-CyS-Lys-Asp(NH_2)-Phe-Phe-Trp-Lys-Thr-Phe-Thr-Ser-CyS-OH

Figure 6.3 Structure of growth hormone-release inhibiting hormone (GH-RIH; somatostatin)

Schally *et al.*, 1976*b*). Recently we found larger and more basic forms of somatostatin in pig hypothalami (Schally *et al.*, 1976*b*). These materials possess different physicochemical properties from somatostatin, are biologically and immunologically active, and may represent precursors of somatostatin. However, the hypothalamus is not the only tissue where somatostatin and somatostatin-like materials are present, since in extracts of the pancreas, stomach and duodenum of the rat we have also found a high concentration of somatostatin, as well as two types of immunoreactive somatostatin (Arimura *et al.*, 1975*b*). Somatostatin has been synthesised by several groups (Rivier *et al.*, 1973; Coy *et al.*, 1973*a*; Yamashiro and Li, 1973; Immer *et al.*, 1974*a*), providing ample amounts of material for a variety of basic and clinical investigations.

Biological Effects of Somatostatin

Synthetic somatostatin has been shown to produce powerful but transient antisecretory effects on various tissues.

Pituitary

Somatostatin inhibits the secretion of pituitary GH in several species of mammals, including humans (Hall *et al.*, 1973; Siler *et al.*, 1973; Besser *et al.*, 1974; Yen *et al.*, 1974), monkeys (Brazeau *et al.*, 1973), dogs (Lovinger *et al.*, 1974) and rats (Brazeau *et al.*, 1973; Coy *et al.*, 1973*a*). Because the basal GH levels are normally low, the elevation of *in vivo* GH levels prior to somatostatin infusion in the above studies was induced by various stimuli such as barbiturates, L-dopa, arginine, insulin, morphine, prostaglandin E_2, exercise, sleep, electrical stimulation of hypothalamus or amygdala, and catecholamine infusion into the third ventricle. Both the reduced (linear) and oxidised (cyclic) forms are active in inhibiting GH levels (Vale *et al.*, 1977; Hall and Gomez-Pan, 1976). A physiological role for somatostatin in the regulation of GH secretion is supported by our observations that passive immunisation with anti-somatostatin elevates basal GH levels and prevents the stress-induced decrease of GH in rats (Arimura *et al.*, 1976*c*).

Somatostatin also inhibits the TRH-induced secretion of TSH (Hall and Gomez-Pan, 1976; Carr *et al.*, 1975; Hall *et al.*, 1973), but not of prolactin, *in vitro* and *in vivo*, and it might play a physiological role in the regulation of TSH secretion (Arimura and Schally, 1976). However, *in vitro* secretion of prolactin from rat pituitaries is inhibited by somatostatin.

Pancreas

Somatostatin suppresses the secretion of glucagon and insulin in rats, cats, baboons (Koerker *et al.*, 1974) and in humans (Yen *et al.*, 1974, Mortimer *et al.*, 1974; Alberti *et al.*, 1973). The inhibition of the release of pancreatic hormones is exerted by a direct action of somatostatin on the pancreas (Iversen, 1974; Curry *et al.*, 1974; Sakurai *et al.*, 1974; Efendic *et al.*, 1974; Koerker *et al.*, 1974). These actions may be of physiological importance since high concentrations of immunoreactive somatostatin are present in the pancreas (Arimura *et al.*, 1975*b*). Moreover, the somatostatin-containing cells (D-cells) are located between α and β islet cells (Luft *et al.*, 1974; Pelletier *et al.*, 1975; Orci and Unger, 1975). Somatostatin affects the exocrine pancreas as well, since it inhibits secretion of pancreatic fluid and bicarbonate (Boden *et al.*, 1975; Konturek *et al.*, 1976*b*).

Stomach

Our observation that administration of somatostatin delayed the rise in blood-sugar following oral glucose (Besser *et al.*, 1974; Mortimer *et al.*, 1974) suggested that somatostatin might act on the gastrointestinal tract. We subsequently found that somatostatin decreases the circulating levels of gastrin in men (Bloom *et al.*, 1974) and in dogs (Konturek *et al.*, 1976*a*). In addition to inhibiting gastrin release, somatostatin also exerts a direct antisecretory effect on both parietal and peptic cells (Gomez-Pan *et al.*, 1975; Konturek *et al.*, 1976*a*), since it inhibits pentagastrin-induced gastric acid and pepsin secretion in cats and dogs. The food-stimulated gastric acid response is also suppressed by somatostatin. This has established for the first time that this hormone can exert exocrine as well as endocrine effects.

Duodenum and other tissues

Somatostatin inhibits the release of secretin and cholecystokinin/pancreozymin from the duodenal mucosa (Boden *et al.*, 1975; Konturek *et al.*, 1976*b*) as well as pancreatic fluid and bicarbonate secretion in response to these hormones. Release of vasoactive intestinal peptide (VIP) and motilin are similarly inhibited by somatostatin (Besser and Mortimer, 1976).

Inhibitory effects of somatostatin on non-neural tissues such as salivary glands, kidney (renin), liver (hepatic glucose, probably on glycogenolysis) have also been reported (Hall and Gomez-Pan, 1976), indicating that this molecule may be a local inhibitory agent.

Tumours

Somatostatin suppresses GH secretion in acromegalics and ACTH secretion by pituitary tumours in Nelson's syndrome (Hall and Gomez-Pan, 1976). Secretion of insulin by insulinomas and glucagon by glucagonomas (Mortimer *et al.*, 1974) can also be inhibited by somatostatin. Similarly, gastrin secretion by a pancreatic tumour has been suppressed in a patient with the Zollinger–Ellison syndrome (Bloom *et al.*, 1974). Recently Bloom (1975) has shown that circulating levels of VIP caused by a pancreatic tumour were reduced in a patient with the Verner–Morrison syndrome after infusion of somatostatin.

CNS effects

Somatostatin potentiates the behavioural effects (hyperactivity) of L-dopa in mice (Plotnikoff *et al.*, 1974); it does not reverse the tremors induced by oxytremorine (Kastin *et al.*, 1976, 1977). There are reports that somatostatin inhibits spontaneous electrical activity of some neuronal systems (Renaud *et al.*, 1975), and that it affects behaviour to the extent that it induces sedation, hypothermia and 'barrel rotation' (Vale *et al.*, 1977). In addition, somatostatin suppresses slow-wave and REM sleep. Both a 'tranquillising effect' and a 'mild euphoria' have been reported in a small number of patients after administration of somatostatin. Since some somatostatin is found outside the hypothalamus, all these findings taken together suggest that somatostatin could also be a neurotransmitter (Plotnikoff *et al.*, 1974; Kastin *et al.*, 1976, 1977).

RIA for Somatostatin

An antiserum to somatostatin was generated in a rabbit using synthetic somatostatin conjugated with human α- and β-globulin as an immunogen. RIA can be carried out using labelled Try^1-somatostatin or *N*-tyrosinyl-somatostatin (Arimura *et al.*, 1975*a*).

Localisation of Somatostatin

The distribution of immunoreactive somatostatin in the rat and guinea-pig brain has been studied utilising immunohistochemical methods (Hökfelt *et al.*, 1974; King *et al.*, 1975). Somatostatin has been found in the external layer of the median eminence around the capillary loops of the hypophyseal portal vessels and in the ventromedial nucleus (Sétáló *et al.*, 1978). Somatostatin in the brain is found principally in neuronal cell bodies in the preoptic and suprachiasmatic area and in the neuronal processes of the rat. A high concentration of immunoreactive somatostatin detected by RIA in the pancreas, duodenum and stomach of the rat (Arimura *et al.*, 1975*b*) suggests that somatostatin could play a role in the regulation not only of the pituitary, but also of the pancreas, duodenum and the stomach. Histoimmunological studies on hypothalamus and pancreas support this view (Hökfelt *et al.*, 1974, 1975*a*; Luft *et al.*, 1974). Luft *et al.* (1974) showed that somatostatin-like immunoreactivity was present in certain islet cells of guinea-pig and rat pancreas by immunohistochemical methods. It was later shown by combined immunocytochemical and histological methods (Polak *et al.*, 1975; Hökfelt *et al.*, 1974, 1975*a*) that pancreatic somatostatin was localised in D-cells within the islets, the function and secretory product of which were unknown previously. The anatomical position of islet D-cells between α and β cells of the pancreas suggests that they might be involved in physiological ('paracrine') control over insulin and glucagon secretion (Unger and Orci, 1977). Somatostatin-secreting pancreatic tumours (somatostatinomas) have also been reported recently (Larsson *et al.*, 1977; Ganda *et al.*, 1977). Somatostatin has also been found in cells similar to D-cells throughout the gastrointestinal mucosa (Polak *et al.*, 1975), in nerve fibres of the intestine, thyroid (Vale *et al.*, 1977) and in spinal cord and fluid. All these studies and the fact that somatostatin has been detected by histoimmunology in a variety of vertebrates such as rat, guinea-pig, chicken, cattle, newt and trout, indicates a broad phylogenetic as well as anatomical distribution (Reichlin *et al.*, 1976; Sétáló *et al.*, 1978).

Mechanism of Action

Somatostatin inhibits the release of GH induced by prostaglandin E_2, dibutyryl cyclic AMP and theophylline (Vale *et al.*, 1972*a*; Labrie *et al.*, 1976), and suppresses cyclic AMP accumulation in rat pituitary gland (Borgeat *et al.*, 1974). Thus inhibitory effects of somatostatin on GH release could be mediated by cyclic AMP. However, somatostatin may also act at a stage following cyclic AMP formation (Vale *et al.*, 1977).

Functional receptors in the three anterior pituitary types (somatotrophs, thyrotrophs and mammotrophs) could be very similar (Labrie *et al.*, 1976). However, somatostatin analogues have been prepared which have dissociated actions on GH, insulin, glucagon and gastric acid (see below), indicating that there may be differences between the receptors in the pituitary cells, stomach and pancreas. The somatostatin-induced inhibition of insulin and glucagon secretion can be reversed by increasing Ca^{2+} concentration (Curry and Bennett, 1976; Bhathena *et al.*, 1976) and thus somatostatin may inhibit Ca^{2+} 'influx' or at least a reduction in calcium concentration may be involved in some stage of the inhibition by somatostatin of the secretory processes.

Somatostatin Analogues

Somatostatin is of little therapeutic value since it has many actions and a short biological half-life. Therefore, attempts have been and continue to be made to produce analogues which would have prolonged activity and which would inhibit the release of some or only one hormone or which would be more selective in their actions. Analogues which would preferentially suppress the secretion of GH might be useful for treatment of gigantism and acromegaly, and perhaps even diabetic retinopathy. Those analogues which selectively inhibit glucagon might find application in the treatment of various types of diabetes, whilst those which inhibit gastrin, gastric acid and pepsin without inhibiting pancreatic bicarbonate secretion could be considered for the treatment of peptic ulcers. The use of solid-phase techniques made possible the synthesis of numerous analogues and provided much information on the structure-function relationship of somatostatin.

Analogues with enhanced activities

Incorporation of D-amino acids into the somatostatin backbone may lead to analogues more powerful than somatostatin. Thus Rivier *et al.* (1975) reported, and we later confirmed (Coy *et al.*, 1976*a*, 1978) that D-Trp^{8}-somatostatin was 2–8 times as active as somatostatin in inhibiting GH release. The increased potency of this analogue was related to conformational effects and not to a prolonged biological half-life. D-Ala^{2}-somatostatin is twice as active as somatostatin (Rivier, J. *et al.*, 1977). Multiple substitutions can sometimes be additive, as in D-Ala^{2}, D-Trp^{8}-somatostatin (Coy *et al.*, 1976*a*), which has a potency twenty times greater than somatostatin on inhibition of GH release. *N*-Tyr, D-Trp^{8}-somatostatin has ten times the potency of somatostatin to inhibit GH release whilst Ala^{2}, D-Trp^{8}, D-Cys^{14}-somatostatin is four times more active than somatostatin in its ability to inhibit pentagastrin-induced gastric acid secretion in cats (Brown *et al.*, 1978). Some of these analogues and their activities are listed in table 6.4.

Analogues with selective activities

That the multiple inhibitory actions of somatostatin could be dissociated was first demonstrated with the synthesis of desAla^{1},Gly^{2},Asn^{5}-somatostatin by Sarantakis *et al.* (1976*a*). Efendic *et al.* (1975) reported that this analogue inhibited insulin and to a lesser extent GH secretion without affecting glucagon release.

Table 6.4 Analogues of somatostatin with more selective biological activities

Analogue of somatostatin	*% Activity relative to somatostatin on:* growth hormone	insulin	glucagon	gastric secretion
Somatostatin	100	100	100	100
D-Cys-14	100	7	22	159
Ala-2-D-Cys-14	150	6	81	193
D-Trp-8-D-Cys-14	1000	10	220	290
Ala-2-D-Trp-8-D-Cys-14-	500	–	–	430
des(Ala-1-Gly-2)-desamino-[Cys-3] dicarba 3, 14-*	50 (?)‡	–	–	100†
des(Ala-1-Gly-2)-desamino-[Cys-3] descarboxy[Cys-14] dicarba 3, 14-*	50 (?)‡	–	–	100†
D-Ala-2-D-Trp-8	2000	–	–	331
D-Trp-8	200–800	–	–	165

*Non-reducible cyclic analogues of somatostatin having both sulphur atoms replaced by methylene groups.
†Reduction of total acid output.
‡Our assays indicate a lower potency.

Grant *et al.* (1976) described an analogue which did not affect insulin or glucagon levels but retained some GH-release inhibiting activity.

Recently Meyers *et al.* (1977) and Brown, M. *et al.* (1977) reported that D-Cys^{14}-somatostatin and D-Trp^{8}, D-Cys^{14}-somatostatin selectively inhibited GH and glucagon secretion more than insulin release. D-Trp^{8}, D-Cys^{14}-somatostatin has a ratio of 22:1 for the selective inhibition of glucagon over insulin secretion and 100:1 for that of GH over insulin release (Meyers *et al.*, 1977). D-Trp^{8}-somatostatin and D-Trp^{8}, D-Cys^{14}-somatostatin are much less potent in suppressing pentagastrin-induced acid than GH secretion (Coy *et al.*, 1976*a*) (table 6.4). Similarly, D-Ala^{2}, D-Trp^{8}-somatostatin, which was twenty times more potent than somatostatin in suppressing GH release, was only three times as potent in inhibiting pentagastrin-induced gastric acid secretion (Brown *et al.*, 1978). Veber *et al.*, (1976) prepared two non-reducible cyclic analogues of the somatostatin ring portion by replacing the sulphur atoms with methylene groups. These analogues were said to retain 50 per cent of the GH-release inhibiting activity of somatostatin and had 100 per cent activity in inhibiting pentagastrin-induced gastric secretion (Hirschmann, 1977). Thus the disulphide bridge is not required for biological activity.

Analogues with prolonged activity

So far no superactive somatostatin analogue with prolonged activity has been reported. Despite its extremely high potency, D-Ala^{2}, D-Trp^{8}-somatostatin does not have prolonged activity *in vivo* (Coy *et al.*, 1978). Acylated Cys^{3}-somatostatin analogues and D-Ala^1-somatostatin retain full GH-release inhibiting activity, but have durations of action identical to somatostatin in rats and humans (Ferland *et al.*, 1976*a*; Evered *et al.*, 1975). Sarantakis *et al.* (1976*b*) have reported a non-reducible cyclic analogue with a shortened ring structure which significantly suppressed plasma GH for four hours after injection, but high doses were required. Since potent analogues of somatostatin with selective activities can be prepared, it is likely that long-acting analogues can be developed too, and that such future analogues may be useful in the treatment of such disorders as acromegaly, diabetic retinopathy, juvenile diabetes, insulinomas, glucagonomas, peptic ulcers and acute pancreatitis.

ENDORPHINS AND ENKEPHALINS

After reports (Hughes, 1975; Pasternak *et al.*, 1975; Terenius and Wahlstrom, 1975) of the presence of endogenous opiate activity in the brain, the pentapeptides Tyr-Gly-Gly-Phe-Met and Tyr-Gly-Gly-Phe-Leu (named Met- and Leu-enkephalin respectively) were isolated from extracts of porcine (Hughes *et al.*, 1975) and calf (Simantov and Snyder, 1976) brain tissue. It was noted that the Met-enkephalin sequence was identical to residues 61–65 of β-lipotrophin, a peptide of unknown function isolated from the pituitary of several animal species (Li *et al.*, 1965; Lohmar and Li, 1967; Graf *et al.*, 1971). Subsequently several other larger C-terminal fragments of β-lipotrophin with opiate activity were isolated (Ling *et al.*, 1976; Li and Chung, 1976) from pituitary and hypothalamic extracts, the most potent being β-endorphin, corresponding to the 61–91 sequence of β-lipotrophin. Both Met-enkephalin and β-endorphin have potent morphine-like effects (Hughes *et al.*, 1975; Li and Chung, 1976; Chang *et al.*, 1976) and bind to opiate receptors (Bradbury *et al.*, 1976; Li *et al.*, 1976). Like the opiate drugs, the opiate peptides are potent stimulators of growth hormone release (Dupont *et al.*, 1977*a*; Rivier, C. *et al.*, 1977), and prolactin release (Rivier, C. *et al.*, 1977; Dupont *et al.*, 1977*b*) and might be mediators in the mechanisms of secretion of hypothalamic releasing hormones. We have also demonstrated behavioural actions of these materials, which seem to be independent of their opiate actions (Kastin *et al.*, 1978).

We have been carrying out extensive structure-activity studies on both Met-enkephalin and the endorphins. Several analogues of the pentapeptide have been synthesised (Coy *et al.*, 1976*c*), for instance D-Ala^2-Met-enkephalin and its amide and alkylamides, which are many times more effective and longer-acting (Walker *et al.*, 1977) than the native sequence.

CONCLUSIONS

A critical survey has been made of the present status of knowledge on hypothalamic regulatory hormones and their synthetic analogues. Technological advances in the methods for measuring ACTH (such as the RIA) and for the separation of similar molecules (such as high-pressure liquid chromatography) have permitted the resumption of work on isolation of corticotrophin releasing factor (CRF) from porcine hypothalamus. Potent preparations of CRF have been obtained and, in agreement with earlier data, an apparent multiplicity of substances capable of releasing ACTH *in vivo* and *in vitro* has again been demonstrated. One of these materials has an MW of about 4000 and is a polypeptide. Thyrotrophin releasing hormone (TRH), the first hypothalamic hormone to have its structure elucidated, is being used more and more routinely in the clinic for diagnostic purposes. Recent research on TRH has consisted mainly of the synthesis of its analogues, some of which are more potent than TRH itself, in the neuropharmacological studies and investigation of the possible function of TRH as a neurotransmitter. TRH releases prolactin in several animal species, including humans, but TRH may not be the principal prolactin releasing hormone, since highly purified hypothalamic preparations free from TRH can release prolactin. In spite of intensive work over more than five years the nature of the physiological prolactin release inhibiting factor (PIF) is still unknown. The work in our laboratory resulted in the isolation from porcine hypothalami of catecholamines and of GABA, both of which inhibit the release of prolactin *in vivo* and *in vitro*. However, several other substances with PIF activity exist in the hypothalamus and the identity of the physiological PIF remains to be established. Work on PIF is complicated by the interference of TRH and other substances with prolactin-releasing activity in the assays. A similar situation exists for MSH release inhibiting factor (MIF), although Pro-Leu-Gly-NH_2 has been shown to have some ability to inhibit MSH release. Our concept that pyroGlu-His-Trp-Ser-Tyr-Gly-Leu-Arg-Pro-Gly-NH_2 is not only the LH-releasing hormone (LH-RH) but also the FSH-releasing hormone (FSH-RH) has not been seriously challenged. Various immunological, biochemical and physiological studies support this view. There has been a great expansion in the field of synthesis of analogues of LH-RH, because of the need to produce better therapeutic agents for the treatment of infertility as well as the attempts to develop new birth-control methods based on antagonists of LH-RH. Extensive physiological and clinical studies have been carried out with superactive, long-acting agonists of LH-RH, principally on D-Leu^6- and D-Ala^6-LH-RH ethylamides, and D-Trp^6-LH-RH. These and related analogues have been shown to cause prolonged liberation of LH and FSH in animals and humans when administered by various routes. However, LH-RH and some of its superactive agonists exert paradoxical antifertility effects in animals when given in large doses and hence caution must be exercised in devising clinical protocols. Potent antagonists of LH-RH have also been developed. D-Phe^2, Phe^3, D-Phe^6-LH-RH inhibits LH and FSH release for several hours and blocks ovulation in rats, golden hamsters, rabbits and some species of monkeys. D-Phe^2, D-Trp^3, D-Phe^6-LH-RH is still more potent and capable of blocking the LH response to LH-RH in humans. The structure of growth hormone-releasing factor (GH-RF) is still unknown in spite of intensive work for over a decade. The assays necessary for the isolation of this substance are complicated because of the presence in hypothalamic extract of several substances inhibiting GH release, principally somatostatin and what appear to be larger molecular forms of it, most probably its precursors. Recently, however, using antisera to somatostatin or columns of Sepharose linked to antisomatostatin-gamma-globulin, we have obtained new evidence for the existence of GH-RF and have purified several fractions with GH-RF activity. In the past few years intensive activity has occurred in the field of the growth hormone-release inhibiting hormone (somatostatin, GH-RIH). We have isolated porcine somatostatin, shown that it has a structure identical to the ovine hormone, and also demonstrated the existence of what appear to be larger and more basic forms of somatostatin. One of these substances, now nearly pure, is most likely pro-somatostatin. Studies with synthetic somatostatin have demonstrated that this hormone inhibits the secretion of GH, TSH, insulin, glucagon, gastrin, gastric HCl and pepsin, secretin and cholecystokinin and other substances. Somatostatin has been found not only in the brain but also in the stomach, gut and pancreas. Thus this hormone may play a role in the regulation of secretions not only from the pituitary, but also from the pancreas, stomach and duodenum. The effects of somatostatin are short-lived, but the synthesis of long-acting analogues might result in the development of useful therapeutic agents. Some analogues, such as D-Trp^8-somatostatin and D-Ala^2, D-Trp^8-somatostatin, are much more active than somatostatin. Dissociation of inhibition of gastric and pancreatic secretion and GH release has also been found with several D-Trp^8- analogues. Three analogues of somatostatin, D-Cys^{14}, Ala^2, D-Cys^{14}-, and D-Trp^8, D-Cys^{14}-somatostatin, suppress the release of growth hormone *in vitro* and glucagon *in vivo*, but have less effect on insulin secretion *in vivo*. The analogues or their derivatives might have greater potential than somatostatin for the possible treatment of diabetes mellitus. Several opiate peptides have been isolated from hypothalamic and pituitary tissue. Together with several superactive analogues, they exert behavioural actions and are potent stimulators of GH and prolactin release *in vivo*. This effect is exerted on CNS centres and not the pituitary, so that these peptides represent neither PRH nor GH-RH.

REFERENCES

Alberti, K. G. M. M., Christensen, N. J., Christensen, S. E., Hansen-Prange, Aa., Iversen, J., Lundbaek, K., Seyer-Hansen, K. and Ørskov, H. (1973). Inhibition of insulin secretion by somatostatin. *Lancet*, **ii**, 1299–1301

Arimura, A., Debeljuk, L. and Schally, A. V. (1974*a*). Blockade of the preovulatory surge of LH and FSH and of ovulation by anti-LH-RH serum in rats. *Endocrinology*, **95**, 323–324

Arimura, A., Kastin, A. J., Schally, A. V., Sato, M., Kumasaka, T., Yaoi, Y., Nishi, N. and Ohjura, K. (1974*b*). Immunoreactive LH-releasing hormone in plasma: mid-cycle elevation in women. *J. Clin. Endocrinol. Metab.*, **38**, 510–513

Arimura, A., Nishi, N. and Schally, A. V. (1976*a*). Delayed implantation caused by administration of sheep immunogamma globulin against LH-RH in the rat. *Proc. Soc. Exp. Biol. Med.*, **152**, 71–75

Arimura, A., Pedroza, E., Vilchez, J. A. and Schally, A. V. (1978). Prevention of implantation by [D-Trp^6]-LH-RH in the rat: comparative study with the effects of large doses of HCG on pregnancy. *Endocrine Res. Commun.*, to be published

Arimura, A., Saito, T. and Schally, A. V. (1967). Assays for corticotrophin-releasing factor (CRF) using rats treated with morphine, chlorpromazine, dexamethasone and nembutal. *Endocrinology*, **81**, 235–245

Arimura, A., Sato, H., Coy, D. H. and Schally, A. V. (1975*a*). Radioimmunoassay for GH-release inhibiting hormone. *Proc. Soc. Exp. Biol. Med.*, **148**, 784–789

Arimura, A., Sato, H., Dupont, A., Nishi, N. and Schally, A. V. (1975*b*). Somatostatin: abundance of immunoreactive hormone in rat stomach and pancreas. *Science*, **189**, 1007–1009

Arimura, A., Sato, H., Kumasaka, T., Worobec, R. B., Debeljuk, L., Dunn, J. D. and Schally, A. V. (1973). Production of antiserum to LH-releasing hormone (LH-RH) associated with gonadal atrophy in rabbits: development of radioimmunoassay for LH-RH. *Endocrinology*, **93**, 1092–1093

Arimura, A. and Schally, A. V. (1976). Increase in basal and thyrotrophin releasing hormone (TRH)-stimulated secretion of thyrotrophin (TSH) by passive immunization with antiserum to somatostatin in rats. *Endocrinology*, **98**, 1069–1072

Arimura, A., Shiino, M., de la Cruz, K. G., Rennels, E. G. and Schally, A. V. (1976*b*). Effect of active and passive immunization with luteinising hormone-releasing hormone and follicle-stimulating hormone levels and the ultrastructure of the pituitary gonadotrophs in castrated male rats. *Endocrinology*, **99**, 291–303

Arimura, A., Smith, W. D. and Schally, A. V. (1976*c*). Blockade of the stress-induced decrease in blood GH by anti-somatostatin serum in rats. *Endocrinology*, **98**, 540–543

Auclair, C., Kelly, P. A., Labrie, F., Coy, D. H. and Schally, A. V. (1977). Inhibition of testicular luteinizing hormone receptor level by treatment with a potent luteinizing hormone-releasing hormone agonist or human chorionic gonadotropin. *Biochem. Biophys. Res. Commun.*, **76**, 855–862

Baba, Y., Matsuo, H. and Schally, A. V. (1971). Structure of the porcine LH-and FSH-releasing hormone. Part II: Confirmation of the proposed structure by conventional sequential analyses. *Biochem. Biophys. Res. Commun.*, **44**, 459–463

Banik, U. K. and Givner, M. L. (1975). Ovulation induction and antifertility effects of an LH-RH analogue (AY 25,205) in cyclic rats. *J. Reprod. Fertil.*, **44**, 87–94

Banik, U. K. and Givner, M. L. (1976). Effects of a luteinizing hormone-releasing hormone analog on mating and fertility in rats. *Fertil. Steril.*, **27**, 1078–1084

Barden, N. and Labrie, F. (1973). Receptor for thyrotropin-releasing hormone in plasma membranes of bovine anterior pituitary gland. *J. Biol. Chem.*, **248**, 7601–7606

Bassiri, R. M. and Utiger, R. D. (1972). The preparation and specificity of antibody to thyrotropin releasing hormone. *Endocrinology*, **90**, 722–727.

Bauer, K. and Lipmann, F. (1976). Attempts towards biosynthesis of the thyrotropin-releasing hormone and studies on its breakdown in hypothalamic tissue preparations. *Endocrinology*, **99**, 230–242

Beattie, C. W., Corbin, A., Foell, T. J., Garsky, V., Rees, R. W. A. and Yardley, J. (1976). Anti-ovulatory/anti-pregnancy effects of [D-Phe^2]-LRH analogs administered early in the rat oestrous cycle. *Contraception*, **13**, 341–353

Besser, G. M. and Mortimer, C. H. (1976). Clinical Neuroendocrinology. In L. Martini and W. F. Ganong (eds.). *Frontiers in Neuroendocrinology*, Raven Press, New York, 227–254

Besser, G. M., Mortimer, C. H., Carr, D., Schally, A. V., Coy, D. H., Evered, D., Kastin, A. J., Tunbridge, W. M. G., Thorner, M. O. and Hall, R. (1974). Growth hormone release inhibiting hormone in acromegaly. *Brit. Med. J.*, **1**, 352–355

Bhathena, S. J., Perrino, P. V., Voyles, N. R., Smith, S. S., Wilkins, S. D., Coy, D. H., Schally, A. V. and Recant, L. (1976). Reversal of somatostatin inhibition of insulin and glucagon secretion. *Diabetes*, **25**, 1031–1040

Birge, C. A., Jacobs, L. S., Hammer, C. T. and Daughaday, W. (1970). Catecholamine inhibition of prolactin secretion by isolated rat adenohypophyses. *Endocrinology*, **86**, 120–130

Bloom, S. R. (1975). Vasoactive intestinal peptide and the Verner–Morrison syndrome. *Gut*, **16**, 399

Bloom, S. R., Mortimer, C. H., Thorner, M. O., Besser, G. M., Hall, R., Gomez-Pan, A., Roy,

V. M., Russell, R. C. G., Coy, D. H., Kastin, A. J. and Schally, A. V. (1974). Inhibition of gastrin and gastric acid secretion by growth hormone release inhibiting hormone. *Lancet,* **ii**, 1106–1109

Boden, G., Sivitz, M. C., Owen, O. E., Essa-Koumar, N. and Landor, J. H. (1975). Somatostatin suppresses secretin and pancreatic exocrine secretion. *Science,* **190**, 163–165

Bøler, J., Enzmann, F., Folkers, K., Bowers, C. Y. and Schally, A. V. (1969). The identity of chemical and hormonal properties of the thyrotropin releasing hormone and pyroglutamyl-histidyl-proline-amide. *Biochem. Biophys. Res. Commun.,* **37**, 705–710

Borgeat, P., Chavancy, G., Dupont, A., Labrie, F., Arimura, A. and Schally, A. V. (1972). Stimulation of adenosine 3′:5′-cyclic monophosphate accumulation in anterior pituitary gland *in vitro* by synthetic luteinizing hormone-releasing hormone. *Proc. Nat. Acad. Sci. USA*, **69**, 2677–2681

Borgeat, P., Labrie, F., Drouin, J., Belanger, A., Immer, H., Sestanj, K., Nelson, V., Götz, M., Schally, A. V., Coy, D. H. and Coy, E. J. (1974). Inhibition of adenosine 3′,5′-monophosphate accumulation in anterior pituitary gland *in vitro* by growth hormone-release inhibiting hormone. *Biochem. Biophys. Res. Commun.,* **56**, 1052–1054

Bower, A., Sr., Hadley, M. E. and Hruby, V. J. (1971). Comparative MSH release inhibiting activities of tocinoic acid (the ring of oxytocin) and L-Pro-Leu-Gly-NH_2 (the side chain of oxytocin). *Biochem. Biophys. Res. Commun.,* **45**, 1185–1191

Bowers, C. Y., Currie, B. L., Johansson, K. N. G. and Folkers, K. (1973). Biological evidence that separate hypothalamic hormones release the follicle stimulating and luteinizing hormones. *Biochem. Biophys. Res. Commun.,* **50**, 20–26

Bowers, C. Y., Friesen, H., Hwang, P., Guyda, H. and Folkers, K. (1971). Prolactin and thyrotropin release in man by synthetic pyroglutamyl-histidyl-proline-amide. *Biochem. Biophys. Res. Commun.,* **45**, 1033–1041

Bowers, C. Y., Weil, A., Chang, J. K., Sievertsson, H., Enzmann, F. and Folkers, K. (1970). Activity-structure relationships of thyrotropin releasing hormone. *Biochem. Biophys. Res. Commun.,* **40**, 683–691

Boyd, A. E., Spencer, E., Jackson, I. M. D. and Reichlin, S. (1976). Prolactin-releasing factor (PRF) in porcine hypothalamic extract distinct from TRH. *Endocrinology*, **99**, 861–871

Bradbury, A. F., Smyth, D. G. and Snell, C. R. (1976). C fragment of lipotropin has a high affinity for brain opiate receptors. *Nature,* **260**, 793–795

Brazeau, P., Vale, W., Burgus, R., Ling, N., Butcher, M., Rivier, J. and Guillemin, R. (1973). Hypothalamic polypeptide that inhibits the secretion of immunoreactive pituitary growth hormone. *Science*, **179**, 77–79

Brown, G. M., van Loon, G. R., Hummel, B. C. W., Grota, L. J., Arimura, A. and Schally, A. V. (1977). Characteristics of antibody produced during chronic treatment with LH-RH *J. Clin. Endocrinol. Metab.,* **44**, 784–790

Brown, M., Rivier, J. and Vale, W. (1977). Somatostatin: analogs with selected biological activities. *Science,* **196**, 1467–1469

Brown, M. P., Coy, D. H., Gomez-Pan, A., Hirst, B. H., Hunter, M., Meyers, C., Reed, J. D., Schally, A. V. and Shaw, B. (1978). Structure–activity relationships of eighteen somatostatin analogues on gastric secretion. *J. Physiol.*, (*Lond.*), **277**, 1–14

Burgus, R., Butcher, M., Amoss, M., Ling, N., Monahan, M., Rivier, J., Fellows, R., Blackwell, R., Vale, W. and Guillemin, R. (1972). Primary structure of the ovine hypothalamic luteinizing hormone releasing factor (LRF). *Proc. Natn. Acad. Sci. USA*, **69**, 278–282

Burgus, R., Dunn, T. F., Desiderio, D., Ward, D. N., Vale, W. and Guillemin, R. (1970). Characterization of ovine hypothalamic hypophysiotropic TSH releasing factor. *Nature,* **226**, 321–325

Carr, D. Gomez-Pan, A., Weightman, D. R., Roy, V. C. M., Hall, R., Besser, G. M., Thorner, M. O., McNeilly, A. S., Schally, A. V., Kastin, A. J. and Coy, D. H. (1975). Growth hormone release inhibiting hormone: actions on thyrotrophin and prolactin secretion after thyrotrophin-releasing hormone. *Brit. Med. J.,* **3**, 67–69

Cehovic, G., Posternak, T. and Charollais, E. (1972). A study of the biological activity and resistance to phosphodiesterase of some derivatives and analogues of cyclic AMP. *Adv. Cycl. Nucleotide Res.*, **1**, 521–524

Celis, M. E., Taleisnik, S. and Walter, R. (1971*a*). Release of pituitary melanocyte-stimulating hormone by the oxytocin fragment, H-Cys-Tyr-Ile-Gln-Asn-OH. *Biochem. Biophys. Res. Commun.*, **45**, 564–569

Celis, M. E., Taleisnik, S. and Walter, R. (1971*b*). Regulation of formation and proposed structure of the factor inhibiting the release of melanocyte stimulating hormone. *Proc. Natn. Acad. Sci. USA*, **68**, 1428–1433

Chang, J. K., Fong, B. T. W., Pert, A. and Pert, C. B. (1976). Opiate receptor affinities and behavioral effects of enkephalin: structure-activity relationship of ten synthetic peptide analogues. *Life Sci.*, **18**, 1473–1482

Chang, J. K., Sievertsson, H., Currie, H., Folkers, K. and Bowers, C. (1971). Synthesis of analogs of the thyrotropin releasing hormone and structure-activity relationships. *J. Med. Chem.*, **14**, 484–487

Cooper, D. M. F., Synetos, D., Christie, R. B. and Schulster, D. (1976). Studies on the nature of corticotrophin releasing hormone and its partial purification from porcine hypothalami. *J. Endocrinol.*, **71**, 171–172

Corbin, A., Beattie, C. W., Rees, R., Yardley, J., Foell, T. J., Chai, S. Y., McGregor, H. Gorsky, V., Sarantakis, D. and McKinley, W. A. (1977). Post-coital contraceptive effects of agonist analogs of luteinizing hormone-releasing hormone. *Fertil. Steril.*, **28**, 471–475

Corbin, A., Beattie, C. W., Yardley, J. and Foell, T. J. (1976). Post-coital contraceptive effects of an agonistic analogue of luteinizing hormone releasing hormone. *Endocrine Res. Commun.*, **3**, 359–376

Coy, D. H., Coy, E. J., Arimura, A. and Schally, A. V. (1973*a*). Solid-phase synthesis of growth hormone-release inhibiting factor. *Biochem. Biophys. Res. Commun.*, **54**, 1267–1273

Coy, D. H., Coy, E. J., Hirotsu, Y. and Schally, A. V. (1974*a*). Synthesis and biological properties of [2-L-β-(pyrazolyl-3) alanine]-luteinizing hormone-releasing hormone. *J. Med. Chem.*, **17**, 140–142

Coy, D. H., Coy, E. J., Hirotsu, Y., Vilchez-Martinez, J. A., Schally, A. V., Van Nispen, J. W. and Tesser, G. I. (1974*b*). Investigation of the role of tryptophan in the luteinizing hormone-releasing hormone. *Biochemistry*, **13**, 3550–3551

Coy, D. H., Coy, E. J., Meyers, G., Drouin, J., Ferland, L., Gomez-Pan, A. and Schally, A. V. (1976*a*). Structure–function studies on somatostatin. Program, 58th Endocrine Society Meeting, San Francisco, June, 1976, Abstract 305, p. 209

Coy, D. H., Coy, E. J. and Schally, A. V. (1975*a*). Structure–activity relationship of LH-FSH releasing hormone. In Marks, N. and Rodnight, R. (eds). *Research Methods in Neurochemistry*, **3**, Plenum Press, New York, 393–406

Coy, D. H., Coy, E. J., Schally, A. V., Vilchez-Martinez, J. A., Hirotsu, Y. and Arimura, A. (1974*c*). Synthesis and biological properties of [D-Ala-6,Des-Gly-NH_2-10]-LH-RH ethylamide, a peptide with greatly enhanced LH- and FSH-releasing activity. *Biochem. Biophys. Res. Commun.*, **57**, 335–340

Coy, D. H., Coy, E. J., Vilchez-Martinez, J. A., de la Cruz, A., Arimura, A. and Schally, A. V. (1976*b*). Suppression of gonadotropin release and ovulation in animals by inhibitory analogs of luteinising hormone-releasing hormone. In F. Labrie, J. Meites and G. Pelletier (eds.). *Hypothalamus and Endocrine Functions*, Plenum Press, New York, 339–354

Coy, D. H., Hirotsu, Y., Redding, T. W., Coy, E. J. and Schally, A. V. (1975*b*). Synthesis and biological properties of the 2-L-beta-(pyrazolyl-l) alanine analogs of luteinising hormone-releasing hormone and thyrotropin-releasing hormone. *J. Med. Chem.*, **18**, 948–949

Coy, D. H., Kastin, A. J., Schally, A. V., Morin, O., Caron, N. G., Labrie, F., Walker, J. M., Fertel, R., Berntson, G. G. and Sandman, C. A. (1976*c*). Synthesis and opiod activities of sterioisomers and other D-amino acid analogs of methionine enkephalin. *Biochem. Biophys. Res. Commun.*, **73**, 632–638

Coy, D. H., Labrie, F., Savary, M., Coy, E. J. and Schally, A. V. (1976*d*). Antagonistic activity of analogs of luteinising hormone-releasing hormone (LH-RH) *in vitro. Mol. Cell. Endocrinol.*, **5**, 201–208

Coy, D. H., Meyers, C., Schally, A. V., Drouin, J., Ferland, L., Beaulieu, M. and Labrie, F. (1978). Synthesis and growth hormone-release inhibiting properties of some stereoisomers of somatostatin. *Mol. Cel. Endocrinol.*, to be published

Coy, D. H., Schally, A. V., Vilchez-Martinez, J. A., Coy, E. J. and Arimura, A. (1975*c*). Stimulatory and inhibitory analogs of LH-RH. In M. Motta, P. G. Crosignani and L. Martini (eds.). *Hypothalamic Hormones,* Academic Press, London, 1–12

Coy, D. H., Vilchez-Martinez, J. A., Coy, E. J., Arimura, A. and Schally, A. V. (1973*b*). A peptide inhibitor of luteinizing hormone-releasing hormone (LH-RH). *J. Clin. Endocrinol. Metab.*, **37**, 331–333

Coy, D. H., Vilchez-Martinez, J. A., Coy, E. J., Nishi, N., Arimura, A. and Schally, A. V. (1975*d*). Polyfluoroalkylamine derivatives of luteinising hormone-releasing hormone. *Biochemistry,* **14**, 1848–1851

Coy, D. H., Vilchez-Martinez, J. A., Coy, E. J. and Schally, A. V. (1976*e*). Analogs of luteinizing hormone-releasing hormone with increased biological activity produced by D-amino acid substitutions in position 6. *J. Med. Chem.*, **19**, 423–425

Coy, D. H., Vilchez-Martinez, J. A. and Schally, A. V. (1977). Structure-function states of LRF. In Loffet, A. (ed.), *Peptides 1976,* Editions de L'Univers de Bruxelles, 463–469

de la Cruz, A., Arimura, A., de la Cruz, K. G. and Schally, A. V. (1976*a*). Effect of administration of anti-serum to luteinizing hormone-releasing hormone on gonadal function during the oestrous cycle in the hamster. *Endocrinology,* **98**, 490–497

de la Cruz, A., Coy, D. H., Vilchez-Martinez, J. A., Arimura, A. and Schally, A. V . (1976*b*). Blockage of ovulation in rats by inhibitory analogs of luteinising hormone-releasing hormone. *Science,* **191**, 195–197

Curry, D. L. and Bennett, L. L. (1976). Does somatostatin inhibition of insulin secretion involve two mechanisms of action? *Proc. Natn. Acad. Sci. USA*, **73**, 248–251

Curry, D. L., Bennett, L. L. and Li, C. H. (1974). Direct inhibition of insulin secretion by synthetic somatostatin. *Biochem. Biophys. Res. Commun.*, **58**, 885–889

Debeljuk, L., Redding, T. W., Arimura, A. and Schally, A. V. (1973). Effect of TRH and triiodothyronine on prolactin release in sheep. *Proc. Soc. Exp. Biol. Med.*, **142**, 421–423

Dular, R., LaBella, F., Vivian, S. and Eddie, L. (1974). Purification of prolactin-releasing and inhibiting factors from beef. *Endocrinology*, **94**, 563–567

Dupont, A., Cusan, L., Garon, M., Labrie, F. and Li, C. H. (1977*a*). β-endorphin: Stimulation of growth hormone release *in vivo. Proc. Nat. Acad. Sci. USA*, **74**, 358–359

Dupont, A., Cusan, L., Labrie, F., Coy, D. H. and Li, C. H. (1977*b*). Stimulation of prolactin release in the rat by intraventricular injection of β-endorphin and methionine-enkephaline. *Biochem. Biophys. Res. Commun.*, **75**, 76–82

Efendic, S., Luft, R. and Grill, V. (1974). Effect of somatostatin on glucose-induced insulin release in isolated perfused rat pancreas and isolated rat pancreatic islets. *FEBS Letts.*, **42**, 169–172

Efendic, S., Luft, R. and Sievertsson, H. (1975). Relative effects of somatostatin and two somatostatin analogues on the release of insulin, glucagon and growth hormone. *FEBS Letts.*, **58**, 302–305

Enjalbert, A., Priam, M. and Kordon, C. (1977). Evidence in favour of the existence of a dopamine-free prolactin inhibiting factor (PIF) in rat hypothalamic extracts. *Eur. J. Pharmacol.*, **41**, 243–244

Evered, D. C., Gomez-Pan, A., Tunbridge, W. M. G., Hall, R., Lind, T., Besser, G. M., Mortimer, C. H., Thorner, M. O., Schally, A. V., Kastin, A. J. and Coy, D. H. (1975). Analogues of growth-hormone release-inhibiting hormone. *Lancet,* **i**, 1250

Failli, A., Sestanj, K., Immer, H. U. and Götz, M. (1977). Synthetic MIF analogues. Part 1: Synthesis by four-component condensation (4CC) and classical methods. *Arzneimittelforsch.*, **27(ii)**, 2286–2289

Fawcett, C. P., Breezley, A. E. and Wheaton, J. E. (1975). Chromatographic evidence for the existence of another species of luteinizing hormone releasing factor (LRF). *Endocrinology*, **96**, 1311–1314

Ferland, L., Labrie, F., Coy, D. H., Arimura, A. and Schally, A. V. (1976*a*). Inhibition by six somatostatin analogs of plasma growth hormone levels stimulated by thiamylal and morphine in the rat. *Mol Cell. Endocrinol.*, **4**, 79–88

Ferland, L., Labrie, F., Savary, M., Beaulieu, M., Coy, D. H., Coy, E. J. and Schally, A. V. (1976*b*). Inhibitory activity of analogues of luteinizing hormone releasing hormone (LH-RH) *in vitro* and *in vivo*. *Clin. Endocrinol.*, **5** (suppl.), 279s–290s

Folkers, K., Enzmann, F., Bøler, J., Bowers, C. Y. and Schally, A. V. (1969). Discovery of modification of the synthetic tripeptide-sequence of the thyrotropin releasing hormone having activity. *Biochem. Biophys. Res. Commun.*, **37**, 123–126

Fujino, M., Fukuda, T., Shinagawa, S., Kobayashi, S., Yamazaki, I., Nakayama, R., Seely, J. H., White, W. F. and Rippel, R. H. (1974). Synthetic analogs of luteinising hormone releasing hormone (LH-RH) substituted in positions 6 and 10. *Biochem. Biophys. Res. Commun.*, **60**, 406–413

Fujino, M., Kobayashi, S., Obayashi, M., Shinagawa, S., Fukuda, T., Kitada, C., Nakayama, R., Yamazaki, I., White, W. F. and Rippel, R. H. (1972). Structure-activity relationships in the C-terminal part of luteinizing hormone releasing hormone (LH-RH). *Biochem. Biophys. Res. Commun.*, **49**, 863–869

Ganda, O. P., Weir, G. C., Soeldner, J. S., Legg, M. A., Chick, W. L., Patel, Y. C., Ebeid, A. M., Gabbay, K. H. and Reichlin, S. (1977). 'Somatostatinoma': a somatostatin-containing tumor of the endocrine pancreas. *N. Engl. J. Med.*, **296**, 963–967

Geiger, R., König, W., Wissman, H., Geisen, K. and Enzmann, F. (1971). Synthesis and characterisation of a decapeptide having LH-RH/FSH-RH activity. *Biochem. Biophys. Res. Commun.*, **45**, 767–773

Gillessen, D., Piva, F., Steiner, H. and Studer, R. O. (1971). Über die Bedeutung des Histidins im 'Thyrotropin-releasing' Hormon (TRH). *Helv. Chim. Acta*, **54**, 1335–1342

Gomez-Pan, A., Reed, J. D., Albinus, M., Shaw, B., Hall, R., Besser, G. M., Coy, D. H., Kastin, A. J. and Schally, A. V. (1975). Direct inhibition of gastric acid and pepsin secretion by growth-hormone release-inhibiting hormone in cats. *Lancet*, **i**, 888–890

Gorbman, A. and Hyder, M. (1973). Failure of mammalian TRH to stimulate thyroid function in the lungfish. *Gen. Comp. Endocrinol.*, **20**, 588–589

Graf, L., Barat, E., Cseh, G. and Sajgo, M. (1971). Amino acid sequence of porcine β-lipotropic hormone. *Biochim. Biophys. Acta*, **229**, 276–278

Grant, N., Clark, D., Garsky, V., Juanakais, I., McGregor, W. and Sarantakis, D. (1976). Dissociation of somatostatin effects. Peptides inhibiting the release of growth hormone but not glucagon or insulin in rats. *Life. Sci.*, **19**, 629–632

Green, J. D. and Harris, G. W. (1947). The neurovascular link between the neurohypophysis and adenohypophysis. *J. Endocrinol.*, **5**, 136–146

Guillemin, R., Hearn, W. R., Cheek, W. R. and Housholder, D. E. (1957). Control of corticotrophin release: further studies with *in vitro* methods. *Endocrinology*, **60**, 488–506

Hall, R., Besser, G. M., Schally, A. V., Coy, D. H., Evered, D., Goldie, D. J., Kastin, A. J., McNeilly, A. S., Mortimer, C. H., Phenekos, C., Tunbridge, W. M. G. and Weightman, D. (1973). Action of growth hormone release inhibitory hormone in healthy men and in acromegaly. *Lancet*, **ii**, 581–584

Hall, R. and Gomez-Pan, A. (1976). The hypothalamic regulatory hormones and their clinical applications. In *Advances in Clinical Chemistry*, Vol. 18, Academic Press, New York, 173–212

l'Hermite, M., Robyn, C., Goldstein, J., Rothenbuchner, G., Birk, J., Loos, U., Bonnyns, M. and Vanhaelst, L. (1974). Prolactin and thyrotropin in thyroid diseases: lack of evidence for a physiological role of thyrotropin-releasing hormone in the regulation of prolactin secretion. *Horm. Metab. Res.*, **6**, 190–195

Hilliard, J., Schally, A. V. and Sawyer, C. H. (1971). Progesterone blockade of the ovulatory response to intrapituitary infusion of LH-RH in rabbits. *Endocrinology*, **88**, 730–736

Hinkle, P. M., Woroch, E. L. and Tashjian, A. H. Jr. (1974). Receptor–binding affinities and biological activities of analogs of thyrotrophin-releasing hormone in prolactin-producing pituitary cells in culture. *J. Biol. Chem.*, **249**, 3085–3090

Hirschmann, R. F. (1977). Some recent developments in the synthesis of biologically-active peptides. In Matthieu, J. (ed.), *Proc. 5th Int. Symp. Med. Chem.*, Paris, July, 1976

Hofmann, K. and Bowers, C. Y. (1970). Polypeptides, XLVII. Effect of the pyrazole-imidazole replacement on the biological activity of thyrotropin-releasing hormone. *J. Med. Chem.*, **13**, 1099–1101

Hökfelt, T., Efendic, S., Hellerström, C., Johansson, O., Luft, R. and Arimura, A. (1975*a*). Cellular localisation of somatostatin in endocrine-like cells and neurons of the rat with special references to the A1-cells of the pancreatic islets and to the hypothalamus. *Acta Endocrinol.*, **80** (Suppl. 200), 5–41

Hökfelt, T., Efendic, S., Johansson, O., Luft, R. and Arimura, A. (1974). Immunohisto-chemical localisation of somatostatin (growth hormone releasing factor) in the guinea-pig brain. *Brain Res.*, **80**, 165–169

Hökfelt, T., Fuxe, K., Johansson, O., Jeffcoate, S. and White, N. (1975*b*). Thyrotrophin releasing hormone (TRH)-containing nerve terminals in certain brain stem nuclei and in the spinal cord. *Neuroscience Letters*, **1**, 133–139

Hsueh, A. J. W., Dufau, M. L. and Catt, K. J. (1976). Regulation of luteinizing hormone receptors in testicular interstitial cells by gonadotropin. *Biochem. Biophys. Res. Commun.*, **72**, 1145–1152

Hughes, J. (1975). Isolation of an endogenous compound from the brain with pharmacological properties similar to morphine. *Brain Res.*, **88**, 295–308

Hughes, J., Smith, T. W., Kosterlitz, H. W., Fothergill, L. A., Morgan, B. A. and Morris, H. R. (1975). Identification of two related pentapeptides from the brain with potent opiate agonist activity. *Nature*, **258**, 577–580

Humphrey, R. R., Windsor, B. L., Reel, J. R. and Edgren, R. A. (1977). The effects of luteinizing hormone releasing hormone (LH-RH) in pregnant rats. *Biol. Reprod.*, **16**, 614–621

Humphries, J., Wan, Y.-P., Folkers, K. and Bowers, C. Y. (1976). Presence of proline in position 3 for potent inhibition of the activity of the luteinizing hormone releasing hormone and of ovulation. *Biochem. Biophys. Res. Commun.*, **72**, 939–944

Immer, H. U., Nelson, V. R., Revesz, C., Sestanj, K. and Götz, M. (1974*a*). Luteinizing hormone-releasing hormone and analogs. Synthesis and biological activity. *J. Med. Chem.*, **17**, 1060–1065

Immer, H. U., Sestanj, K., Nelson, V. R. and Götz, M. (1974*b*). Synthesis of somatostatin. *Helv. Chim. Acta*, **57**, 730–734

Iversen, J. (1974). Inhibition of pancreatic glucagon release by somatostatin: *in vitro*. *Scand. J. Clin. Lab. Invest.*, **33**, 125–129

Jackson, I. M. D. and Reichlin, S. (1974). Thyrotropin-releasing hormone (TRH): distribution in hypothalamic and extrahypothalamic brain tissues of mammalian and sub-mammalian chordates. *Endocrinology*, **95**, 854–862

Jacobs, L., Snyder, P., Wilber, J., Utiger, R. and Daughaday, W. (1971). Increased serum prolactin after administration of synthetic thyrotropin releasing hormone (TRH) in man. *J. Clin. Endocrinol. Metab.*, **33**, 996–998

Jaramillo-Jaramillo, C., Charro-Salgado, A., Perez-Infante, V., Bordiu-Obanza, E., Cano-Iglesias, F., Fernandez-Cruz, A., Coy, D. H. and Schally, A. V. (1978*a*). Clinical studies with D-Trp^6-LH-RH in men with hypogonadotropic hypogonadism. *Fertil. Steril.*, to be published

Jaramillo-Jaramillo, C., Charro-Salgado, A., Perez-Infante, V., Lopez del Campo, G., Botella-Llusia, J., Coy, D. H. and Schally, A. V. (1978*b*). Clinical studies with D-Trp^6-LH-RH in anovulatory women. *Fertil. Steril.*, to be published

Johnson, E. S., Gendrich, R. L. and White, W. F. (1976*a*). Delay of puberty and inhibition of reproductive processes in the rat by a gonadotropin-releasing hormone agonist analog. *Fertil. Steril.*, **27**, 853–860

Johnson, E. S., Seely, J. H., White, W. F. and de Sombre, E. F. (1976*b*). Endocrine-dependent rat mammary tumor regression: use of a gonadotropin releasing hormone analog. *Science*, **194**, 329–330

Jones, M. T., Gillham, B. and Hillhouse, E. W. (1977). The nature of corticotrophin factor from rat hypothalamus *in vitro. Fed. Proc.*, **36**, 2104–2109

Kamberi, I. A., Mical, R. S. and Porter, J. C. (1971). Effect of anterior pituitary perfusion and intraventricular injection of catecholamines on prolactin release. *Endocrinology*, **88**, 1012–1020

Kastin, A. J., Miller, L. H., Sandman, C. A., Schally, A. V. and Plotnikoff, N. P. (1977). CNS and pituitary effects of hypothalamic peptides and MSH. In Youdim, M.B.H., Lovenberg, W., Sharman, D. F. and Lagnado, J. R. (eds.), *Essays in Neurochemistry and Neuropharmacology*, **1**, 139–176, John Wiley and Sons, London

Kastin, A. J., Plotnikoff, N. P., Schally, A. V. and Sandman, C. A. (1976). Endocrine and CNS effects of hypothalamic peptides and MSH. In Ehrenpreis, S. and Kopin, I. J. (eds.), *Reviews of Neuroscience*, **2**, 111–148, Raven Press, New York

Kastin, A. J., Sandman, C. A., Schally, A. V. and Ehrensing, R. A. (1978). Clinical effects of hypothalamic–pituitary peptides upon the central nervous system. In Klawans, H. L. (ed.), *Clinical Neuropharmacology*, Volume 3, Raven Press, New York, 133–152

Kastin, A. J. and Schally, A. V. (1967). MSH activity in pituitaries of rats treated with hypothalamic extracts from various animals. *Gen. Comp. Endocrinol.*, **8**, 344–347

Kastin, A. J., Schally, A. V., Gual, C., Glick, S. and Arimura, A. (1972). Clinical evaluation in men of a substance with growth hormone-releasing activity in rats. *J. Clin. Endocrinol. Metab.*, **35**, 326–329

Kerdelhué, B., Catin, S., Kordon, C. and Jutisz, M. (1976). Delayed effects of *in vivo* LH-RH immunoneutralisation on gonadotropins and prolactin secretion in the female rat. *Endocrinology*, **98**, 1539–1549

Kerdelhué, B., Jutisz, M., Gillessen, D. and Studer, R. O. (1973). Obtention of antisera against a hypothalamic decapeptide (luteinizing hormone follicle stimulating hormone releasing hormone) which stimulates the release of pituitary gonadotrophins and development of its radioimmunoassay. *Biochim. Biophys. Acta*, **297**, 540–548

King, J. C., Gerall, A. A., Fishback, J. B., Elkind, K. E. and Arimura, A. (1975). Growth hormone-release inhibiting hormone (GH-RIH) pathway of the rat hypothalamus revealed by the unlabelled antibody peroxidase–antiperoxidase method. *Cell Tiss. Res.*, **160**, 423–430

Koch, Y., Baram, T., Hazum, E. and Fridkin, M. (1977*a*). Resistance to enzymic degradation of LH-RH analogues possessing increased biological activity. *Biochem. Biophys. Res. Commun.*, **74**, 488–491

Koch, Y., Chobsieng, P., Zor, U., Fridkin, M. and Lindner, H. R. (1973). Suppression of gonadotropin secretion and prevention of ovulation in the rat by antiserum to synthetic gonadotropin-releasing hormone. *Biochem. Biophys. Res. Commun.*, **55**, 623–629

Koch, Y., Goldhaber, G., Fireman, I., Zor, U., Shani, J. and Tal, E. (1977*b*). Suppression of prolactin and thyrotropin secretion in the rat by antiserum to thyrotropin-releasing hormone. *Endocrinology*, **100**, 1476–1478

Koerker, D. J., Ruch, W., Chideckel, E., Palmer, J., Goodner, C. J. and Gale, C. C. (1974). Somatostatin: hypothalamic inhibitor of the endocrine pancreas. *Science,* **184**, 482–484

König, W., Sandow, J. and Geiger, R. (1975). Structure-function relationships of LH-RH/FSH-RH. In R. Walter and J. Meienhofer (eds.), *Peptides: Chemistry, Structure and Biology,* Ann Arbor Science Pub. Inc., Ann Arbor, Michigan, 883–888

Konturek, S. J., Tasler, J., Cieszkowski, M., Coy, D. H. and Schally, A. V. (1976*a*). Effect of growth hormone release-inhibiting hormone on gastric secretion, mucosal blood flow , and serum gastrin. *Gastroenterology,* **70**, 737–741

Konturek, S. J., Tasler, J., Obtulowicz, W., Coy, D. H. and Schally, A. V. (1976*b*). Effect of growth hormone-release inhibiting hormone (TRH)-stimulated secretion of thyrotropin (TSH) by passive immunization with antiserum to somatostatin in rats. *J. Clin. Invest.,* **58**, 1–6

Krulich, L., Dhariwal, A. P. S. and McCann, S. M. (1968). Stimulatory and inhibitory effects of purified hypothalamic extracts on growth hormone release from rat pituitary *in vitro. Endocrinology,* **83**, 783–790

Labrie, F., Pelletier, G., Drouin, J., Belanger, A., Ferland, L., Lemay, A., Lemaire, S. and Beaulieu, M. (1976). Mechanism of action of hypothalamic hormones in the anterior pituitary gland. In Charro-Salgado, A. L., Fernandez-Durango, R. and Lopez del Campo, J. G. (eds.), *Basic Applications and Clinical Uses of Hypothalamic Hormones,* Excerpta Medica, Amsterdam, 100–110

Larsson, L-I., Holst, J. J., Kuhl, C., Lundqvist, G., Hirsch, M. A., Ingemansson, S., Lindkaer-Jensen, S., Rehfeld, J. F. and Schwartz, T. W. (1977). Pancreatic somatostatinoma. *Lancet,* **i**, 666–668

Levitan, D., Beitins, I. Z., Milton, G., Barnes, A. and McArthur, J. W. (1977). Insensitivity of Bonnet Monkeys to (D-Ala, Des Gly) LH-RH ethylamide, a potent new luteinizing hormone releasing hormone analogue in rats and mice. *Endocrinology*, **100**, 918–922

Li, C. H., Barnafi, L., Chretien, M. and Chung, D. (1965). Isolation and amino-acid sequence of β-LPH from sheep pituitary glands. *Nature*, **208**, 1093–1094

Li, C. H. and Chung, D. (1976). Isolation and structure of an untriakontapeptide with opiate activity from camel pituitary glands. *Proc. Natn. Acad. Sci. USA,* **78**, 1145–1148

Li, C. H., Lemaire, S., Yamashiro, D. and Doreen, B. A. (1976). The synthesis and opiate activity of beta-endorphin. *Biochem. Biophys. Res. Commun.,* **71**, 19–25

Ling, N., Burgus, R. and Guillemin, R. (1976). Isolation, primary structure, and synthesis of α-endorphin and γ-endorphin, two peptides of hypothalamic-hypophyseal origin with morphinomimetic activity. *Proc. Natn. Acad. Sci. USA,* **73**, 3942–3946

Lohmar, P. and Li, C. H. (1967). Isolation of bovine β-lipotropic hormone. *Biochim. Biophys. Acta,* **147**, 381–383

Lovinger, R., Boryczka, A. T., Schackelford, R., Kaplan, S. L., Ganong, W. F. and Grumbach, M. M. (1974). Effect of synthetic somatotrophin release inhibiting factor on the increase in plasma growth hormone elicited by L-dopa in the dog. *Endocrinology*, **95**, 943–946

Luft, R., Efendic, S., Hokfelt, T., Johansson, O. and Arimura, A. (1974). Immunohistochemical evidence for the localization of somatostatin-like immuno-reactivity in a cell population of the pancreatic islets. *Med. Biol.,* **52**, 428–430

MacLeod, R. M. (1969). Influence of norepinephrine and catecholamine-depleting agents on the synthesis and release of prolactin and growth hormone. *Endocrinology,* **85**, 916–923

Martin, J. B., Renaud, L. P. and Brazeau, P. (1975). Hypothalamic peptides: new evidence for 'peptidergic' pathways in the CNS. *Lancet*, **ii**, 393–395

Matsuo, H., Arimura, A., Nair, R. M. G. and Schally, A. V. (1971*a*). Synthesis of the porcine LH and FSH releasing hormone by the solid-phase method. *Biochem. Biophys. Res. Commun.,* **45**, 822–827

Matsuo, H., Baba, Y., Nair, R. M. G., Arimura, A. and Schally, A. V. (1971*b*). Structure of the porcine LH- and FSH-releasing hormone. Part I. The proposed amino acid sequence. *Biochem. Biophys. Res. Commun.,* **43**, 1334–1339

Mayar, M. Q., Tarnavsky, G. K. and Reeves, J. J. (1977). Ovarian growth and uptake of iodinated D-Leu[6] desGly NH_2^{10}-LH-RH-Ethylamide in HCG treated prepubertal rats. *Proc. Western Sect. Am. Soc. Anim. Sci.*, **28**, 182–184

Meites, J. and Clemens, J. A. (1972). Hypothalamic control of prolactin secretion. *Vit. Horm.*, **30**, 165–221

Meyers, C., Arimura, A., Gordin, A., Fernandez-Durango, R., Coy, D. H., Schally, A. V., Drouin, J., Ferland, L., Beaulieu, M. and Labrie, F. (1977). Somatostatin analogs which inhibit glucagon and growth hormone more than insulin release. *Biochem. Biophys. Res. Commun.*, **74**, 630–636

Monahan, M. W., Amoss, M. S., Anderson, H. A. and Vale, W. (1973). Synthetic analogs of the hypothalamic luteinizing hormone releasing factor with increased agonist or antagonist properties. *Biochemistry*, **12**, 4616–4620

Mortimer, C. H., McNeilly, A. S., Rees, L. H., Lowry, P. J., Gilmore, D. and Dobbie, H. G. (1976). Radioimmunoassay and chromatographic similarity of circulating endogenous gonadotrophin releasing hormone and hypothalamic extracts in man. *J. Clin. Endocrinol. Metab.*, **43**, 882–888

Mortimer, C. H., Tunbridge, W. M. G., Carr, D., Yeomans, L., Lind, T., Coy, D. H., Bloom, S. R., Kastin, A., Mallinson, C. N., Besser, G. M., Schally, A. V. and Hall, R. (1974). Effects of growth-hormone release-inhibiting hormone on circulatory glucagon, insulin and growth hormone in normal, diabetic, acromegalic, and hypopituitary patients. *Lancet*, **i**, 697–701

Moss, R. L. (1977). Role of hypophysiotropic neurohormones in mediating neural and behavioral events. *Fed. Proc.*, **36**, 1978–1983

Moss, R. L. and McCann, S. M. (1973). Induction of mating behavior in rats by luteinising hormone-releasing factor. *Science*, **181**, 177–179

Nair, R. M. G., Barrett, J. F., Bowers, C. Y. and Schally, A. V. (1970). Structure of porcine thyrotropin releasing hormone. *Biochemistry*, **9**, 1103–1106

Nair, R. M. G., Kastin, A. J. and Schally, A. V. (1971*a*). Isolation and structure of hypothalamic MSH release inhibiting hormone. *Biochem. Biophys. Res. Commun.*, **43**, 1376–1381

Nair, R. M. G., Redding, T. W. and Schally, A. V. (1971*b*). Inactivation of thyrotropin-releasing hormone by human plasma. *Biochemistry*, **10**, 3621–3624

Neill, J. D., Patton, J. M., Dailey, R. A., Tsou, R. C. and Tindall, G. T. (1977). Luteinizing hormone releasing hormone (LH-RH) in pituitary stalk blood of rhesus monkeys: Relationship to level of LH release. *Endocrinology*, **101**, 430–434

Nillius, S. J., Fries, H. and Wide, L. (1975). Successful induction of follicular maturation and ovulation by prolonged treatment with LH-releasing hormone in women with anorexia nervosa. *Am. J. Obstet. Gynecol.*, **122**, 921–928

Nillius, S. J. and Wide, L. (1972). Variations in LH and FSH response to LH-releasing hormone during the menstrual cycle. *J. Obstet. Gynaecol. Brit. Commun.*, **79**, 865–873

Nishi, N., Arimura, A., de la Cruz, K. G. and Schally, A. V. (1976). Termination of pregnancy by sheep anti-LH-RH gamma globulin in rats. *Endocrinology*, **98**, 1024–1030

Oliver, C., Charvet, J. P., Codaccioni, J. L., Vague, J. and Porter, J. C. (1974). TRH in human CSF. *Lancet*, **i**, 873

Orci, L. and Unger, R. H. (1975). Functional subdivision of islets of Langerhans and possible role of D-cells. *Lancet*, **ii**, 1243–1244

Pasternak, G. W., Goodman, R. and Snyder, S. H. (1975). An endogenous morphine-like factor in mammalian brain. *Life Sci.*, **16**, 1765–1769

Pedroza, E., Vilchez-Martinez, J. A., Fishback, J., Arimura, A. and Schally, A. V. (1977). Binding capacity of LH-RH and its analogue for pituitary receptor sites, *Biochem. Biophys. Res. Commun.*, **79**, 234–238

Pelletier, G., Leclerc, R., Arimura, A. and Schally, A. V. (1975). Immunohistochemical localization of somatostatin in the rat pancreas. *J. Histochem. Cytochem.*, **23** 699–701

Plotnikoff, N. P., Kastin, A. J. and Schally, A. V. (1974). Growth hormone release inhibiting hormone: Neuropharmacological studies. *Pharm. Biochem. Behavior,* **2**, 693–696

Polak, J. M., Pearse, A. G. E., Grimelius, L., Bloom, S. R. and Arimura, A. (1975). Growth-hormone release-inhibiting hormone in gastrointestinal and pancreatic D-cells. *Lancet,* **i**, 1220–1222

Redding, T. W. and Schally, A. V. (1971). The distribution of radioactivity following the administration of labeled thyrotrophin-releasing hormone (TRH) in rats and mice. *Endocrinology*, **89**, 1075–1081

Redding, T. W. and Schally, A. V. (1977). The purification of a growth hormone-releasing factor from porcine hypothalamic tissue. Program 59th Meeting, Endocrine Society, Chicago, June, 1977, Abstract No. 350, 231

Rees, L. H., Cook, D. M., Kendall, J. W., Allen, C. F., Kramer, R. M., Ratcliffe, J. G. and Knight, R. A. (1971). A radioimmunoassay for rat plasma ACTH. *Endocrinology*, **89**, 254–261

Rees, R. W. A., Foell, T. J., Chai, S. and Grant, N. (1974). Synthesis and biological activities of analogs of the luteinizing hormone-releasing hormone (LH-RH) modified in position 2. *J. Med. Chem.*, **17**, 1016–1019

Reeves, J. J., Tarnavsky, G. K., Becker, S. R., Coy, D. and Schally, A. V. (1977). Uptake of iodinated luteinizing hormone releasing hormone analogs of the pituitary. *Endocrinology,* **101**, 540–547

Reichlin, S., Saperstein, R., Jackson, I. M. D., Boyd, A. E. (III) and Patel, Y. (1976). Hypothalamic hormones. *Ann. Rev. Physiol.*, **38**, 389–424

Renaud, L. P., Martin, J. B. and Brazeau, P. (1975). Depressant action of TRH, LH-RH and somatostatin on activity of central neurones. *Nature,* **255**, 233–235

Rippel, R. H. and Johnson, E. S. (1976). Inhibition of HCG-induced ovarian and uterine weight augmentation in the immature rat by analogs of GnRH. *Proc. Soc. Exp. Biol. Med.*, **152**, 432–436

Rivier, C., Vale, W., Ling, N., Brown, M. and Guillemin, R. (1977). Stimulation *in vivo* of the secretion of prolactin and growth hormone by β-endorphin. *Endocrinology*, **100**, 238–241

Rivier, J., Brazeau, P., Vale, W., Ling, N., Burgus, R., Gilon, C., Yardley, J. and Guillemin, R. (1973). Synthèse totale par phase solide d'un tétradécapeptide ayant les propriétés chimiques et biologiques de la somatostatine. *C. R. Acad. Sci. Paris.*, **276**, 2737–2740

Rivier, J., Brown, M., Rivier, C., Ling, N. and Vale, W. (1977). Hypothalamic hypophysiotropic hormones: a review on the design of synthetic analogues. In Loffet, A. (ed.), *Peptides 1976*, Editions de l'Univers de Bruxelles, 427–451

Rivier, J., Brown, M. and Vale, W. (1975). D-Trp^8-somatostatin: an analog of somatostatin more potent than the native molecule. *Biochem. Biophys. Res. Commun.*, **65**, 746–751

Rivier, J., Vale, W., Monahan, M., Ling, N. and Burgus, R. (1972). Synthetic thyrotropin-releasing factor analogs. Part 3: Effect of replacement or modification of histidine residue on biological activity. *J. Med. Chem.*, **15**, 479–482

Saffran, M. and Schally, A. V. (1955). Release of corticotrophin by anterior pituitary tissue *in vitro. Canad. J. Biochem.*, **33**, 408–415

Sakurai, H., Dobbs, R. and Unger, R. H. (1974). Somatostatin-induced changes in insulin and glucagon secretion in normal and diabetic dogs. *J. Clin. Invest.*, **54**, 1395–1402

Sandow, J., von Rechenberg, W., Konig, W., Han, M., Jerabek, G. and Fraser, H. (1977). Physiological studies with highly active analogues of LH-RH. In *Proc. 2nd Europ. Col. Hypothal. Hormones,* Tubingen, July, 1976, Verlag Chemie, Hamburg

Sarantakis, D., McKinley, W. A., Juanakais, I., Clark, D. and Grant, N. H. (1976*a*). Structure activity studies on somatostatin. *Clin. Endocrinol.*, **5** (suppl.), 275–278

Sarantakis, D., Teichman, J., Lien, E. L. and Fenichel, R. L. (1976*b*). A novel cyclic undecapeptide, WY-40, 770, with prolonged growth hormone release inhibiting activity. *Biochem. Biophys. Res. Commun.*, **73**, 336–342

Schalch, D. S., Gonzales-Barcena, D., Kastin, A. J., Schally, A. V. and Lee, L. A. (1972). Abnormalities in the release of TSH in response to thyrotrophin-releasing hormone (TRH) in patients with disorders of the pituitary, hypothalamus and basal ganglia. *J. Clin. Endocrinol. Metab.*, **35**, 609–615

Schally, A. V., Anderson, R. N., Lipscomb , H. S., Long, J. M. and Guillemin, R. (1960). Evidence for the existence of two corticotrophin-releasing factors, α and β. *Nature*, **188**, 1192–1193

Schally, A. V. and Arimura, A. (1978). Physiology and nature of hypothalamic regulatory hormones. In Martini, L. and Besser, G. M. (eds.), *Clinical Neuroendoc.*, Academic Press, New York, 2–4

Schally, A. V., Arimura, A., Baba, Y., Nair, R. M. G., Matsuo, H., Redding, T. W., Debeljuk, L. and White, W. F. (1971*a*). Isolation and properties of the FSH and LH-releasing hormone. *Biochem. Biophys. Res. Commun.*, **43**, 393–399

Schally, A. V., Arimura, A., Bowers, C. Y., Kastin, A. J., Sawano, S. and Redding, T. W. (1968). Hypothalamic neurohormones regulating anterior pituitary function. In Astwood, E. B. (ed .), *Recent Progress in Hormone Research*, **24**, Academic Press, New York, 497–588

Schally, A. V., Arimura, A., Carter, W. H., Redding, T. W., Geiger, R., König, W., Wissman, H., Jaeger, G., Sandow, J., Yanaihara, N., Yanaihara, T., Hashimoto, T. and Sakagami, M. (1972*a*). Luetinising hormone-releasing hormone (LH-RH) activity of some synthetic polypeptides. 1. Fragments shorter than decapeptide. *Biochem. Biophys. Res. Commun.*, **48**, 366–375

Schally, A. V., Arimura, A. and Kastin, A. J. (1973*a*). Hypothalamic regulatory hormones. *Science*, **179**, 341–350

Schally, A. V., Arimura, A., Kastin, A. J., Matsuo, H., Baba, Y., Redding, T. W., Nair, R. M. G., Debeljuk, L. and White, W. F. (1971*b*). Gonadotrophin-releasing hormone: one polypeptide regulates secretion of luetinizing and follicle-stimulating hormones. *Science*, **173**, 1036–1038

Schally, A. V., Arimura, A., Redding, T. W., Chihara, K., Gordin, A., Huang, W. Y. and Saffran, M. (1977*a*). Purification of corticotrophin releasing factor from porcine hypothalami. Program, 59th Meeting, Endocrine Society, Chicago, June, 1977, Abstract, 77, p. 95

Schally, A. V., Arimura, A., Redding, T. W., Debeljuk, L., Carter, W., Dupont, A. and Vilchez-Martinez, J. A. (1976*a*). Re-examination of porcine and bovine hypothalamic fractions for additional luteinizing hormone and follicle stimulating hormone-releasing activities. *Endocrinology*, **98**, 380–391

Schally, A. V., Arimura, A., Wakabayashi, I., Redding, T. W., Dickerman, E. and Meites, J. (1972*b*). Biological activity of a synthetic decapaptide corresponding to the proposed growth hormone-releasing hormone. *Experientia*, **28**, 205–206

Schally, A. V., Baba, Y., Nair, R. M. G. and Bennett, C. (1971*c*). The amino acid sequence of a peptide with growth hormone-releasing activity isolated from porcine hypothalamus. *J. Biol. Chem.*, **246**, 6647–6650

Schally, A. V., Bowers, C. Y., Redding, T. W. and Barrett, J. F. (1966). Isolation of thyrotropin releasing factor (TRF) from porcine hypothalamus. *Biochem. Biophys. Res. Commun.*, **25**, 165–169

Schally, A. V., Dupont, A., Arimura, A., Redding, T. W., Nishi, N., Linthicum, G. L. and Schlesinger, D. H. (1976*b*). Isolation and structure of somatostatin from porcine hypothalami. *Biochemistry*, **15**, 509–514

Schally, A. V., Dupont, A., Arimura, A., Takahara, J., Redding, T. W., Clemens, J. and Shaar, C. (1976*c*). Purification of a catecholamine-rich fraction with prolactin release-inhibiting factor (PIF) activity from porcine hypothalami. *Acta Endocrinol.*, **82**, 1–14

Schally, A. V. and Kastin, A. J. (1966). Purification of a bovine hypothalamic factor which elevates pituitary MSH levels in rats. *Endocrinology*, **79**, 768–772

Schally, A. V., Kastin, A. J. and Arimura, A. (1971*d*). Hypothalamic follicle stimulating hormone (FSH) and luteinising hormone (LH)-regulating hormone: structure, physiology and clinical studies. *Fertil. Steril.*, **22**, 703–721

Schally, A. V., Kastin, A. J. and Coy, D. H. (1976*d*). LH-releasing hormone and its analogues: recent basic and clinical investigations. *Internat. J. Fertil.*, **21**, 1–30

Schally, A. V., Nair, R. M. G., Redding, T. W. and Arimura, A. (1971*e*). Isolation of the luteinising hormone and follicle-stimulating hormone-releasing hormone from porcine hypothalami. *J. Biol. Chem.*, **246**, 7230–7236

Schally, A. V., Redding, T. W., Arimura, A., Dupont, A. and Linthicum, G. L. (1977*b*). Isolation of gamma-amino butyric acid from pig hypothalami and demonstration of its prolactin release-inhibiting (PIF) activity *in vivo* and *in vitro*. *Endocrinology*, **100**, 681–691

Schally, A. V., Redding, T. W., Takahara, J., Coy, D. H. and Arimura, A. (1973*b*). Lack of growth hormone-releasing activity of (Pyro)Glu-Ser-Gly-NH_2. *Biochem. Biophys. Res. Commun.*, **55**, 556–562

Schally, A. V., Saffran, M. and Zimmerman, B. (1958). A corticotrophin-releasing factor: partial purification and amino acid composition. *Biochem. J.*, **70**, 97–103

Schally, A. V., Sawano, S., Arimura, A., Barrett, J., Wakabayashi, I. and Bowers, C. Y. (1969). Isolation of growth hormone-releasing hormone (GH-RH) from porcine hypothalami. *Endocrinology*, **84**, 1493–1506

Sétáló, G., Flerko, B., Arimura, A. and Schally, A. V. (1978). Brain cells as producers of releasing and inhibiting hormones. In Bourne, H. R. and Danielli, J. F. (eds.), *Int. Rev. Cytol.*, **55**, Academic Press, New York, to be published

Sétáló, G., Vigh, S., Schally, A. V., Arimura, A. and Flerko, B. (1975). LH-RH containing neural elements in the rat hypothalamus. *Endocrinology*, **96** 135–142

Sievertsson, H., Castensson, S. and Lindgren, O. (1974). Studies on tetrapeptides related to thyrotropin releasing hormone and luteinizing hormone releasing hormone. *Acta Pharmaceutica Suecica*, **11**, 67–76

Sievertsson, H., Chang, J.-K., Folkers, K. and Bowers, C. Y. (1972). On the role of the histidine moiety in the structure of the thyrotropin-releasing hormone. *J. Med. Chem.*, **15**, 219–221

Siler, T. M., Vandenberg, G., Yen, S. S. C., Brazeau, P., Vale, W. and Guillemin, R. (1973). Inhibition of growth hormone release in humans by somatostatin. *J. Clin. Endocrinol. Metab.*, **37**, 632–634

Simantov, R. and Snyder, S. H. (1976). Isolation and structure identification of a morphine-like peptide 'enkephalin' in bovine brain. *Life Sci.*, **18**, 781–788

Smith, J. R., la Hann, T. R., Chesnut, R. M., Carino, M. A. and Horita, A. (1977). Thyrotropin releasing hormone: stimulation of colonic activity following intracerebroventricular administration. *Science*, **196**, 660–661

Szabo, M. and Frohman, L. A. (1976). Dissociation of prolactin-releasing activity from thyrotropin-releasing hormone in porcine stalk median eminence. *Endocrinology*, **98**, 1451–1459

Takahara, J., Arimura, A. and Schally, A. V. (1974*a*). Suppression of prolactin release by a purified porcine PIF preparation and catecholamines infused into a rat hypophyseal portal vessel. *Endocrinology*, **95**, 462–465

Takahara, J., Arimura, A. and Schally, A. V. (1974*b*). Effect of catecholamines on the TRH-stimulated release of prolactin and growth hormone from sheep pituitaries *in vitro*. *Endocrinology*, **95**, 1490–1494

Takahara, J., Arimura, A. and Schally, A. V. (1975). Assessment of GH releasing hormone activity in Sephadex-separated fractions of porcine hypothalamic extracts by hypophyseal portal vessels. *Acta Endocrinol.*, 78, 428–434

Tashjian, A., Barowski, N. and Jensen, D. (1971). Thyrotropin releasing hormone: direct evidence for stimulation of prolactin production by pituitary cells in culture. *Biochem. Biophys. Res. Commun.*, **43**, 516–523

Terenius, L. and Wahlstrom, A. (1975). Search for an endogenous ligand for the opiate receptor. *Acta Physiol. Scand.*, **94**, 74–81

Tilders, F. J. H., Mulder, A. H. and Smelik, P. G. (1975). On the presence of an MSH release inhibiting system in the rat neurointermediate lobe. *Neuroendocrinology*, **18** , 125–130

Tsou, R. C., Dailey, R. A., McLanahan, C. S., Parent, A. D., Tindall, G. T. and Neill, J. D. (1977). Luteinizing hormone releasing hormone (LH-RH) levels in pituitary and stalk plasma during the preovulatory gonadotropin surge of rabbits. *Endocrinology*, **101**, 534–539

Unger, R. and Orci, L. (1977). Possible roles of the pancreatic D-cell in the normal and diabetic states. *Diabetes*, **26**, 241–244

Vale, W., Brazeau, P., Grant, G., Nussey, A., Burgus, R., Rivier, J., Ling, N. and Guillemin, R. (1972*a*). Premières observations sur le mode d'action de la somatostatine, un facteur hypothalamique qui inhibe la sécrétion de l'hormone de croissance. *C. R. Acad. Sci. Paris*, **275**, 2913–2916

Vale, W., Grant, G., Amoss, M., Blackwell, R. and Guillemin, R. (1972*b*). Culture of enzymatically dispersed anterior pituitary cells: functional validation of a method. *Endocrinology*, **91**, 562–572

Vale, W., Grant, G., Rivier, J., Monahan, M., Amoss, M., Blackwell, R., Burgus, R. and Guillemin, R. (1972*c*). Synthetic polypeptide antagonists of the hypothalamic luteinizing hormone releasing factor. *Science*, **176**, 933–934

Vale, W., Rivier, C. and Brown, M. (1977). Regulatory peptides of the hypothalamus. *Ann. Rev. Physiol.*, **39**, 473–527

Veber, D. F., Strachan, R. G., Bergstrand, S. J., Holly, F. W., Homnick, C. F., Hirschmann, R., Torchiana, M. L. and Saperstein, R. (1976). Non-reducible cyclic analogues of somatostatin. *J. Amer. Chem. Soc.*, **98**, 2367–2369

Vecsei, P. (1974). Glucocorticoids cortisol, corticosterone and compound S., In Jaffe, B. M. and Behrman (eds.), *Methods of Hormone Radioimmunoassay*, Academic Press, London, 393–415

Vilchez-Martinez, J. A., Arimura, A. and Schally, A. V. (1974*a*). Influence of estradiol benzoate on pituitary responsiveness to LH-RH at different stages of the estrous cycle in rats (38206). *Proc. Soc. Exp. Biol. Med.*, **146**, 859–862

Vilchez-Martinez, J. A., Coy, D. H., Arimura, A., Coy, E. J., Hirotsu, Y. and Schally, A. V. (1974*b*). Synthesis and biological properties of [Leu-6]-LH-RH and [D-Leu-6, Desgly-$NH_2{}^{10}$]-LH-RH ethylamide. *Biochem. Biophys. Res. Commun.*, **59**, 1226–1232

Vilchez-Martinez, J. A., Coy, D. H., Coy, E. J., Schally, A. V. and Arimura, A. (1975). Anti-luteinizing (LH)-releasing activity of several analogues of LH-releasing hormone. *Fertil. Steril.*, **26**, 554–559

Walker, J. M., Berntson, G. G., Sandman, C. A., Coy, D. H., Schally, A. V. and Kastin, A. J. (1977). An analog of enkephalin having prolonged opiate-like effects *in vivo*. *Science*, **196**, 85–87

Wilber, J. F. and Flouret, G. (1971). Thyrotropin releasing hormone (TRH) analogues: stimulation of thyrotropin (TSH) secretion *in vitro*. Program, 53rd Meeting, The Endocrine Society, San Francisco, June, 1971, A-87

Yamashiro, D. and Li, C. H. (1973). Synthesis of a peptide with full somatostatin activity. *Biochem. Biophys. Res. Commun.*, **54**, 882–887

Yanaihara, N., Yanaihara, C., Sakagami, M., Tsuji, K., Hashimoto, T., Kaneko, T., Oka, H., Schally, A. V., Arimura, A. and Redding, T. W. (1973). Synthesis and biological evaluation of LH and FSH releasing hormone and its analogs. *J. Med. Chem.*, **16**, 373–377

Yen, S. S. C., Siler, T. M. and De-Vane, G. W. (1974). Effect of somatostatin in patients with acromegaly: suppression of growth hormone, prolactin, insulin and glucose levels. *New Engl. J. Med.*, **290**, 935–938

Youdaev, N. A., Outecheva, Z. F., Novikova, T. E., Chvatchkin, Y. P. and Smirnova, A. P. (1973). A hypothalamic peptide stimulating the release of growth hormone. *Doklady Akad. Nauk. URSS*, **210**, 731–732

Zimmerman, E. A. (1977). Localisation of hormone secreting pathways in the brain by immunohistochemistry and light microscopy: a review. *Fed. Proc.*, **36**, 1964–1967

Discussion

Carpenter (Manchester)
What is the time course of the inhibition of gastric acid secretion by somatostatin and could you speculate on the mechanism of its action?

Schally
I think the onset is immediate. Somatostatin is less successful at inhibiting histamine-induced gastric secretion and the antagonistic action is only seen in some experiments. We suspect the mechanism involves messengers such as prostaglandins and 3′–5′ cyclic AMP but we are not sure.

Morris (Manchester)
What is the danger of antibody generation with the long-term use of LH-RH analogues for contraception?

Schally
Only one of our many thousands of patients from all over the world who have been injected with LH-RH has developed antibodies. This was a hypogonadal patient who was injected three or four times a day with 500 μg LH-RH for 400 days. Therefore I think the danger is very low and if antibodies did develop you could change analogues.

Plumber (London)
The human pituitary is reported to produce lipotrophins rather than MSH. Is there any evidence to suggest that there are lipotrophin releasing factors or release-inhibiting factors?

Schally
I did not mention my interest in obesity control. There may be a lipid mobilising factor (LMF) which inhibits lipotrophin secretion. We have published a paper suggesting an LMF exists but we have been unable to isolate it so far.

Lowe (Alderley Park)
In your list of actions of somatostatin you did not mention platelets. Do you believe that somatostatin has no effect on platelets?

Schally
I was alarmed to hear the reports that somatostatin may lower platelet count. All the work was carried out on baboons and clinical work in the USA subsequently stopped. I feel these baboons died from causes other than somatostatin. Perhaps Professor Besser has some comments.

Besser
There is no doubt that both linear and cylic somatostatin can decrease platelet agglutination in humans. The original experiments in baboons were not very well controlled as the animals had been treated with many different agents. In humans the platelet count does not go down but agglutination indices fall. Our own view is that it is unimportant because its effect is similar to aspirin and no one really worries about this when using aspirin.

Morley (Alderley Park)
You mentioned that you did not think that the decapeptide with CRF activity was actually physiological CRF. Was this comment just based on its low activity?

Schally
We have been able to isolate many CRFs and I do not know which fraction is the physiological one. The one you mention is certainly much less potent than some others, but it is possible that the body releases different CRFs in response to different stresses.

Hollingsworth (Manchester)
Is it possible to develop a low molecular weight non-peptide analogue of the hypothalamic hormones which would be more potent orally?

Schally
Many peptides (e.g. LH-RH analogues and TRH analogues) can be given by mouth. Ultimately we hope to give somatostatin analogues by mouth and the contraceptive peptides also. However, it is not necessary to administer the peptides by mouth since the intranasal route is extremely effective and I feel this may be the way in the future.

Schnieden (Manchester)
With regard to the contraceptive peptide which you are using in Mexico, is there any evidence of cross-antagonism against other hypothalamic peptides?

Schally
The work is very new but we feel that the results point only to one thing, an unquestionable block of LH and FSH release. We hope that this is a specific effect, unlike the effect of TRH. However, we do not yet have an answer.

7

Evaluation of clinical disturbances of the hypothalamus

G. M. Besser*

INTRODUCTION

Endocrine disturbances of hypothalamic function may affect those peptide hormones normally associated with either the anterior or posterior lobes of the pituitary gland. Alternatively such diseases may occur as a result of primary pituitary pathology (figure 7.1). The clinical management involves initial confirmation of the pituitary secretory disorder and then if possible the differentiation of the site of the lesion either at the hypothalamic or the pituitary level.

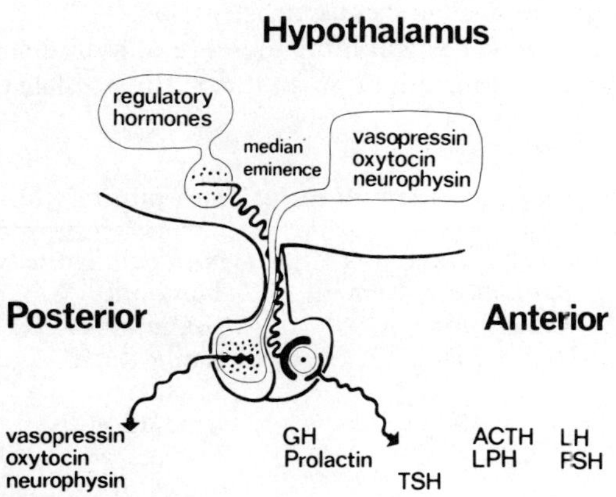

Figure 7.1 Diagrammatic representation of hypothalamic–pituitary relationships. Disease of either the hypothalamus or pituitary may result in apparent abnormalities of pituitary hormone secretion (after Besser (1974), with permission)

*Department of Endocrinology, University of London, at St. Bartholomew's Hospital Medical College, London, EC1A 7BE

DISORDERS OF ANTERIOR PITUITARY HORMONE SECRETION

Gonadotrophins

Reduced or absent gonadal function is a common concomitant of pituitary or hypothalamic diseases. However, care must be taken to avoid error by excluding hyperprolactinaemia since this is even commoner in patients with such diseases and produces gonadal failure despite adequate gonadotrophin secretion. With a progressive lesion causing hypopituitarism from whatever cause, LH is almost always the first hormone to disappear; FSH secretion usually remains until a late stage in the disease.

Gonadotrophin deficiency is usually simple to establish since in the presence of low circulating gonadal steroid concentrations patients with normal gonadotrophin secretory reserve should, because of the negative feedback control mechanisms, show elevated gonadotrophin levels in the blood. Low levels, or levels within the range seen in patients with normal blood gonadal steroid concentrations, indicate gonadotrophin deficiency. If any doubt about the diagnosis remains, clomiphene may be used. Clomiphene, an anti-oestrogen, appears to bind to gonadal steroid receptors within the hypothalamus and pituitary and to stimulate secretion of the gonadotrophin-releasing hormone and of LH and FSH. A dose of 3 mg kg^{-1} day^{-1} is given in divided doses for 10 days (Anderson *et al.,* 1972) up to a maximum of 200 mg/day. Blood for serum LH and FSH is obtained on days 0, 4, 7 and 10. Normal subjects show a definite rise in gonadotrophin levels usually above the normal range. In hypogonadism due to hypothalamic or pituitary disease the response is absent or impaired. Absent responses may also be seen in prepubertal subjects and in patients with severe anorexia nervosa in whom clomiphene responsiveness may return long before spontaneous menstruation.

However, these tests which establish the presence of hypothalamic or pituitary endocrine failure do not distinguish between them. The possible causes are shown in table 7.1.

Table 7.1 Principal causes of failure of secretion of pituitary hormones

Pituitary disease	*Pituitary stalk lesions*	*Hypothalamic disease*
Tumours	Tumour involvement	Tumours
Granulomas	Head injury	Granulomas
Infarction	Meningitis	Hamartoma
Apoplexy		Vascular
Irradiation		Irradiation
Trauma		Trauma
		Congenital deficiency of a releasing hormone (RH): commonly of Gn-RH producing 'isolated gonadotrophin deficiency' or of GH-RH producing 'isolated GH deficiency' in the presence of otherwise normal pituitary function

The gonadotrophin-releasing hormone (Gn–RH) has considerably facilitated the investigation of patients with hypothalamic-pituitary-gonadal dysfunction. The releasing hormone produces a dose-dependent release of both LH and FSH. With a 100 μg dose intravenously, the peak LH and FSH response occurs between 20 and 30 min and this is the basis of the standard test (Besser *et al.*, 1972*a*; Mortimer *et al.*, 1973*a*).

Patients with isolated gonadotrophin deficiency usually show an LH and FSH response after Gn–RH, demonstrating that they are deficient in the releasing hormone and not in gonadotrophin. Some patients require a larger dose (500 μg) or several applications of Gn–RH to produce a response.

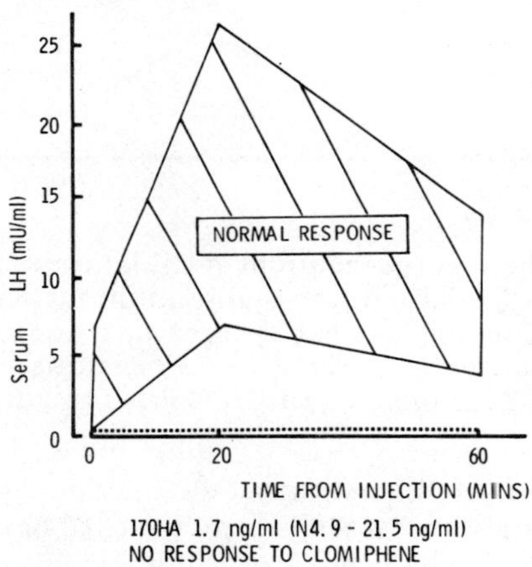

Figure 7.2 Serum LH response to 100 μg intravenous Gn-RH in normal subjects (hatched area) and in a patient with hypogonadism due to a hypothalamic tumour (dotted line). The hypogonadal patient showed no gonadotrophin responses to clomiphene but a brisk testosterone (measured as 17OH-androgens, 17OHA) responses to exogenous gonadotrophin. Despite an intact pituitary there is no LH response to Gn-RH and there was similarly no FSH response

In patients who have organic hypothalamic diseases and who are hypogonadal, impaired or absent LH and FSH responses are usually seen (figure 7.2), but they may be entirely normal if the pituitary gonadotrophs contain hormone. Since many patients with hypothalamic disease but intact pituitary glands fail to show LH/FSH responses to Gn–RH unless this is repeatedly given, it is clear that Gn–RH promotes synthesis as well as release of 'readily releasable' gonadotrophin. Indeed in such patients long-term treatment with Gn–RH promotes normal gonadal function in both men and women (Mortimer *et al.*, 1974, 1976).

In one series of 31 patients with functionless pituitary tumours 25 were clinically hypogonadal. Despite this there was some FSH and LH secretion after Gn–RH in all except one patient. However, in 19 patients the LH response was impaired in contrast to the 28 patients who had a normal FSH response (Mortimer *et al.*, 1973*b*). This suggests that in these patients the ability to secrete LH is lost before FSH. Before puberty, FSH secretion in response to Gn–RH is similar to that of adults, but LH responses are less. Transition through puberty is marked by increasing LH responses.

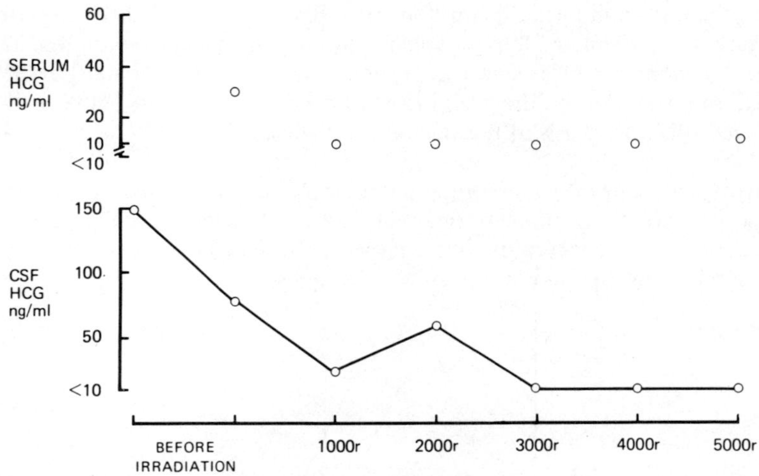

Figure 7.3 Human chorionic gonadotrophin (HCG) concentration in serum and after irradiation of a hypothalamic germinoma in the region of the pineal. This male patient had panhypopituitarism but maintenance of testicular function and a normal circulating testosterone level due to the interstitial cell stimulating activity of the HCG secreted by the tumour. The HCG acted as a 'tumour marker'

It is evident that the Gn–RH test cannot distinguish between primarily hypothalamic and pituitary disease because absent, impaired or normal responses may be seen in either group. It does however establish the 'functional residual capacity' of the pituitary for gonadotrophin secretion, that is the amount of readily available gonadotrophin in the gonadotroph cells. If the response is absent or impaired when Gn-RH is given initially, normal gonadotrophin responses can be obtained after regular injections of this material if the pituitary cells are intact, indicating that Gn-RH can promote synthesis as well as release of gonadotrophins.

Excessive secretion of gonadal steroids, which usually causes precocious puberty, is only rarely associated with tumours of the posterior region of the hypothalamus. It has been suggested that this condition is due to melatonin deficiency or excessive secretion of the gonadotrophin releasing hormone, but this has never been satisfactorily demonstrated. Care must be taken to check for the presence of chorionic gonadotrophin (HCG) rather than LH (HCG may cross-react in an LH radioimmunoassay). This is because, particularly in males, a germinoma of the pineal may secrete HCG (normally only secreted by the placenta) and cause precocious puberty or maintenance of sexual function despite deficiency of all other pituitary hormones. Excessive secretion of LH and FSH due to pituitary tumours is not encountered.

Prolactin

Hypothalamic dopamine, acting as a prolactin inhibiting factor (PIF), tonically inhibits pituitary prolactin secretion (Macleod and Lehmeyer, 1974; Takahara *et al.,* 1974). Hyperprolactinaemia will occur during treatment with drugs which de-

Table 7.2 Drugs which elevate prolactin levels

Dopamine receptor blocking agents	*Dopamine depleting or synthesis-inhibiting agents*	*Agents acting directly on pituitary cells*
Phenothiazines	Reserpine	Oestrogens
Butyrophenones	Alpha-methyl dopa	Thyrotrophin releasing hormone
Pimozide	Alpha-methyl paratyrosine	
Benzamides: metoclopramide sulpiride		

plete the brain of dopamine or which block pituitary dopamine receptors, or with a pituitary stalk or hypothalamic lesion since PIF will not reach the pituitary; there is therefore release from the normal hypothalamic inhibition of pituitary prolactin secretion. However, small or large pituitary tumours may secrete prolactin in excess, and in the absence of obvious radiological changes in the pituitary it is important to be able to identify them by endocrine investigation.

Stimulation of prolactin release by TRH is thought to occur by a direct action on the pituitary. Until recently the secretagogue actions of neuroleptics such as the phenothiazines, metoclopramide and sulpiride (table 7.2), and inhibition of secretion by L-dopa or dopamine, were thought to occur at the hypothalamic level and observations of the differences between the responses to TRH and the neuroleptics were used in attempts to distinguish hypothalamic from pituitary causes of hyperprolactinaemia. However, recent evidence shows that all these agents work at the pituitary level (figure 7.4) even though they may also work within the mid-brain

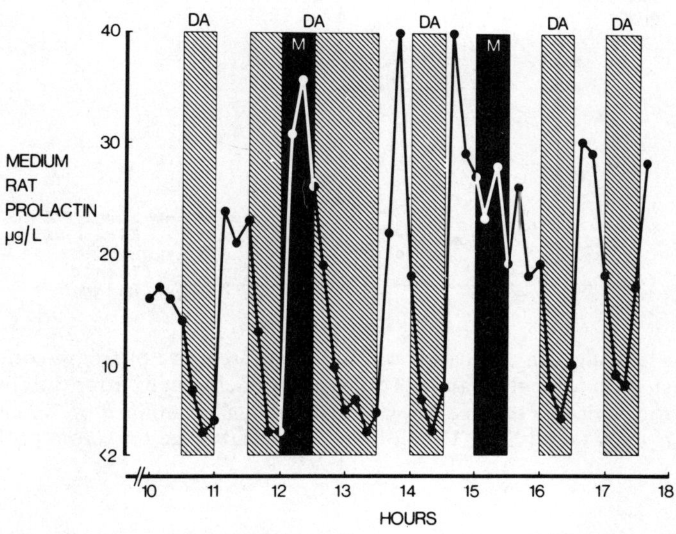

Figure 7.4 Prolactin secretion from a perfused column of isolated rat pituitary cells. Prolactin secretion is inhibited by a direct action of dopamine (DA, 10^{-7} molar) on the cells. This inhibition is blocked by metoclopramide (M, 10^{-5} molar). Thus metoclopramide can act at the pituitary level to inhibit the prolactin-inhibiting activity of dopamine

(Thorner and Besser, 1977). This suggests that the above-mentioned diagnostic pharmacological dissection is invalid.

Hyperprolactinaemia is a common clinical endocrine problem causing infertility and amenorrhoea or menstrual abnormalities in women and impotence in men. The hypogonadism may be accompanied by galactorrhoea. The cause of the gonadal dysfunction in hyperprolactinaemia appears to be a peripheral blockade of the actions of the gonadotrophins at the gonadal level since in the human (unlike the rat) gonadotrophin secretion is not reduced when prolactin rises (Thorner and Besser, 1977). The presence of secretion-inhibiting dopamine receptors on the prolactin secreting cells of the pituitary is of great clinical significance since long-acting dopamine agonists such as bromocriptine, lergotrile, lisuride and piribedil may be used to lower the circulating prolactin levels and normalise the gonadal function in any cause of hyperprolactinaemia (figure 7.5). Bromocriptine is the only agent to have had extensive clinical trial and is highly successful (Besser *et al.*, 1972*b*; Thorner *et al.*, 1974; Thorner and Besser, 1977).

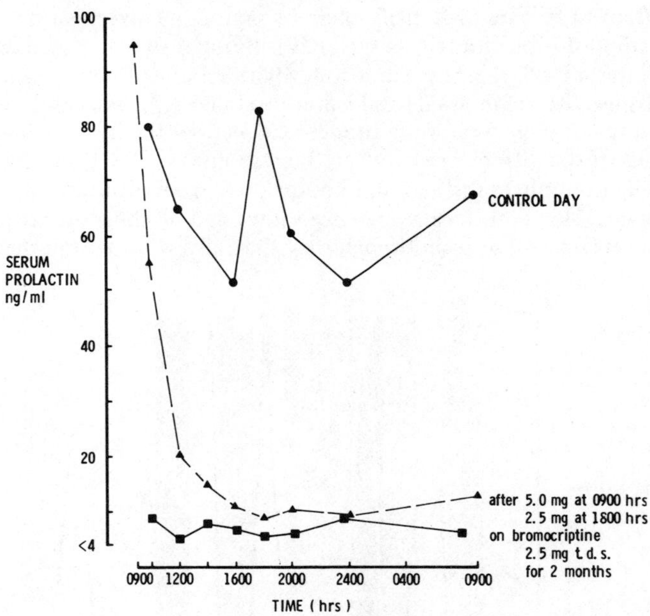

Figure 7.5 Changes in circulating serum immunoreactive prolactin concentration in a woman with idiopathic amenorrhoea and galactorrhoea after the first 5 mg dose of bromocriptine (dotted line) and later when maintained on 7.5 mg of bromocriptine daily (solid line). Upper limit of normal range for serum prolactin 18 ng/ml

Growth Hormone (GH)

Excess secretion causing acromegaly or giantism is usually associated with pituitary tumours, but it is still far from clear whether the primary defect is hypothalamic or pituitary. Evidence in favour of a hypothalamic cause is the abnormal hypothalamic glucostatic control of GH in acromegaly: many patients show a rise in GH with hyperglycaemia, while normal subjects show a fall. However, primary defects in the pituitary are suggested by the abnormal GH release in response to thyrotrophin releasing hormone (TRH) and Gn-RH, an effect not seen in normals (figure 7.6).

Dopamine and other dopamine agonists produce a rise in GH in normal subjects (figure 7.7) but there is frequently a paradoxical fall in GH in acromegaly and giantism (figure 7.8) (Liuzzi *et al.*, 1972; Chiodini *et al.*, 1974). There must

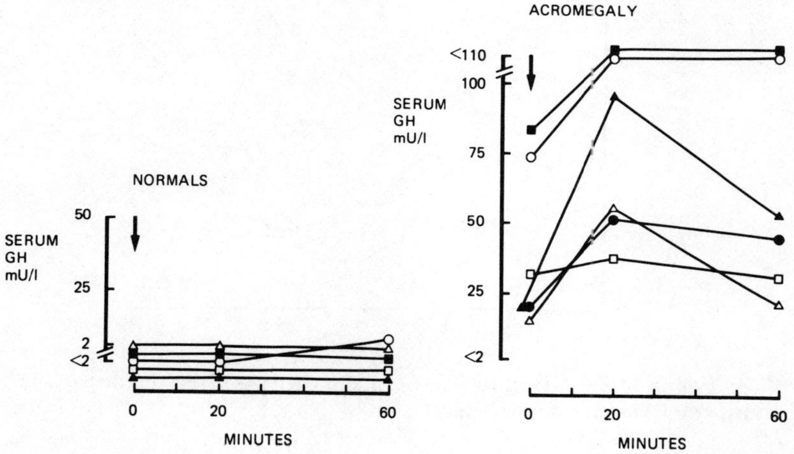

Figure 7.6 Lack of serum GH response to 200 μg TRH administered intravenously at arrow in normal subjects (left); the acromegalic patients (right), however, show a rise in GH levels. Similar effects are often also seen after Gn-RH

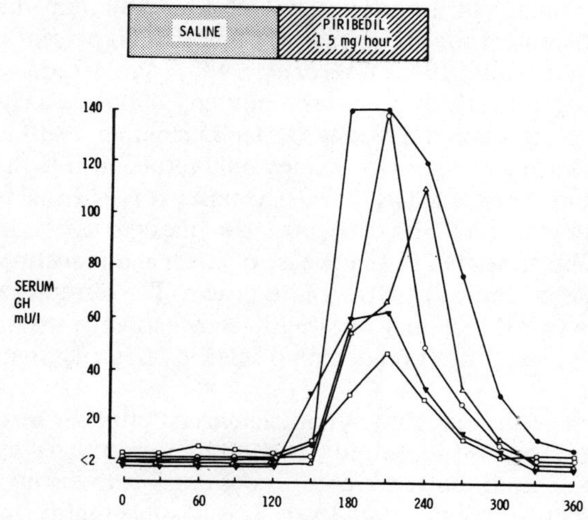

Figure 7.7 Rise in serum GH in five normal subjects given infusions of the dopamine agonist piribedil (1.5 mg/h intravenously)

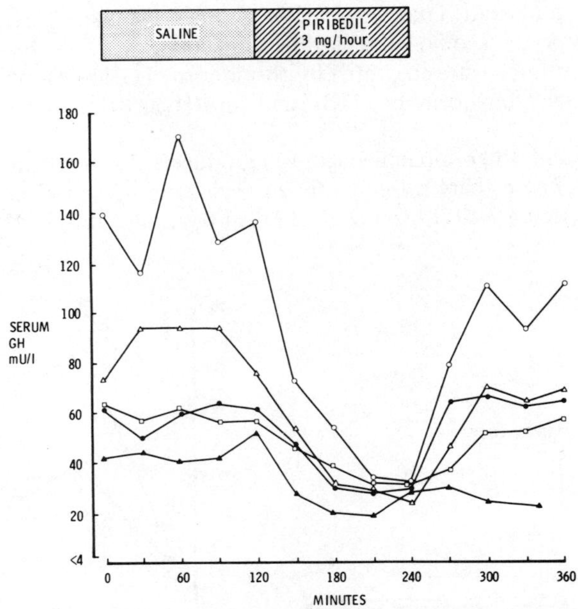

Figure 7.8 Fall in serum GH in five acromegalic patients given the dopamine agonist piribedil (3 mg/h intravenously)

be an abnormality of the dopaminergic mechanisms either in the median eminence (outside the blood-brain barrier for dopamine) or more probably in the pituitary itself in acromegaly, as well as altered receptors for TRH and Gn-RH.

The inhibitory action of dopamine agonists on GH in acromegaly is of great clinical value since the use of the dopamine agonist bromocriptine has provided the first medical treatment for this disease. We find that 80 percent of patients respond well (Thorner *et al.,* 1975; Wass *et al.,* 1977) (figure 7.9).

GH deficiency occurs early in progressive pituitary failure, usually shortly after LH, and the growth retardation is readily treated in children if sufficient quantities of human GH are available. Since GH therapy only results in growth acceleration in cases of short stature when true GH deficiency exists, it is essential to diagnose this condition with great accuracy. The best technique is to assess the GH rise following the combined stimuli to GH release of stress and reduction in blood-sugar following administration of intravenous insulin. The test has the additional advantage that the ACTH secretory reserve can be consistently assessed. Less satisfactory tests involve administration of amino acids, particularly arginine (Edwards and Besser, 1974).

Unfortunately it seems that both hypothalamic and pituitary mechanisms are involved in all the recognised stimuli to GH release—stress, hypoglycaemia, exercise, sleep and amino acids—so at present it is not possible to distinguish by simple hormonal assessment between the two levels. It is possible that further elucidation of the mechanism of the GH response to dopamine may help in this respect.

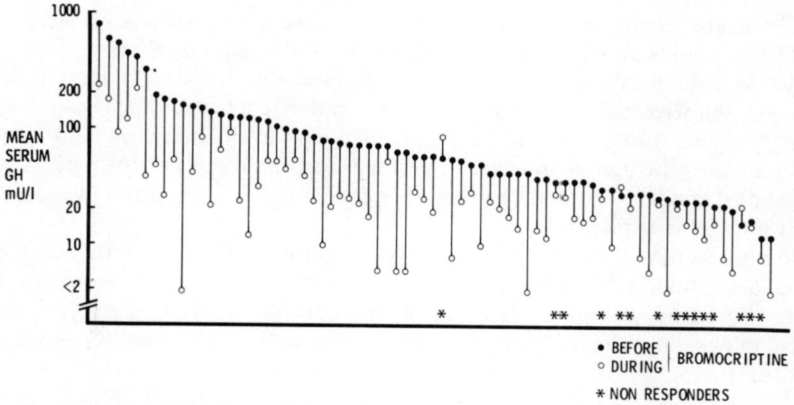

Figure 7.9 Change in serum GH levels (mean of four samples through one day) before (solid symbols) and after (open symbols) 3 to 25 months of treatment with the dopamine agonist bromocriptine (20 to 60 mg/day). The asterisks indicate patients considered as non-responders (after Wass *et al.* (1977), with permission)

Adrenocorticotrophic Hormone (ACTH) and Lipotrophin (LPH)

These hormones are normally secreted together. Neither α- nor β-MSH is secreted in man. α-MSH does not exist in the human post-natally and β-MSH is a sequence within the LPH molecule.

Pituitary-dependent Cushing's disease could be primarily the result of over-stimulation of pituitary ACTH release from the hypothalamus. Evidence in favour of this is obtained from the occasional patient whose disease may be controlled by dopamine agonists or serotonin antagonists. Alternatively the defect could reside primarily in the pituitary–evidence for which is the absence of hyperplastic cells around the microdenomata found in the pituitaries of some patients. The situation is far from clear and clinical evaluation does not help. Overt hypothalamic disease does not result in ACTH over-secretion but many patients with Cushing's disease eventually develop an enlarged pituitary fossa unless pituitary irradiation is given.

Four main stimulation tests have been used to investigate patients with suspected adrenal insufficiency secondary to hypothalamic or pituitary disease: the insulin-induced hypoglycaemia test (Greenwood *et al.,* 1966); the lysine-vasopressin test (Gwinup, 1965; Landon *et al.,* 1965); the metyrapone test (Liddle *et al.,* 1959); and the pyrogen test (Jenkins and Elkington, 1964). Hypoglycaemia and bacterial pyrogens stimulate ACTH secretion only if there is an entirely intact hypothalamic-pituitary portal system with a normal secretory reserve capable of response to stress. Lysine-vasopressin, however, appears to have a direct effect on anterior corticotrophic pituitary cells causing ACTH release, but may also produce a non-specific stress response which acts via the hypothalamus. More recently, glucagon has been introduced as a test of ACTH and growth hormone reserve (Mitchell *et al.,* 1970; Cain *et al.*, 1972).

The relative value of these tests has been assessed by Jacobs and Nabarro (1969) in a series of 102 patients with hypothalamic–pituitary disease. They concluded that the insulin test was clinically the most useful. Properly performed it was safe, sensitive and reliable and had fewer side-effects than lysin-vasopressin or pyrogen. It can also be used as a stimulus to growth hormone release and the result answers the most important question as to whether or not the patient can respond adequately to stress and therefore determines if corticosteroid replacement therapy is required.

If patients have a normal plasma cortisol response to hypoglycaemia they will respond adequately to stress and therefore do not require corticosteroid replacement. In patients who have been treated with corticosteroids, this test is extremely useful in deciding which require corticosteroid cover during surgery (Plumpton *et al.*, 1969).

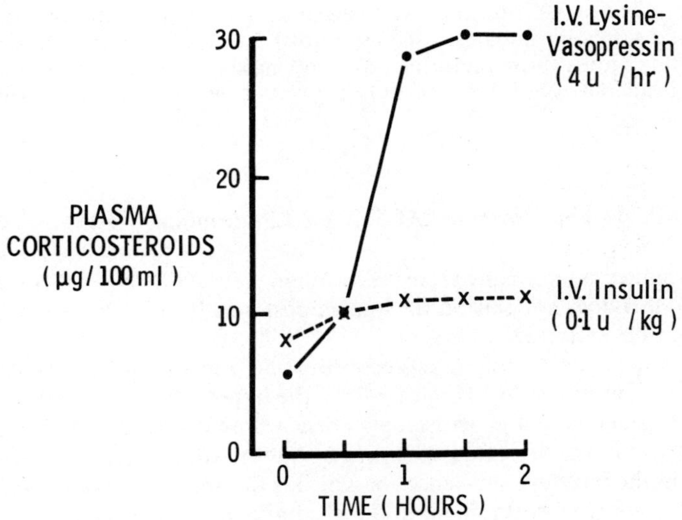

Figure 7.10 Plasma corticosteroid responses in a patient with a craniopharyngioma: impaired response to the stress of insulin-induced hypoglycaemia (dotted line) but a normal response to lysine vasopressin (solid line) acting as a corticotrophin-releasing hormone directly on the pituitary cell (after Edwards and Besser (1974), with permission)

Some patients with clinical hypopituitarism have a normal corticosteroid response to lysine-vasopressin but an inadequate response to hypoglycaemia (figure 7.10). This suggests that their anterior pituitary cells are capable of synthesising but not of releasing ACTH and indicates the presence of a hypothalamic or stalk lesion. The metyrapone test merely tests the responsiveness of ACTH to a fall in cortisol and hence tests the sensitivity to operation of the negative feedback mechanism for ACTH release, not its responsiveness to stress. The pyrogen test is most unpleasant for the patient. The result of the hypoglycaemia test is the clinically relevant one.

Thyrotrophin (TSH)

Assessment of the hypothalamic-pituitary-thyroid axis has been extended beyond simple measurement of circulating thyroid hormone levels by the introduction of the thyrotrophin releasing hormone, TRH. Administration of TRH to normal subjects is followed by a rise in serum TSH levels which are maximal at about 20 min and then gradually fall to pre-TRH levels in about 2 to 3 h. The releasing hormone can be given either orally or intravenously, but much larger doses are required when it is given orally. The test can be performed at any time during the day but should not be done at night. The patients do not need to be fasting or resting and TRH can be given at the same time as other endocrine function tests, such as the insulin tolerance test and the Gn-RH test (Besser *et al.*, 1971; Mortimer *et al.*, 1973*b*).

The response to TRH has now replaced the TSH stimulation test in the differential diagnosis of primary and secondary hypothyroidism, and also the T_3 suppression test used for the diagnosis of thyrotoxicosis. Before doing the TRH test, basal thyroid function should be assessed by measuring the serum thyroxine and the resin uptake of ^{125}I-labelled triiodothyronine. The test has been described by Ormston *et al.* (1971) and Hall *et al.* (1972).

The TSH response to TRH may be normal, absent, impaired or delayed. The response to TRH will be impaired in patients on pharmacological doses of corticosteroids or replacement therapy with triiodothyronine or thyroxine. Primary hypothyroidism is associated with an excessive response and in thyrotoxicosis the response is absent. The typical response in hypothalamic disease is a delayed rise in TSH, whether or not hypothyroidism is present (Hall *et al.*, 1972). Although the delayed response is most commonly found in patients with hypothalamic disease, it is not diagnostic and may also be seen in patients with pituitary disease. Patients with pituitary (secondary) hypothyroidism typically show an impaired or absent response, but this too may be normal. In acromegaly, the response to TRH may appear impaired even though the patient is euthyroid. The frequency of deviation from the typical patterns is so high that the TRH test is often of little value in investigating pituitary and hypothalamic diseases. It does however indicate how much functioning pituitary thyrotroph tissue remains. Diagnosis is further helped since any hypothyroid patient who fails to show an excessive TSH response to TRH will not have primary thyroid disease and the defect must lie in the hypothalamic-pituitary system.

DISORDERS OF POSTERIOR PITUITARY HORMONE SECRETION

Both oxytocin and vasopressin are synthesised in the cell bodies of the supraoptic and paraventricular nuclei of the hypothalamus, and are then transported as neurosecretory granules via the hypothalamic-neurohypophyseal tract to the posterior pituitary where they are stored before being released. In addition to the octapeptides, these granules also contain a binding protein called neurophysin (Ginsburg and Ireland, 1963; Hollenberg and Hope, 1968). Specific neurophysins exist for both oxytocin and vasopressin, and these proteins are released into the circulation at the same time as the octapeptides (Fawcett *et al.*, 1968; Cheng and Freisen, 1971).

The development of radioimmunoassays for neurophysin (Legros *et al.*, 1969; Cheng and Friesen, 1971; Martin *et al.*, 1972) has shown that neurophysin is released into the circulation in response to dehydration, hypertonic saline, parturition and suckling, and suggests that specific assays for neurophysin may be useful in the investigation of posterior pituitary function. Such assays are much easier to establish and perform than radioimmunoassays for oxytocin and vasopressin.

Most of the investigations of vasopressin secretion are based on indirect parameters of neurohypophyseal function, such as changes in the rate of urine flow, osmolality or free water clearance, rather than direct measurement of the hormone released. This is because of the major difficulties involved in measuring the hormone. The assay of choice is the ethanol anaesthetised rat but this assay has a limited sample capacity and requires considerable practice before reliable results can be obtained. Radioimmunoassays have been developed for vasopressin (Klein *et al.*, 1966; Beardwell and Wright, 1968; Robertson *et al.*, 1970; Edwards *et al.*, 1970, 1972), but there are still major technical problems and they must therefore be regarded at present as research procedures.

There is very little information about pathophysiological abnormalities in oxytocin secretion. The only aspect which has been looked at carefully is the outcome of pregnancy in patients with diabetes insipidus. All the evidence suggests that labour is usually normal in these patients and this is in keeping with the data that the foetal rather than maternal posterior pituitary may play the dominant role in parturition (Chard *et al.*, 1971). One patient with a pancreatic carcinoma has been shown to have ectopic oxytocin secretion (Marks *et al.*, 1968).

Cranial Diabetes Insipidus

The major disease which results from impaired function of the neurohypophysis is diabetes insipidus (DI) in which there is inadequate secretion of the antidiuretic hormone, arginine vasopressin (AVP). This results in polyuria which, in the presence of a normal thirst centre and an unimpaired level of consciousness, precipitates polydipsia. The disease has been classified into primary and secondary groups, of which the former consists of familial and idiopathic cases. Familial DI usually has a simple dominant mode of inheritance. The secondary group comprises cases resulting from trauma caused by neurosurgery or a head injury, or from primary or secondary neoplasia. Other rare causes include sarcoidosis and other granulomata, histocytosis and basal meningitis. In general the higher the stalk lesion the greater is the chance of DI developing.

Diabetes insipidus has to be distinguished from other causes of polyuria and polydipsia. These can be divided into those in which vasopressin secretion is decreased, such as in psychogenic polydipsia, and those in which the renal tubules do not respond normally to vasopressin. This latter group includes nephrogenic diabetes insipidus, patients with diabetes mellitus and some patients with chronic renal failure.

The simplest, least unpleasant and most important procedure for the diagnosis of DI is the water-deprivation test (Dashe *et al.*, 1963; Price and Lauener, 1966). This is safe providing the period of water deprivation does not exceed eight hours and consideration is given to stopping the test if the patient loses more than three

Table 7.3 Results of water deprivation tests in eight normal subjects (after Edwards and Besser, 1974, with permission)

Time	*Urine specimen*	*Mean urine osmolality* ±1 s.d. (mOsm/kg)	*Range of urine osmolality* (mOsm/kg)	*Plasma specimen*	*Mean plasma osmolality* ±1 s.d. (mOsm/kg)	*Range of plasma osmolality* (mOsm/kg)
0–1 h	U_1	440 ± 265.5	155–889	P_1	285 ± 2.5	280–288
3–4 h	U_2	724 ± 118.5	562–880	P_2	285 ± 2.6	281–289
6–7 h	U_3	915 ± 118.3	774–1160	P_3	287 ± 2.3	283–289
7–8 h	U_4	962 ± 100.0	828–1158	P_4	291 ± 2.5	286–294

Table 7.4 Water deprivation tests in six patients with cranial diabetes insipidus* (after Edwards and Besser, 1974, with permission)

Time	*Urine specimen*	*Mean urine osmolality* ±1 s.d. (mOsm/kg)	*Range of urine osmolality* (mOsm/kg)	*Plasma specimen*	*Mean plasma osmolality* ±1 s.d. (mOsm/kg)	*Range of plasma osmolality* (mOsm/kg)
0–1 h	U_1	116 ± 3.14	83–162	P_1	297 ± 5.5	289–305
3–4 h	U_2	156 ± 67.4	84–245	P_2	307 ± 4.1	301–313
6–7 h	U_3	221 ± 78.0	94–304	P_3	308 ± 7.4	303–319
7–8 h	U_4	221 ± 57.9	207–337	P_4	306 ± 3.5	302–310

*In one patient the test was stopped after 4 h and in another after 6 h

per cent of body weight (Edwards and Besser, 1974). If the patient fails to concentrate urine during the test the ability to respond to exogenous vasopressin should be tested.

The urine and plasma osmolalities measured during an 8 h water deprivation in eight normal subjects are shown in table 7.3. At the end of the test a normal subject produces a urine with an osmolality greater than 800 mOsm/kg and the plasma osmolality does not rise above 294 mOsm/kg. This is in marked contrast to the results in the six patients with cranial diabetes insipidus shown in table 7.4. In these the urine osmolality does not exceed the plasma osmolality, and the plasma osmolality at the end of the test is greater than 300 mOsm/kg.

Patients with nephrogenic DI show similar responses to water deprivation as patients with cranial DI. However, unlike those with cranial DI they do not concentrate their urine after endogenous vasopressin.

Patients with psychogenic polydipsia can usually be easily distinguished from those with diabetes insipidus. The plasma osmolality at the start of the test is usually normal or low, unlike the patients with DI. The urine osmolality rise may be impaired, as is the response to exogenous vasopressin (de Wardener and Herxheimer, 1957).

Partial defects in vasopressin secretion may be difficult to identify with the routine water deprivation test, but may be further investigated using a procedure suggested by Miller *et al.*, (1970). The patient is deprived of water until the osmolality of urine samples voided hourly reaches a plateau, with the osmolality not increasing by more than 30 mOsm/kg between two consecutive collections. Vasopressin is then injected subcutaneously and urine collected for one further hour. The period of dehydration required to reach a constant urine osmolality ranges from 4 to 18 h.

In normal subjects the urine osmolality does not increase more than five per cent after the injection of vasopressin, whereas patients with partial diabetes insipidus show a further increase in urine osmolality after vasopressin (9–67 per cent). In psychogenic polydipsia, patients show a similar response to normal subjects.

The Syndrome of Inappropriate Vasopressin Secretion

This syndrome can result either from ectopic secretion of vasopressin (ADH) by certain tumours (usually an oat-cell carcinoma of the bronchus), or from alterations in its secretion, metabolism or antidiuretic actions by a variety of drugs and other disease processes (most commonly chest infections, cerebral tumours, infections or trauma, and chlorpropamide therapy). The syndrome can be diagnosed if a patient with normal renal and adrenal function has hyponatraemia with decreased serum osmolality with an inappropriate urinary osmolality (where the urine is not maximally dilute). In some cases a continuing renal salt loss also occurs (Schwartz *et al.*, 1957). The clinical presentation is dependent on the severity of the hyponatraemia. Despite the increase in total body water there is usually no oedema.

If diagnostic uncertainty exists the patient may be given one litre of water to drink over 20 min. Normal subjects dilute their urine to less than the osmolality of their plasma forming free water in the urine and excrete the water load. In inappropriate ADH secretion patients dilute their plasma further, fail to render their urine less concentrated than plasma, do not form free water and do not excrete the water load (Edwards and Besser, 1974).

REFERENCES

Anderson, D. C., Marshall, J. C., Young, J. L. and Fraser, T. R. (1972). Stimulation tests on pituitary–Leydig cell function in normal male subjects and hypogonadal men. *Clin. Endocrinol.*, **1**, 127–140

Beardwell, C. G. and Wright, A. D. (1968). The radioimmunoassay of arginine-vasopressin. In C. Gual (ed.). *International Congress Series,* **157**, Excerpta Medica, Amsterdam, 48

Besser, G. M. (1974). Hypothalamus as an endocrine organ: Part I. *Brit. Med. J.*, **3**, 560–564

Besser, G. M., McNeilly, A. S., Anderson, D. C., Marshall, J. C., Harsoulis, P., Hall, R., Ormston, B. J., Alexander, L. and Collins, W. P. (1972*a*). Hormonal responses to synthetic luteinizing hormone and follicle stimulating hormone-releasing hormone in man. *Brit. Med. J.*, **3**, 267–271

Besser, G. M., Parke, L., Edwards, C. R. W., Forsythe, I. A. and McNeilly, A. S. (1972*b*). Galactorrhoea–successful treatment with reduction of plasma prolactin levels by Bromergocryptine. *Brit. Med. J.*, **3**, 669–672

Besser, G. M., Ratcliffe, J. G., Kilborn, J. B., Ormston, B. J. and Hall, R. (1971). Interaction between thyrotrophin, corticotrophin and growth hormone secretion in man. *J. Endocrinol.*, **51**, 699–706

Cain, J. P., Williams, G. H. and Dluhy, R. G. (1972). Glucagon-initiated human growth hormone release: a comparitive study. *Can. Med., Ass. J.*, **107**, 617–622

Chard, T., Hudson, C. N., Edwards, C. R. W. and Boyd, N. R. H. (1971). Release of oxcytocin and vasopressin by the human foetus during labour. *Nature,* **234**, 352–354

Cheng, K. W. and Friesen, H. G. (1971). A radioimmunoassay for vasopressin binding proteins--neurophysin. *Endocrinology,* **88**, 608–619

Chiodini, P. G., Liuzzi, A., Botalla, L., Cremascoli, G. and Silvestrini, F. (1974). Inhibitory effect of dopaminergic stimulation on GH release in acromegaly. *J. Clin. Endocrinol. Metab.*, **38**, 200–206

Dashe, A. M., Cramm, R. E., Crist, C. A., Habener, J. F. and Solomon, D. H. A. (1963). A water deprivation test for the differential diagnosis of polyuria. *J. Am. Med. Ass.*, **185**, 699–703

Edwards, C. R. W. and Besser, G. M. (1974). Diseases of the hypothalamus and pituitary gland. In R. I. S. Bayliss (ed.). *Clinics in Endocrinology and Metabolism.* **3**, 475–505

Edwards, C. R. W., Chard, T., Kitau, M. J. and Forsling, M. L. (1970). The development of a radioimmunoassay and a plasma extraction method for vasopressin. *J. Endocrinol.*, **48**, xi–xii

Edwards, C. R. W., Chard, T., Kitau, M. J., Forsling, M. L. and Landon, J. (1972). The development of a radioimmunoassay for arginine-vasopressin: production of antisera and labelled hormone; separation techniques; specificity and sensitivity of the assay in aqueous solution. *J. Endocrinol.*, **52**, 279–288

Fawcett, C. T., Powell, A. E. and Sachs, H. (1968). Biosynthesis and release of neurophysin. *Endocrinology,* **83**, 1299–1310

Ginsburg, M. and Ireland, M. (1963). The hormone-binding protein of the neurohypophysis. *J. Physiol.*, **169**, 15P

Greenwood, F. C., Landon, J. and Stamp, T. C. (1966). The plasma sugar, free fatty acid, cortisol, and growth hormone response to insulin. Part I: in control subjects. *J. Clin. Invest.*, **45**, 429–436

Gwinup, G. (1965). Studies on the mechanism of vasopressin-induced steroid secretion in man. *Metabolism,* **14**, 1282–1286

Hall, R., Ormston, B. J., Besser, G. M., Cryer, R. J. and McKendrick, M. (1972). The thyrotrophin-releasing hormone test in diseases of the pituitary and hypothalamus. *Lancet,* **i**, 759–763

Hollenberg, M. D. and Hope, D. B. (1968). The isolation of the native hormone-binding proteins from bovine pituitary posterior lobes. *Biochem. J.,* **106**, 557–564

Jacobs, H. A. and Nabarro, J. D. (1969). Tests of hypothalamic–pituitary–adrenal function in man. *Q. J. Med.,* **38**, 475–491

Jenkins, J. S. and Elkington, S. E. (1964). Metyrapone and pyrogen in the assessment of pituitary–adrenal function after removal of pituitary adenoma. *Lancet,* **ii**, 991–994

Klein, L. A., Roth, J. and Petersen, M. J. (1966). Radioimmunoassay of arginine vasopressin (antidiuretic hormone). *S. Forum,* **17**, 240–242

Landon, J., James, V. H. T. and Stoker, D. J. (1965). Plasma-cortisol reponse to lysine-vasopressin. *Lancet,* **ii**, 1156–1159

Legros, J. J., Franchimont, P. and Hendrick, J. C. (1969). Dosage radioimmunologique de la neurophysine dans le sérum des femmes normales et des femmes enceintes. *C. R. Soc. Biol.,* **163**, 2773–2777

Liddle, G. W., Este, P. H. I., Kendall, J. W., Williams, W. C. and Towns, A. W. (1959). Clinical application of a new test of pituitary reserve. *J. Clin. Endocrinol. Metab.,* **19**, 875–894

Liuzzi, A., Chiodini, P. G., Botalla, L., Cremoscoli, G., Muller, E. E. and Silvestrini, F. (1972). Inhibitory effect of L-dopa on GH release in acromegalic patients. *J. Clin. Endocrinol. Metab.,* **35**, 941–943

Macleod, R. M. and Lehmeyer, J. E. (1974). Studies on the mechanism of the dopamine-mediated inhibition of prolactin secretion. *Endocrinology,* **94**, 1077–1085

Marks, L. J., Berde, B., Klein, L. A., Roth, J., Goonan, S. R., Blumen, D. and Nabseth, D. C. (1968). Inappropriate vasopressin secretion and carcinoma of the pancreas. *Am. J. Med.,* **45**, 967–974

Martin, M. J., Chard, T. and Landon, J. (1972). The development of a radioimmunoassay for bovine neurophysin. *J. Endocrinol.,* **52**, 481–495

Miller, M., Dalakos, T., Moses, A. M., Fellerman, H. and Streeten, D. H. P. (1970). Recognition of partial defects in antidiuretic hormone secretion. *Ann. Int. Med.,* **73**, 721–729

Mitchell, M. L., Byrne, M. J., Sanchez, Y. and Sawin, C. T. (1970). Detection of growth-hormone deficiency. The glucagon stimulation test. *New Engl. J. Med.,* **282**, 539–541

Mortimer, C. H., Besser, G. M., McNeilly, A. S., Marshall, J. C., Harsoulis, P., Tunbridge, W. M. G., Gomez-Pan, A. and Hall, R. (1973*a*). Luteinizing hormone and follicle stimulating hormone-releasing hormone test in patients with hypothalamic pituitary-gonadal dysfunction. *Brit. Med. J.,* **4**, 73–77

Mortimer, C. H., Besser, G. M., McNeilly, A. S., Tunbridge, W. M. G., Gomez-Pan, A. and Hall, R. (1973*b*). Interaction between secretion of the gonadotrophins, prolactin, growth hormone, thyrotrophin and corticosteroids in man: the effects of LH/FSH-RH, TRH and hypoglycaemia alone and in combination. *Clin. Endocrinol.,* **2**, 317–326

Mortimer, C. H., McNeilly, A. S. and Besser, G. M. (1976). Gonadotrophin releasing hormone therapy. *Ann. Biol. Anim. Bioch. Biophys.,* **16**, 235–243

Mortimer, C. H., McNeilly, A. S., Murray, M. A. F., Fisher, R. A. F. and Besser, G. M. (1974). Gonadotrophin-releasing hormone therapy in hypogonadal males with hypothalamic or pituitary dysfunction. *Brit. Med. J.,* **4**, 617–621

Ormston, B. J., Garry, R., Cryer, R. J., Besser, G. M. and Hall, R. (1971). Thyrotrophin releasing hormone as a thyroid-function test. *Lancet,* **ii**, 10–14

Plumpton, F. S., Besser, G. M. and Cole, P. V. (1969). Corticosteroid treatment and surgery. Part I: An investigation of the indications for steroid cover. *Anaesthesia,* **24**, 3–11

Price, J. D. E. and Lauener, R. W. (1966). Serum and urine osmolalities in the differential diagnosis of polyuric states. *J. Clin. Endocrinol. Metab.,* **26**, 143–148

Robertson, G. L., Klein, L. A., Roth, J. and Gorden, P. (1970). Immunoassay of plasma vasopressin in man. *Proc. Natn., Acad. Sci., Washington,* **66**, 1298–1305

Schwartz, W. B., Bennett, W., Curelop, S. and Bartter, F. C. (1957). A syndrome of renal sodium loss and hyponatremia probably resulting from inappropriate secretion of antidiuretic hormone. *Am. J. Med.,* **23**, 529–542

Takahara, J., Arimura, A. and Schally, A. V. (1974). Suppression of prolactin release by a purified porcine PIF preparation and catecholamines infused into a rat hypophyseal portal vessel. *Endocrinology*, **95**, 462–465

Thorner, M. O. and Besser, G. M. (1977). Clinical significance of dopaminergic mechanisms in the hypothalamus and pituitary. In J. L. H. O'Riordan (ed.). *Recent Advances in Endocrinology and Metabolism.* Churchill Livingston, London, 1–17

Thorner, M. O., Chait, A., Aitken, M., Benker, G., Bloom, S. R., Mortimer, C. H., Sanders, P., Stuart Mason, A. and Besser, G. M. (1975). Bromocryptine treatment of acromegaly. *Brit. Med. J.,* **1**, 299–303

Thorner, M. O., McNeilly, A. S., Hagan, C. and Besser, G. M. (1974). Long-term treatment of galactorrhoea and hypogonadism with bromocryptine. *Brit. Med. J.,* **2**, 419–422

de Wardener, M. E. and Herxheimer, A. W. (1957). The effect of a high water intake on the kidney's ability to concentrate the urine in man. *J. Physiol.,* **139**, 42–52

Wass, J. A. H., Thorner, M. O., Morris, D. V., Rees, L. H., Stuart Mason, A., Jones, A. E. and Besser, G. M. (1977). Long-term treatment of acromegaly with bromocryptine. *Brit. Med. J.,* **1**, 875–878

Discussion

Foy (Bradford)
I believe that a large number of acromegalics are also diabetic. Does the treatment which you describe for acromegaly also ameliorate or cure the diabetes?

Besser
Of 73 patients with acromegaly whom we treated with bromocriptine, 23 were diabetics. Of these 15 were cured, and in 5 others glucose tolerance was substantially improved. Thus the answer to your question is yes, it makes a dramatic difference. It is interesting that insulin levels are almost unchanged and the effect is due to a reduction of the anti-insulin action of growth hormone and somatamedin.

Daly (Ware)
There have been clinical reports of increased serum prolactin levels during treatment with histamine H_2 receptor antagonists. Do you have any comments on the mechanisms involved or on the role of histamine receptors in this system?

Besser
This is currently under investigation. In our experiments we have not seen any rise in prolactin levels. It may be a question of dose or duration of treatment but we have no evidence to support the report to which you refer.

Flack (Harlow)
Firstly, is metoclopramide really as extensively prescribed as aspirin? Secondly, I imagine that you and many other clinical investigators are using metoclopramide because it is the most specific of the available dopamine antagonists. Also it does not have other highly undesirable CNS effects such as those seen with chlorpromazine and similar drugs. Thirdly, on the question of your experiments with the normal volunteers who showed hyperprolactinaemia and impotence after metoclopramide, is it not true that after the dopamine receptor blocker the androgen levels did not change at all? If you are suggesting that these patients were infertile we should need to see some data on this.

Besser
There is no doubt that metoclopramide is prescribed very widely in general practice and that it has undesirable effects which were not anticipated. Hyperprolactinaemia in the male is not necessarily associated with low testosterone but it is associated with impotence. Hyperprolactinaemia in the male is usually associated with testosterone levels within or at the lower end of the normal range and responds to treatment fairly quickly. Somehow, giving dopamine agonists which lower prolactin levels dramatically cures the impotence. I do believe that the effects of metoclopramide on these people were related to the hyperprolactin-

aemia, or at least to the administration of the dopamine receptor blocker. Finally, metoclopramide does have central sedative effects, even though these may not have been well documented.

Martini
In Italy metoclopramide is sold more widely than aspirin.

Besser
In countries outside Britain, where sulpiride is among the most commonly prescribed drugs, similar effects to those of metoclopramide are produced since sulpiride is just a chemical modification of metoclopramide.

Fraser (Edinburgh)
You mentioned possible future therapy using long-acting LH-RH stimulator analogues. It has been shown that the LH response to repeated injection of these analogues can, in fact, decline with time. This might be due to the effect on pituitary LH-RH receptors or feedback of gonadal steroids but it is obviously an important one.

Besser
This point is very important. If you give a continuous infusion of these analogues you deliver a great deal of LH-RH to the pituitary and you can deplete it of LH and FSH. Initially LH and FSH levels are high and then they fall off and you get much lower levels than you started with. Similarly, with TRH you can get a surge of TSH followed by very low levels, and if you give LH-RH too frequently the same thing happens. I am absolutely certain that if the administration of long-acting analogues is too frequent you will obtain anti-fertility effects and Professor Schally has just shown us some evidence for that. In our treatment regime we give 500 μg LH-RH every 8 h because we know it only works for 5 to 6 h. The 'off' period is to enable the hormones to shift from the non-releasable to the readily releasable pool. This is because LH-RH only seems to release from the readily releasable pool. One must be very careful in using the long-acting analogues too frequently and a 12 h gap between the end of one injection and the next injection might be appropriate.

Milson (Newcastle upon Tyne)
In a recent trial in Newcastle using the beta-adrenoceptor antagonist tolamolol it was found that there were increased levels of prolactin in some patients. I wondered if you had experienced any similar changes with beta-antagonists?

Besser
We have not studied this aspect very intensively and certainly have not noticed changes in prolactin levels. One has to be very careful to ensure that there are adequate controls because prolactin secretion is very sensitive to stress; it is also secreted in pulses.

8
The hypothalamus as an endocrine target organ

L. Martini*

INTRODUCTION

The concept that the brain may be a target for hormone action originated with the pioneer studies of Berthold, who, in 1849, reported the striking changes in aggressive and sexual behaviour occurring in cockerels after castration and concluded that testicular secretions may influence the brain.

It is now abundantly clear that ovarian, testicular, adrenal and thyroid hormones may influence several aspects of brain function and of behaviour. More recently, data have appeared which indicate that some pituitary and hypothalamic hormones may also modify cerebral functions. In the last few years several laboratories have obtained important information on: (1) the anatomical distribution of hormones in the brain and in the anterior pituitary; (2) the existence and the intracellular localisation of specific binding sites for several hormones; and (3) the occurrence of important metabolic transformations of steroid hormones in several neuroendocrine tissues.

The data referring to sex steroids will be summarised in this chapter. For other hormones the reader is referred to the recent reviews by Moss (1976), Celotti *et al.* (1978), McEwen (1978) and Piva *et al.* (1978).

TOPOGRAPHY OF SEX STEROID RECEPTORS IN THE CENTRAL NERVOUS SYSTEM AND IN THE ANTERIOR PITUITARY

Oestrogens

In the rat, mouse, hamster, guinea-pig and other vertebrates putative receptor sites for oestradiol and other oestrogens are found in the anterior pituitary (Notides, 1970) and in a number of brain regions including the basomedial hypothalamus, medial preoptic area, corticomedial amygdala, midbrain central grey, ventral

*Department of Endocrinology, University of Milan, Via Andrea del Sarto 21, 20129 Milano, Italy

hippocampus, spinal cord and cerebellum (McEwen *et al.*, 1972; Keefer *et al.*, 1973; Morrell *et al.*, 1975; Stumpf *et al.*, 1975; Krieger *et al.*, 1976; Fox, 1977). Oestrogen receptors isolated from the brain or the anterior pituitary are similar to those found in the uterus, having a sedimentation coefficient in sucrose density gradients of 8S at low ionic strength, a high affinity ($K_d \approx 10^{-10}$M) for oestradiol-17β, and a preference for active oestrogens such as oestradiol-17β and diethylstilboestrol (Notides, 1970; Mowles *et al.*, 1971; Plapinger and McEwen, 1973; Armstrong and Villee, 1977). The cell nuclear receptor is reported to have a sedimentation coefficient of 5–7S (Mowles *et al.*, 1971; Vertes and King, 1971), whilst the uterine nuclear receptor is generally believed to sediment at 5S (Jensen and de Sombre, 1973).

Androgens

Binding of labelled testosterone by the anterior pituitary and by the brain of the rat is less intense than that of oestradiol. However, the same brain regions are usually labelled by the two hormones (Sar and Stumpf, 1975). Autoradiographic studies reveal that only 15 per cent of rat pituitary cells (mostly basophils) are labelled after the injection of labelled testosterone (Sar and Stumpf, 1973*a*). In the medial preoptic area of the rat a majority of the cells are labelled by both testosterone and oestradiol (Sar and Stumpf, 1973*b*). This indicates either an overlap of the cells binding the two hormones or a local conversion of testosterone into oestradiol (see below). The brain of the male rat is known to contain oestrogen receptors, with a topography closely resembling that of the female rat brain; however, the number of oestrogen-binding sites seems to be lower in males than in females (Maurer and Woolley, 1974; Barley *et al.*, 1977).

Several laboratories have presented evidence for the presence in the brain and in the anterior pituitary of soluble macromolecules binding testosterone and 5α-androstan-17β-ol-3-one (dihydrotestosterone, DHT), a functional metabolite of testosterone (Ginsburg *et al.*, 1971; Jouan *et al.*, 1973; Naess *et al.*, 1975; Kato, 1976; Armstrong and Villee, 1977; Lieberburg *et al.*, 1977). It is interesting to note that apparently specific receptors for 5α-androstan-3α, 17β-diol (3α-diol), another important metabolite of testosterone and DHT formed in the brain and in the anterior pituitary, are lacking in central structures (Mercier *et al.*, 1976).

Progesterone

The evidence for specific uptake and binding of progesterone in the brain and in the anterior pituitary has remained conflicting and unclear until very recently. Several investigators have reported equivocal uptake of progesterone by the central structures of the rat, mainly in the hypothalamus and in the midbrain region (Seiki *et al.*, 1969; Whalen and Luttge, 1971; Wade *et al.*, 1973; Sar and Stumpf, 1973*c*; Atger *et al.*, 1974), while others have claimed the existence of specific progesterone binding sites in the central tissues of rats and guinea-pigs (Laumas and Farooq, 1966; Seiki and Hattori, 1971, 1973). It must be recalled that the study of specific progesterone binding sites is complicated both by the demonstration that progesterone and corticoids may bind to the same cytosol proteins, and by the fact that progesterone may undergo a variety of metabolic transformations in the brain and in the anterior pituitary (see below).

However, the results of recent experiments seem to have clarified the stituation. Kato (1975) and Seiki *et al.* (1977) have reported that progesterone-binding components are present in the cytosol of the hypothalamus and of the anterior pituitary of the rat. High-affinity and low-capacity progesterone binding components have also been found in the hypothalamus of guinea-pigs and rabbits by Iramain *et al.* (1973). Finally Kato and Onouchi (1977) have detected specific progesterone binding components in hypothalamic and anterior pituitary cytosol from oestrogen-primed immature and mature female rats using the synthetic progestin R 5020 which binds specifically to progesterone receptors. Little binding was detected in similar preparations of the cerebral cortex, the amygdala or the reticular formation. It is interesting that 5α-pregnan-3, 20-dione (dihydroprogesterone, DHP), a metabolite formed from progesterone in the brain and in the anterior pituitary (see below), is also specifically linked by the same binding component.

METABOLISM OF SEX STEROIDS IN THE CENTRAL NERVOUS SYSTEM AND IN THE ANTERIOR PITUITARY

Oestrogens

Oestrogens undergo considerable metabolism in the CNS and in the anterior pituitary. Three major processes which are typical of oestrogen metabolism in other organs (for example the liver and placenta) have also been clearly demonstrated in neuroendocrine structures. These are: (1) hydrolysis of oestrone sulphate; (2) interconversion of oestrone and oestradiol; and (3) hydroxylation of position 2 in the steroid nucleus to give 2 hydroxy-oestrogens or catecholoestrogens (see figure 8.1).

The anterior pituitary, the hypothalamus and the cerebral cortex of the ewe (Payne *et al.*, 1973; Kazama and Longcope, 1974) and of the rat (Kishimoto, 1973) possess sulphatase activity, which allows for the formation of free oestrone from oestrone sulphate. In foetal ewes the sulphatase activity is higher in the anterior pituitary than in the hypothalamus (Jenkin *et al.*, 1975). In the sheep castration decreases the deconjugation of oestrone sulphate in the hypothálamus and in the cerebral cortex, but does not affect its hydrolysis in the anterior pituitary (Payne *et al.*, 1973).

A 17β-oxidoreductase is present both in the CNS (hypothalamus, limbic system and cerebral cortex) and in the anterior pituitary of all animal species (rat, sheep, rhesus monkey) so far studied (Luttge and Whalen, 1970; Flores *et al.*, 1973; Jenkin *et al.*, 1975). This permits the interconversion of oestrone and oestradiol, a process which is more active in the anterior pituitary (Payne *et al.*, 1973) than in CNS structures (Reddy *et al.*, 1974*a*). The conversion of oestradiol to oestrone is much more active in the foetal than in the adult anterior pituitary (Jenkin *et al.*, 1975) and castration does not affect such a process (Payne *et al.*, 1973; Jenkin *et al.*, 1975).

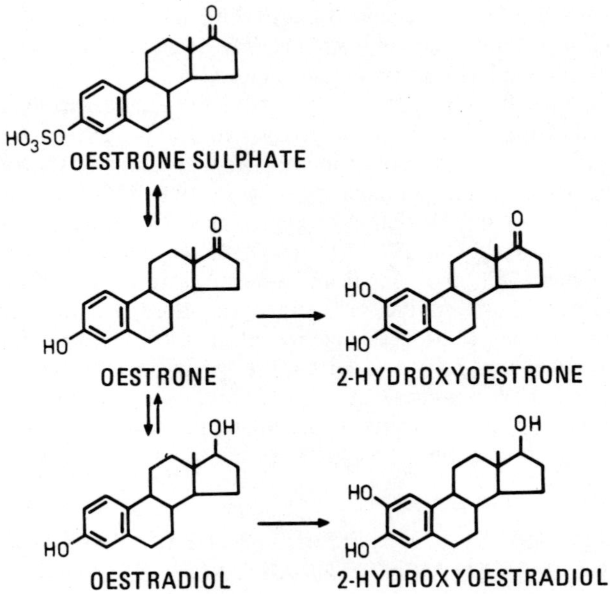

Figure 8.1 Metabolism of oestrogens

Sufficient data are available to indicate that hydroxylation of oestrogens in position 2, leading to the formation of catecholestrogens, occurs in neuroendocrine structures. Depending on the precursor used, 2-hydroxy-oestrone or 2-hydroxy-oestradiol is formed. In normal adult animals this process has been shown to occur in the hypothalamus, but it is minimal in the cerebral cortex and in the anterior pituitary (Fishman and Norton, 1975; Fishman, 1976). Ovariectomy increases the yields in all tissues but especially in the anterior pituitary (Fishman and Norton, 1975; Fishman, 1976). Catecholoestrogens are also formed from oestradiol—in descending order by the hypothalamus, the cerebral cortex and the anterior pituitary of the human foetus (Fishman, 1976; Fishman *et al.*, 1976). The biosynthesis of catecholoestrogens in specific tissues involved in the control of neuroendocrine phenomena may have important physiological implications. The data to support this statement are still rather scanty since catecholoestrogens are very unstable compounds. However, preliminary evidence is available which indicates that they may affect gonadotrophin secretion. Catecholoestrogens have been reported to induce an increase in plasma gonadotrophins in immature male rats (Naftolin *et al.*, 1975*a*) and to lower gonadotrophin secretion when administered into the amygdala of miniature pigs (Parvizi and Ellendorff, 1975). These data are not necessarily in conflict because of the species differences and routes of administration.

Catecholoestrogens may affect gonadotrophin secretion either by their ability to bind to oestrogen receptors or by an interference with catecholamine metabolism (data are available in support of each of these hypotheses). Catecholoestrogens bind to hypothalamic and pituitary oestrogen receptors (Davies *et al.*, 1975), and this binding confers antioestrogenic properties to these biologically inert molecules

(Gordon *et al.*, 1964). On the other hand, catecholoestrogens are excellent competitive inhibitors of the O-methylation of catecholamines by catechol-O-methyltransferase (Ball *et al.*, 1972; Bolt and Kappus, 1976), and it is well known that brain catecholamines are involved in the control of gonadotrophin secretion (Fuxe and Hökfelt, 1969; Weiner, 1975).

It is obvious that the metabolic processes described in the preceding paragraphs may occur not only on oestrogens which have arrived in central tissues from the periphery, but also on those formed *in situ* by the local aromatisation of androgens (see below).

Androgens

5α-reduction and aromatisation

Testosterone and other androgens undergo extensive metabolic conversion in the CNS and in the anterior pituitary. Two major enzymatic systems have been identified: the 5α-reductase pathway and the aromatising pathway. For clarity these metabolic processes will be discussed separately. However, it must be understood that they operate concomitantly, or even in an additive fashion.

The biochemical processes through which testosterone and other androgens (such as androstenedione) may be transformed into DHT and 3α-diol in the central structures of male animals are summarised in figure 8.2, which also shows the enzymes involved.

The anterior pituitary of adult male animals converts testosterone into DHT and 3α-diol with yields which are second only to those recorded in the prostate and in the seminal vesicles (Kniewald *et al.*, 1971; Massa *et al.*, 1972*a*; Cresti and Massa, 1976; Juneja *et al.*, 1976; Martini, 1976). In adult animals some transfor-

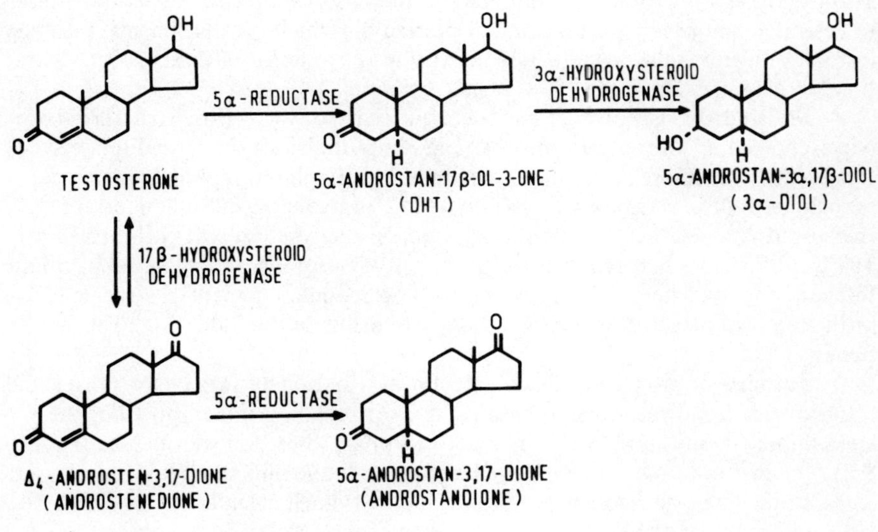

Figure 8.2 The 5α-reduction of androgens

mations of testosterone into these metabolites occur also in descending order in: the hypothalamus (Kniewald *et al.*, 1971; Massa *et al.*, 1972*a*; Cresti and Massa, 1976; Juneja *et al.*, 1976; Martini, 1976); in the midbrain (Denef *et al.*, 1973); in the limbic structures (amygdala and hippocampus) (Massa *et al.*, 1972*a*; Juneja *et al.*, 1976; Martini, 1976); in the cerebellum (Rommerts and Van der Molen, 1971); and in the cerebral cortex (Kniewald *et al.*, 1971; Massa *et al.*, 1972*a*; Cresti and Massa, 1976; Juneja *et al.*, 1976; Martini, 1976). In general these metabolic processes are similar to those occurring in the peripheral androgen-dependent structures (Wilson and Gloyna, 1970). However, the following two major differences have been reported: the formation of 3α-diol is much greater in central structures than in peripheral target organs (Juneja *et al.*, 1976; Martini, 1976); and 3β-diol, which is a regular metabolite of testosterone at the periphery (Wilson and Gloyna, 1970) is usually not formed in neuroendocrine tissues (Rommerts and Van der Molen, 1971).

In the brain and the anterior pituitary, as in all other mammalian structures, the 5α-reduction of testosterone is an irreversible process, and consequently DHT cannot be enzymatically converted back to testosterone (Cresti and Massa, 1976). However, 3α-diol can easily revert to DHT in the anterior pituitary and the hypothalamus under both *in vitro* and *in vivo* conditions (Cresti and Massa, 1976; Pilven *et al.*, 1976). These observations may have physiological significance because of the different biological effects of each metabolite (see below).

In the adult male rat castration significantly enhances the 5α-reductase activity of the anterior pituitary (Kniewald *et al.*, 1971; Massa *et al.*, 1972*a*; Juneja *et al.*, 1976; Martini, 1976). It is possible that this reflects the changes in the composition of pituitary cell population which follow the operation, since the gonadotrophs increase in size and proliferate after gonadectomy. If so, this observation would provide indirect evidence in favour of a specific localisation of the 5α-reductase in the gonadotrophs (Lloyd and Karavolas, 1975; Celotti *et al.*, 1976). This hypothesis is further substantiated by the observation that the administration to castrated animals of testosterone or oestradiol (which restores normal pituitary histology) restores the 5α-reductase activity in the anterior pituitary to precastration levels (Kniewald *et al.*, 1971; Massa *et al.*, 1972*a*).

Age-related changes of the 5α-reductase activity of the central structures have been reported. In the anterior pituitary and hypothalamus the 5α-reductase system seems to be more active in the neonatal and prepubertal periods than after sexual maturation (Massa *et al.*, 1972*a*, 1975). In neonatal and prepubertal rats the 5α-reductase activity is also very elevated in the cerebral cortex (Massa *et al.*, 1972*a*, 1975), a structure in which this activity is very low following sexual maturation (see above). The progessive decrease of 5α-reductase in central structures from birth to sexual maturity may play an important role in the induction of male puberty.

The process of aromatisation, which converts androgens into oestrogens, is the result of a series of reactions. It entails the hydroxylation of carbon 19 of the steroid nucleus and, ultimately, its loss to form an 18-carbon steroid (see figure 8.3). The loss of two hydrogens which results in the A ring may require enzymatic steps, or may be a consequence of the instability which attends the loss of C 19. Using either testosterone or androstenedione as the substrate, this process has been shown to occur, *in vitro* or *in vivo*, in the hypothalamus and in the limbic system (Naftolin *et al.*, 1975*b*, 1976; Lieberburg and McEwen, 1975, 1977). In adult animals the aromatising enzymes do not seem to be present in the anterior pituitary, which then metabolises androgens exclusively through the 5α-reductase pathway.

It is relevant to emphasise that the aromatisation process occurs in the hypothalamus and in the amygdala of foetal and neonatal animals, with yields higher than those found in adult animals (Reddy *et al.*, 1974*b*; Naftolin *et al.*, 1975*b*,

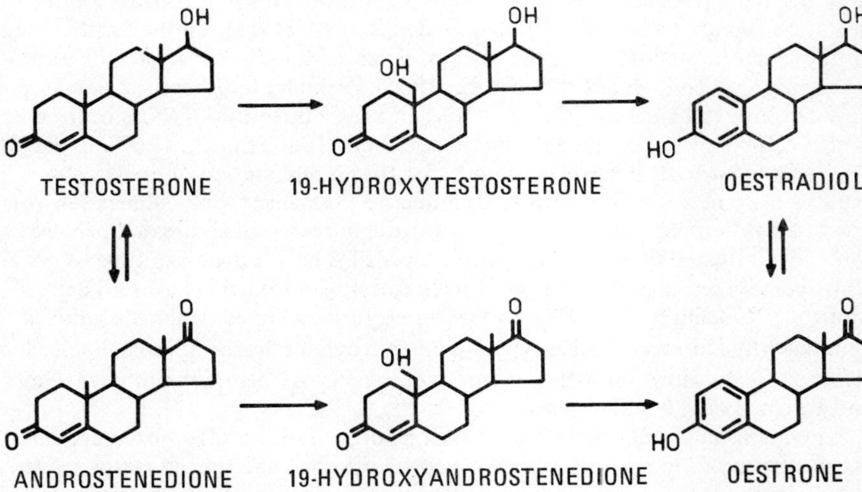

Figure 8.3 The aromatisation of androgens

1976). This may be relevant for the sexual differentiation of the brain, since both these structures have been suggested as taking part in such a process (Raisman and Field, 1971). During foetal and neonatal life the aromatising system is also present in the anterior pituitary and in the cerebral cortex (Weisz and Gibbs, 1974; Naftolin *et al.*, 1975*b*). This contrasts with the situation in adulthood and is reminiscent of the broad distribution of 5α-reductase in the central structures during the neonatal and prepubertal periods (see above).

Using androstenedione as the substrate, a very consistent sex difference has been reported to exist, the hypothalamus of the male having a greater aromatising activity than that of the female (Reddy *et al.*, 1973). Again, this may be important for the sexual differentiation of the brain.

Physiological significance of the processes of 5α-reduction and of aromatisation of androgens

The observation that the brain and the anterior pituitary convert testosterone and other androgens into different types of metabolites has led to the suggestion that these might be important for the regulation of several physiological processes such as the sexual organisation of the brain during the foetal and/or the neonatal period, the control of sexual and other types of behaviour and the control of gonadotrophin secretion.

Sexual organisation of the brain during the foetal and/or the neonatal period It has been proposed that the typical cyclic release of gonadotrophins necessary for ovulation occurs in adult female mammals because the brain of the female is not exposed during the foetal and/or the neonatal period to the action of testosterone and of other androgens. Conversely, the non-cyclical nature of gonadotrophin secretion found in adult males would be due to the fact that, during a critical phase of foetal and/or neonatal life, the brain of these animals is exposed to the action of endogenous androgens (Barraclough, 1967). A similar hypothesis is usually put forward for explaining the "organisation" in a male direction of those centres which control male sexual behaviour in adulthood (Barraclough, 1967). In support of such a hypothesis it has been shown that the administration of exogenous androgens to neonatal female animals induces a permanent "masculinisation" of the brain structures which control gonadotrophin secretion and sexual behaviour. Animals so treated do not show oestrous cyclicity, have a male-like type of gonadotrophin secretion and may exhibit male patterns of sexual behaviour (Barraclough, 1967). On the contrary, castration performed in neonatal male animals permanently induces a female type of gonadotrophin release and sexual behaviour (Gorski, 1971). However, whether similar mechanisms operate in primates (Goy and Resko, 1972) is still unresolved.

The question is still open whether such an organisational effect of androgens is brought about by the transformation of testosterone and related steroids into the corresponding 5α-reduced metabolites or into oestrogens. The presence of high levels of 5α-reductase activity in the brain of neonatal animals might suggest that the 5α-reduction is the metabolic pathway responsible for these effects. However, the majority of the data favour oestrogens as the metabolites of testosterone involved in the sexual differentation of the brain. The data which may be quoted in support of such a hypothesis are: (1) testosterone and other aromatisable androgens (such as androstenedione) masculinise the nervous centres of neonatally-treated female rats (McDonald and Doughty, 1974; Naftolin *et al.,* 1975*b*), while on the other hand DHT (a non-aromatisable androgen) is totally ineffective (McDonald and Doughty, 1972, 1974); (2) 19-hydroxytestosterone, an intermediate in the aromatisation process (see figure 8.3), is a more potent masculinising agent than testosterone when administered to neonatal female rats (McDonald and Doughty, 1974); (3) oestradiol and other oestrogens (Gorski, 1963; Doughty *et al.,* 1975) given to neonatal females have a masculinising effect at doses much lower than those of testosterone, especially if administered directly into the hypothalamus (Döcke and Dörner, 1975); and (4) pretreatment of new-born females with anti-oestrogens blocks the masculinising action of testosterone (McDonald and Doughty, 1972).

If one accepts the view that neonatal aromatisation of androgens is essential for the masculinisation of brain centres, it may appear puzzling that the CNS of neonatal female animals should not be masculinised by the physiological presence of oestrogens. This may occur either because circulating oestrogens are unable to initiate such a process and local intracerebral aromatisation is a necessary step to trigger it; or it may be that the brain of the neonatal female is protected from the effects of circulating oestrogens by the presence of α-foetoprotein (a specific oestrogen-binding protein) which is found in high concentrations in the blood of new-born animals and which may prevent the penetration of active oestrogens through the blood-brain barrier (Raynaud *et al.*, 1971).

Control of male sexual behaviour It is possible that oestrogens are the metabolites of androgens responsible for the induction of male sexual behaviour in adult animals. It is universally accepted that testosterone can initiate or restore all aspects of male sexual behaviour in orchidectomised animals of all the species studied so far (Beyer, 1976). Other aromatisable androgens (androstenedione, androstenediol, 19-hydroxytestosterone and 19-hydroxyandrostenedione) are also able to restore full patterns of male sexual activity (Parrott and McDonald, 1975; Beyer, 1976). Oestrogens *per se* can restore most aspects of copulatory behaviour in several species of animals (Sodersten, 1973; Christiansen and Clemens, 1974). Oestradiol, however, is ineffective in restoring male sexual behaviour in castrated rabbits (Agmo and Sodersten, 1975) and guinea-pigs (Alsum and Goy, 1974).

However, DHT as well as other 5α-reduced non-aromatisable androgens (such as 5α-androstenedione, androsterone, 3α-diol, 3β-diol) have very little effect in restoring male sexual behaviour if given alone to orchidectomised rats and mice (Beyer *et al.*, 1973; Feder *et al.*, 1974). Large species differences seem to exist, since DHT may stimulate male sexual behaviour in guinea-pigs (Alsum and Goy, 1974), hamsters (Whalen and de Bold, 1974), rabbits (Agmo and Sodersten, 1975) and Rhesus monkeys (Goy and Resko, 1972). Also, in those species in which DHT is ineffective if given alone, male sexual behaviour can be induced if this steroid is administered in combination with low doses of oestradiol which are in themselves ineffective in producing behavioural effects (Larsson *et al.*, 1976; Luttge *et al.*, 1975).

Control of gonadotrophin secretion Systemically administered testosterone totally abolishes LH secretion, but exerts only minor effects on FSH release in the majority of the species studied (Swerdloff *et al.*, 1972; Zanisi *et al.*, 1973; Stewart-Bentley *et al.*, 1974). Following systemic administration DHT has been shown in several species of animals and in man to be either as effective as or more active than testosterone in suppressing LH secretion (Swerdloff *et al.*, 1972, 1973; Zanisi *et al.*, 1973; Stewart-Bentley *et al.*, 1974). In the animal species in which it has been tested, systemically administered 3α-diol has always been found to be more active than either testosterone or DHT in inhibiting LH release (Zanisi *et al.*, 1973; Eik-Nes, 1975; Verjans and Eik-Nes, 1977). These facts have led to the hypothesis that the formation of DHT and of 3α-diol in neuroendocrine structures might be essential for testosterone to exert its negative feedback effect on LH secretion.

The situation seems to be quite different with regard to the negative feedback effect which androgens exert on FSH release. The bulk of the available evidence indicates that, following systemic administration, DHT is less effective than testosterone in suppressing the release of this gonadotrophin in castrated experimental animals and in man (Zanisi *et al.*, 1973; Stewart-Bentley *et al.*, 1974). Systemically administered 3α-diol is practically ineffective (Zanisi *et al.*, 1973; Eik-Nes, 1975;

Verjans and Eik-Nes, 1977). These data seem to assign to the 5α-reduced metabolites of testosterone only a secondary role in the inhibition of FSH release. Consequently, it is possible that testosterone intervenes in the negative control of the secretion of FSH, either acting as such, or after having been converted (in the CNS, in the plasma or in the liver) into oestrogenic molecules. This is supported by the observation that oestrogens can indeed effectively suppress the release of FSH (David *et al.,* 1965; Martini *et al.,* 1968; Kulin and Reiter, 1972).

Progesterone

5α-reduction

Evidence is rapidly accumulating which indicates that several nervous structures and the anterior pituitary of female animals contain a 5α-reductase-3α-hydroxysteroid-dehydrogenase system. However, it is premature to say whether these enzymes are identical to those found in the male. Progesterone is the most obvious candidate as the physiological substrate of these central enzymes. However, it must also be remembered that other 3-keto-Δ_4-steroids (corticoids, androgens, and so on) of ovarian or adrenal origin may undergo such a conversion.

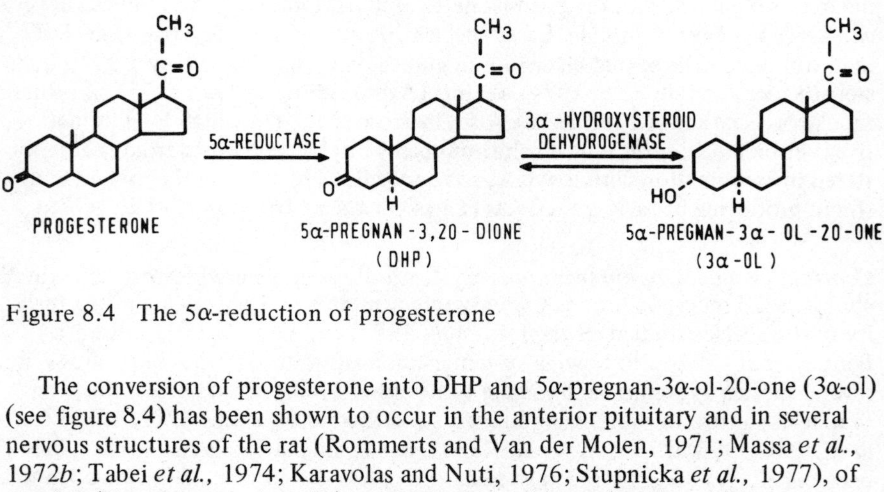

Figure 8.4 The 5α-reduction of progesterone

The conversion of progesterone into DHP and 5α-pregnan-3α-ol-20-one (3α-ol) (see figure 8.4) has been shown to occur in the anterior pituitary and in several nervous structures of the rat (Rommerts and Van der Molen, 1971; Massa *et al.,* 1972*b*; Tabei *et al.,* 1974; Karavolas and Nuti, 1976; Stupnicka *et al.,* 1977), of the dog (Kawahara *et al.,* 1975), of subhuman primates (Tabei and Heinrichs, 1974), and of the human foetus (Mickan, 1972). In the female rat, as in the male, the 5α-reductase-3α-hydroxysteroid-dehydrogenase system is present in higher concentrations in the anterior pituitary than in other CNS structures. Among these, the hypothalamus has the highest converting ability, followed by the amygdala and the cerebral cortex. The thalamus, midbrain, cerebellum, medulla oblongata and pineal gland are also able to convert progesterone into DHP and 3α-ol. These facts have been demonstrated *in vitro* and confirmed by a recent *in vivo* study (Karavolas *et al.,* 1976).

Other progestogens, like 17α-OH-progesterone and 20α-OH-progesterone, are also good substrates for the 5α-reductase and the 3α-hydroxysteroid-dehydrogenase of the anterior pituitary and of the hypothalamus of female animals (Nowak and Karavolas, 1974; Tabei *et al.*, 1974; Nakamura and Tanabe, 1975; Cheng and Karavolas, 1975). This process may be of physiological importance since in some species, such as the rat and rabbit, 17α-OH-progesterone and 20α-OH-progesterone are believed to exert major facilitatory or positive feedback effects on gonadotrophin release (Neill and Smith, 1974).

In the female rat the 5α-reductase activity of the anterior pituitary shows conspicuous changes during the different phases of the oestrous cycle, peak activities being recorded in the morning of oestrus (Massa *et al.*, 1972*b*; Karavolas *et al.*, 1976; Stupnicka *et al.*, 1977). Only minor oestrus-linked modifications of the 5α-reductase activity have been found in the hypothalamus. Oestrous cyclicity does not influence the 5α-reductase of the cerebral cortex and of the amygdala (Massa *et al.*, 1972*b*; Karavolas *et al.*, 1976; Stupnicka *et al.*, 1977). As in males, castration induces a significant increase of the 5α-reductase activity of the anterior pituitary (Massa *et al.*, 1972*b*; Karavolas *et al.*, 1976; Stupnicka *et al.*, 1977), which may be due to the increase in the number of gonadotrophic cells (in which the enzyme is probably localised). In agreement with this hypothesis, the 5α-reductase activity has been shown to be particularly elevated in *in vitro* preparations in which the gonadotrophic component has been increased by velocity sedimentation (Lloyd and Karavolas, 1975; Karavolas and Nuti, 1976). The ovariectomy-induced increase of the 5α-reductase in the anterior pituitary may be returned to normal by *in vivo* administration of either oestrogens or testosterone, that is by two hormones which normalise pituitary histology (Massa *et al.*, 1972*b*; Karavolas *et al.*, 1976; Stupnicka *et al.*, 1977). Progesterone, which does not affect gonadotrophin secretion in the absence of oestrogens, is totally ineffective in this respect (Stupnicka *et al.*, 1977). Ovariectomy and the *in vivo* administration of sex steroids do not exert major effects on the 5α-reductase activity of the hypothalamus, amygdala or cerebral cortex (Massa *et al.*, 1972*b*; Karavolas *et al.*, 1976; Stupnicka *et al.*, 1977).

The 5α-reductase activity of the anterior pituitary, of the hypothalamus and of the cerebral cortex of female rats is higher in the neonatal and prepubertal period and shows a progressive decline with advancing age (Massa *et al.*, 1975). Before sexual maturation the 5α-reductase activity of the female pituitary is even higher than that of the male (Massa *et al.*, 1975). The significance of this phenomenon for the induction of female puberty is still obscure. It is possible that this sex difference is brought about by the neonatal hormonal environment, since neonatal androgenisation decreases the level of the enzymes in the anterior pituitary of females, and neonatal orchidectomy has the opposite effect on the enzyme of the anterior pituitary of the male (Massa *et al.*, 1975).

Physiological significance of the process of 5α-reduction of progestogens

The observations described above suggest that the 5α-reduced metabolites of progesterone might play a role in the expression of the multiple activities of the hormone. In agreement with this hypothesis, it has recently been reported that DHP facilitates ovulation in PMS-treated immature rats (Sanyal and Todd, 1972; Shridharan *et al.*, 1974), and in hamsters in which ovulation has been blocked by pentobarbitone (Bosley and Leavitt, 1972). Moreover, DHP may facilitate LH and

FSH secretion if given systemically to castrated oestrogen-primed rats (Zanisi *et al.*, 1975; Zanisi and Martini, 1975; Karavolas and Nuti, 1976; Stupnicka *et al.*, 1977; Martini, 1977). No facilitatory effects on ovulation have been reported for 3α-ol (Bosley and Leavitt, 1972; Shridharan *et al.*, 1974), but this steroid stimulates LH and FSH release in oestrogen-primed castrated female rats (Zanisi and Martini, 1975; Zanisi *et al.*, 1975; Martini, 1977). These results suggest that the 5α-reduced derivatives of progesterone might be the mediators of the positive feedback of the hormone on gonadotrophin secretion.

Additional data supporting the possibility that progesterone might also operate on central structures following conversion into 5α-reduced compounds may be derived from behavioural studies. DHP induces sexual receptivity in castrated oestrogen-primed rats (Meyerson, 1972), hamsters (Bosley and Leavitt, 1972), guinea-pigs (Wade and Feder, 1972; Czaja *et al.*, 1974) and mice (Gorzalka and Whalen, 1974). The fact that DHP is not as active as progesterone in some behavioural experiments does not contradict the hypothesis because of the low 5α-reducing activity of the nervous structures (see above). Behavioural effects have also been reported after the administration of 3α-ol; however, the activity of this compound is lower than that of DHP (Czaja *et al.*, 1974). DHP and 3α-ol as well as other members of the 5α-pregnane family have been shown to exert depressant effects on the CNS (P'an and Laubach, 1964), although of lower magnitude than those induced by 5-pregnane derivatives. It is therefore possible that the process of 5α-reduction might play a role in the appearance of the sedative-anaesthetic activity of progesterone.

ACKNOWLEDGEMENTS

The experimental work performed in the author's laboratory and presented in this chapter was supported by grants from the Ford Foundation, New York, and the Consiglio Nazionale delle Ricerche, Rome, through the programme *"Biology of Reproduction"*. All such support is gratefully acknowledged.

REFERENCES

Agmo, A. and Sodersten, P. (1975). Sexual behaviour in castrated rabbits treated with testosterone, oestradiol, dihydrotestosterone or oestradiol in combination with dihydrotestosterone. *J. Endocr.*, **67**, 327–332

Alsum, P. and Goy, R. W. (1974). Action of esters of testosterone, dihydrotestosterone or oestradiol on sexual behaviour in castrated male guinea-pigs. *Hormones Behav.*, **5**, 207–217

Armstrong, E. G., Jr. and Villee, C. A. (1977). Characterisation and comparison of oestrogen and androgen receptors of calf anterior pituitary. *J. Steroid Biochem.*, **8**, 285–292

Atger, M., Baulieu, E. E. and Milgrom, E. (1974). An investigation of progesterone receptors in guinea-pig vagina, uterine cervix, mammary glands, pituitary and hypothalamus. *Endocrinology*, **94**, 161–167

Ball, F., Knuppen, R., Haupt, M. and Breuer, H. (1972). Interaction between oestrogens and catecholamines. III. Studies on the methylation of catecholestrogens, catecholamines and other catechols by the catechol-*o*-methyltransferase of human liver. *J. Clin. Endocr. Metab.*, **34** 736–746

Barley, J., Ginsburg, M., MacLusky, N. Y., Morris, I. D. and Thomas, P. J. (1977). Sex differences in the distribution of cytoplasmic oestrogen receptors in rat brain and pituitary: effects of gonadectomy and neonatal androgen treatment. *Brain Res.*, 309–318

Barraclough, C. A. (1967). Modifications in reproductive function after exposure to hormones during the prenatal and early postnatal period. In L. Martini and W. F. Ganong (eds.). *Neuroendocrinology,* Vol. 2, Academic Press, New York, 61–99

Berthold, A. A. (1849). Transplantation of the testes. *Arch. Anat. Physiol. Wiss. Med.*, **16**, 42–46

Beyer, C. (1976). Neuroendocrine mechanisms in sexual behaviour. In F. Naftolin, K. Y. Ryan and I. Y. Davies (eds.). *Subcellular Mechanisms in Reproductive Neuroendocrinology.* Elsevier Scientific Publishing Company, Amsterdam, 471–485

Beyer, C., Larsson, K., Perez-Palacios, G. and Morali, G. (1973). Androgen structure and male sexual behaviour in the castrated rat. *Hormones Behav.*, **4**, 99–108

Bolt, H. M. and Kappus, H. (1976). Interaction by 2-hydroxyestrogens with enzymes of drug metabolism. *J. Steroid Biochem.*, **7**, 311–313

Bosley, C. G. and Leavitt, W. W. (1972). Specificity of progesterone action during the preovulatory period in the cyclic hamster. *Fed. Proc.*, **31**, 257–258

Celotti, F., Farina, J., Cresti, L., Massa, R. and Martini, L. (1976). 5alpha-reductase activity (5alpha-R) in rat pituitary homografts under the kidney capsule. *Program 5th Intern. Congr. Endocr.*, Excerpta Medica, Amsterdam, 44–45

Celotti, F., Massa, R. and Martini, L. (1978). Metabolism of sex steroids in the central nervous system. In L. J. De Groot (ed.). *Metabolic Basis of Endocrinology*. Grune and Stratton, New York, to be published

Cheng, Y. J. and Karavolas, H. J. (1975). Properties and subcellular distribution of Δ^4-steroid (progesterone) 5alpha-reductase in rat anterior pituitary. *Steroids,* **26**, 57–72

Christiansen, L. W. and Clemens, L. G. (1974). Intrahypothalamic implants of testosterone or oestradiol and resumption of masculine sexual behaviour in long-term castrated male rats. *Endocrinology*, **95**, 984–990

Cresti, L. and Massa, R. (1976). Metabolism of androgens in various testosterone-dependent tissues. *Program 5th Intern. Congr. Endocr.*, Excerpta Medica, Amsterdam, 46–47

Czaja, J. A., Goldfoot, D. A. and Karavolas, H. J. (1974). Comparative facilitation and inhibition of lordosis in the guinea-pig with progesterone, 5alpha-pregnane 3.20-dione or 3alpha-hydroxy-5alpha-pregnan-20-one. *Hormones Behav.*, **5**, 261–274

David, M. A., Fraschini, F. and Martini, L. (1965). Parallélisme entre le contenu hypophysaire en FSH et le contenu hypothalmique en FSH-RF (FSH-releasing factor). *C. R. Acad. Sci.*, **261**, 2249–2251

Davies, I. J., Naftolin, F. and Ryan, K. J. (1975). The affinity of catechol for oestrogen receptors in the pituitary and anterior hypothalamus of the rat. *Endocrinology,* **97**, 554–557

Denef, C., Magnus, C. and McEwen, B. S. (1973). Sex differences and hormonal control of testosterone metabolism in rat pituitary and brain. *J. Endocr.*, **59**, 605–621

Döcke, F. and Dörner, G. (1975). Anovulation in adult female rats after neonatal intracerebral implantation of oestrogen. *Endokrinologie,* **65**, 375–377

Doughty, C., Booth, J. E., McDonald, P. G. and Parrott, R. F. (1975). Inhibition, by the anti-oestrogen MER-25, of the defeminization induced by the synthetic oestrogen RU 2858. *J. Endocr.*, **67**, 459–460

Eik-Nes, K. B. (1975). Production and secretion of 5alpha-reduced testosterone (DHT) by male reproductive organs. *J. Steroid Biochem.*, **6**, 337–339

Feder, H. H., Naftolin, F. and Ryan, K. J. (1974). Male and female sexual responses in male rats given estradiol benzoate and 5alpha-androstan-17beta-ol-3-one propionate. *Endocrinology,* **94**, 136–141

Fishman, J. (1976). Estrogen metabolism by neuroendocrine tissues. In F. Naftolin, K. J.

Ryan and I. J. Davies (eds.). *Subcellular Mechanisms in Reproductive Neuroendocrinology.* Elsevier Scientific Publishing Company, Amsterdam, 357–362

Fishman, J., Naftolin, F., Davies, I. J., Ryan, K. J. and Petro, Z. (1976). Catecholestrogen formation by the human foetal brain and pituitary. *J. Clin. Endocr. Metab.*, **42**, 177–180

Fishman, J. and Norton, B. (1975). Catecholestrogen formation in the central nervous system of the rat. *Endocrinology,* **96**, 1054–1058

Flores, F., Naftolin, F., Ryan, K. J. and White, R. J. (1973). Estrogen formation by the isolated perfusal Rhesus monkey brain. *Science,* **180**, 1074–1075

Fox, T. O. (1977). Estradiol and testosterone binding in normal and mutant mouse cerebellum: biochemical and cellular specificity. *Brain Res.*, **128**, 263–273

Fuxe, K. and Hökfelt, T. (1969). Catecholamines in the hypothalmus and in the pituitary gland. In W. F. Ganong and L. Martini (eds.). *Frontiers in Neuroendocrinology,* Vol. 1, Oxford University Press, New York, 47–96

Ginsburg, M., Greenstein, B. D., MacLusky, N. J., Morris, I. D. and Thomas, P. J. (1971). Dihydrotestosterone binding in brain and pituitary cytosol of rats. *J. Endocr.*, **61**, XXIV

Gordon, S., Cantrall, E. W., Cekleniak, W. P., Albers, H. J., Mauer, S., Stolar, S. M. and Bernstein, S. (1964). Steroid and lipid metabolism. The hypocholesteremic effect of estrogen metabolites. *Steroids,* **4**, 267–271

Gorski, R. A. (1963). Modification of ovulatory mechanisms by postnatal administration of estrogen to the rat. *Amer. J. Physiol.*, **205**, 842–844

Gorski, R. A. (1971). Gonadal hormones and the perinatal development of neuroendocrine function. In L. Martini and W. F. Ganong (eds.). *Frontiers in Neuroendocrinology,* Vol. 2. Oxford University Press, New York, 237–290

Gorzalka, B. B. and Whalen, R. E. (1974). Genetic regulation of hormone action: selective effects of progesterone and dihydroprogesterone (5-alpha-pregnane-3, 20-dione) on sexual receptivity in mice. *Steroids,* **23**, 499–505

Goy, R. W. and Resko, J. A. (1972). Gonadal hormones and behaviour of normal and pseudohermaphroditic non-human female primates. *Recent Progr. Hormone Res.*, **28**, 707–733

Iramain, C. A., Danzo, B. J., Stratt, C. A. and Toft, D. O. (1973). Program 4th Intern. Congr. Soc. Psychoneuroendocrinology, p.5.

Jenkin, G., Henville, A. and Heap, R. B. (1975). Metabolism of oestrone sulphate and binding by the brain and pituitary of foetal and adult sheep. *J. Endocr.*, **64**, 22–23P

Jensen, E. V. and de Sombre, E. R. (1973). Estrogen–receptor interaction. *Science,* **182**, 126–134

Jouan, P., Samperez, S. and Thieulant, M. L. (1973). Testosterone 'receptors' in purified nuclei of rat anterior hypophysis. *J. Steroid Biochem.*, **4**, 65–74

Juneja, H. S., Motta, M., Massa, R., Zanisi, M. and Martini, L. (1976). Feedback control of gonadotropin secretion in the male. In P. O. Hubinont and M. L'Hermite (eds.). *Sperm Action,* Karger, Basel, 162–173

Karavolas, H. J., Hodges, D. and O'Brien, D. (1976). Uptake of ^{3}H-progesterone and ^{3}H-5-alpha-dihydroprogesterone by rat tissues *in vivo* and analysis of accumulated radioactivity: accumulation of 5-alpha-dihydroprogesterone by pituitary and hypothalamic tissues. *Endocrinology,* **98**, 164–175

Karavolas, H. J. and Nuti, K. M (1976). Progesterone metabolism by neuroendocrine tissues. In F. Naftolin, K. J. Ryan and I. J. Davies (eds.). *Subcellular Mechanisms in Reproductive Neuroendocrinology,* Elsevier Scientific Publishing Company, Amsterdam, 305–326

Kato, J. (1975). The role of hypothalamic and hypophyseal 5-alpha-dihydrotestosterone, estradiol and progesterone receptors in the mechanism of feedback action. *J. Steroid Biochem.*, **6**, 979–987

Kato, J. (1976). Cytosol and nuclear receptors for 5-alpha-dihydrotestosterone and testosterone in the hypothalamus and hypophysis, and testosterone receptors isolated from neonatal female rat hypothalamus. *J. Steroid Biochem.*, **7**, 1179–1187

Kato, J. and Onouchi, T. (1977). Specific progesterone receptors in the hypothalamus and anterior hypophysis of the rat. *Endocrinology,* **101**, 920–928

Kawahara, F. S., Berman, M. L. and Green, O. C. (1975). Conversion of progesterone [1-2-^{3}H] to 5-beta-pregnane-3. 20-dione by brain tissue. *Steroids,* 25, 459–463

Kazama, N. and Longcope, C. (1974). *In vivo* studies on the metabolism of estrone and estradiol-17-beta by the brain. *Steroids,* 23, 469–481

Keefer, D. A., Stumpf, W. E. and Sar, M. (1973). Estrogen-topographical localisation of estrogen-concentrating cells in the rat spinal cord following ^{3}H-estradiol administration. *Proc. Soc. exp. Biol., Med.,* **143**, 414–417

Kishimoto, Y. (1973). Estrone sulphate in rat brain: uptake from blood and metabolism *in vivo. J. Neurochem.,* **20**, 1489–1492

Kniewald, Z., Massa, R. and Martini, L. (1971). Conversion of testosterone into 5-alpha-androstan-17-beta-ol-3-one at the anterior pituitary and hypothalamic level. In V. H. T. James and L. Martini (eds.). *Hormonal Steroids,* Excerpta Medica, Amsterdam, 784–791

Krieger, M. S., Morrell, J. I. and Pfaff, D. W. (1976). Autoradiographic localisation of estradiol-concentrating cells in the female hamster brain. *Neuroendocrinology,* **22**, 193–205

Kulin, H. E. and Reiter, E. O. (1972). Gonadotrophin suppression by low-dose estrogen in men: evidence for differential effects upon FSH and LH. *J. Clin. Endocr. Metab.,* **35**, 836–839

Larsson, K., Sodersten, P., Beyer, C., Morali, G. and Perez-Palacios, G. (1976). Effects of estrone, estradiol and estriol combined with dihydrotestosterone on mounting and lordosis behavior in castrated male rats. *Hormones Behav.,* 7, 379–390

Laumas, K. R. and Farooq, A. (1966). The uptake *in vivo* of [1, 2-^{3}H] progesterone by the brain and genital tract of the rat. *J. Endocr.,* **36**, 95–96

Lieberburg, I., MacClusky, N. J. and McEwen, B. S. (1977). 5-alpha-dihydrotestosterone (DHT) receptors in rat brain and pituitary cell nuclei. *Endocrinology,* **100**, 598–607

Lieberburg, I. and McEwen, B. S. (1975). Estradiol-17-beta: a metabolite of testosterone recovered in cell nuclei from limbic areas of male adult rat brains. *Brain Res.,* **85**, 165–170

Lieberburg, I. and McEwen, B. S. (1977). Brain cell nuclear retention of testosterone metabolites, 5-alpha-dihydrotestosterone and estradiol-17-beta in adult rats. *Endocrinology,* **100**, 588–597

Lloyd, R. V. and Karavolas, H. J. (1975). Uptake and conversion of progesterone and testosterone to 5-alpha-reduced products by enriched gonadotropic and chromophobic rat anterior pituitary cell fractions. *Endocrinology,* 97, 517–526

Luttge, W. G., Hall, N. R., Wallis, C. J. and Campbell, J. C. (1975) Stimulation of male and female sexual behaviour in gonadectomised rats with estrogen and androgen therapy and its inhibition with concurrent antihormone therapy. *Physiol. Behav.,* **14**, 65–73

Luttge, W. G. and Whalen, R. E. (1970). Regional localisation of estrogenic metabolites in the brain of male and female rats. *Steroids,* **15**, 605–612

McDonald, P. G. and Doughty, C. (1972). Comparison of the effect of neonatal administration of testosterone and dihydrotestosterone in female rat. *J. Reprod. Fertil.,* **30**, 55–62

McDonald, P. G. and Doughty, C. (1974). Effect of neonatal administration of different androgens in the female rat: correlation between aromatisation and the induction of sterilisation. *J. Endocr.,* **61**, 95–103

McEwen, B. S. (1978). Distribution and binding of hormones in different CNS areas. In L. J. De Groot (ed.). *Metabolic Basis of Endocrinology*. Grune and Stratton, New York, to be published

McEwen, B. S., Zigmond, R. E. and Gerlach, G. H. (1972). Sites of steroid binding and action in the brain. In G. H. Bourne (ed.). *Structure and Function of Nervous Tissue,* vol. 5, Academic Press, New York, 205–291

Martini, L. (1976). Androgen reduction by neuroendocrine tissues: physiological significance. In F. Naftolin, K. J. Ryan and I. J. Davies (eds.). *Subcellular Mechanisms in Reproductive*

Neuroendocrinology, Elsevier Scientific Publishing Company, Amsterdam, 327–345

Martini, L. (1977). Recent views on the control of anterior pituitary function. *Acta Endocr.*, Suppl. **214**, 19–32

Martini, L., Fraschini, F. and Motta, M. (1968). Neural control of anterior pituitary functions. *Recent Progr. Hormone Res.*, **24**, 439–496

Massa, R., Justo, S. and Martini, L. (1975). Conversion of testosterone into 5-alpha-reduced metabolites in the anterior pituitary and in the brain of maturing rats. *J. Steroid Biochem.*, **6**, 567–571

Massa, R., Stupnicka, E., Kniewald, Z. and Martini, L. (1972*a*). The transformation of testosterone into dihydrotestosterone by the brain and the anterior pituitary. *J. Steroid Biochem.*, **3**, 385–399

Massa, R., Stupnicka, E. and Martini, L. (1972*b*). Metabolism of progesterone in the anterior pituitary, the hypothalamus and the uterus of female rats. *Program 4th Intern. Congr. Endocr.*, Excerpta Medica, Amsterdam, 118

Maurer, R. A. and Woolley, D. E. (1974). Demonstration of nuclear ^{3}H-estradiol binding in hypothalamus and amygdala of female, androgenised-female and male rats. *Neuroendocrinology,* **16**, 137–147

Mercier, L., LeGuellec, C., Thieulant, M. L., Samperez, S. and Jouan, P. (1976). Androgen and estrogen receptors in the cytosol from male rat anterior hypophysis: further characteristics and differentiation between androgen and estrogen receptors. *J. Steroid Biochem.*, 7, 779–785

Meyerson, B. (1972). Latency between intravenous injection of progestins and the appearance of estrus behaviour in estrogen-treated ovariectomised rats. *Hormones Behav.*, **3**, 1–9

Mickan, H. (1972). Metabolism of 4-^{14}C-progesterone and 4-^{14}C-testosterone in brain of the previable human fetus. *Steroids,* **19**, 659–665

Morrell, J. I., Kelley, D. B. and Pfaff, D. W. (1975). Sex steroid binding in the brains of vertebrates. Studies with light microscopic autoradiography. In K. M. Knigge, D. E. Scott, H. Kobayashi and S. Ishii (eds.). *Brain-Endocrine Interactions. Part II: The Ventricular System in Neuroendocrine Mechanisms,* Karger, Basel, 230–256

Moss, R. L. (1976). Unit responses in preoptic and arcuate neurons related to anterior pituitary function. In L. Martini and W. F. Ganong (eds.). *Frontiers in Neuroendocrinology*, Vol. 4, Raven Press, New York, 95–128

Mowles, T. F., Ashkanazy, B., Mix, E., Jr. and Sheppard, H. (1971). Hypothalamic and hypophyseal estradiol-binding complexes. *Endocrinology,* **89**, 484–491

Naess, O., Attramadal, A. and Aakvaag, A. (1975). Androgen-binding proteins in the anterior pituitary, hypothalamus, preoptic area and brain cortex of the rat. *Endocrinology,* **96**, 1–9

Naftolin, F., Morishita, H., Davies, I. J., Todd, R. and Ryan, K. J. (1975*a*). 2-Hydroxyestrone-induced rise in serum luteinising hormone in the immature male rat. *Biochem. Biophys. Res. Commun.*, **64**, 905–910

Naftolin, F., Ryan, K. J. and Davies, I. J. (1976). Androgen aromatisation by neuroendocrine tissues. In F. Naftolin, K. J. Ryan, and I. J. Davies (eds.). *Subcellular Mechanisms in Reproductive Neuroendocrinology,* Elsevier Scientific Publishing Company, Amsterdam, 347–355

Naftolin, F., Ryan, K. J., Davies, I. J., Reddy, V. V., Flores, F., Petro, Z., Kuhn, M., White, R. S., Takaoka, Y. and Wolin, L. (1975*b*). The formation of estrogens by central neuroendocrine tissues. *Recent Progr. Hormone Res.*, **31**, 295–319

Nakamura, T. and Tanabe, Y. (1975). *In vitro* metabolism of steroid hormones by chicken brain. *Acta Endocr.*, 75, 410–416

Neill, J. D. and Smith, M. S. (1974). Pituitary-ovarian inter-relationships in the rat. In V. H. T. James and L. Martini (eds.). *Current Topics in Experimental Endocrinology*, Vol. 2, Academic Press, New York, 73–106

Notides, A. C. (1970). Binding affinity and specificity of the estrogen receptor of the rat uterus and anterior pituitary. *Endocrinology,* 87, 987–992

Nowak, F. V. and Karavolas, H. J. (1974). Conversion of 20-alpha-hydroxypregnen-4-en-3-one to 20-alpha-hydroxy-5-alpha-pregnan-3-one and 5-alpha-pregnane-3-alpha, 20-alpha-diol by rat medial basal hypothalamus. *Endocrinology,* **94**, 994–997

P'an, S. V. and Laubach, G. D. (1964). Steroid central depressants. In R. I. Dorfman (ed.). *Methods in Hormone Research,* Vol. 3, Academic Press, New York, 415–475

Parrott, R. F. and McDonald, P. G. (1975). Sexual behavior of male rats implanted in the brain with 19-hydroxytestosterone. *J. Endocr.*, **64**, 37–38P

Parvizi, N. and Ellendorff, F. (1975). 2-hydroxy-oestradiol-17-beta as a possible link in steroid brain interaction. *Nature*, **256**, 59–60

Payne, A. H., Lawrence, C. C., Foster, D. L. and Jaffe, R. B. (1973). Intranuclear binding of 17-beta-estradiol and estrone in female ovine pituitaries following incubation with estrone sulfate. *J. Biol. Chem.*, **248**, 1598–1602

Pilven, A., Thieulant, M. L., Ducouret, B., Samperez, S. and Jouan, P. (1976). Rapid and intensive conversion of 5-alpha-androstane-3-alpha, 17-beta-diol into 5-alpha-dihydrotestosterone in the male rat anterior pituitary: *in vivo* and *in vitro* studies. *Steroids,* **28**, 349–359

Piva, F., Motta, M. and Martini, L. (1978). Long, short and ultra-short feedback loops. In L. J. De Groot (ed.). *Metabolic Basis of Endocrinology.* Grune and Stratton, New York, to be published

Plapinger, L. and McEwen, B. S. (1973). Autogeny of estradiol-binding sites in rat brain. Part I: Appearance of presumptive adult receptors in cytosol and nuclei. *Endocrinology,* **93**, 1119–1128

Raisman, G. and Field, P. M. (1971). Anatomical considerations relevant to the interpretation of neuroendocrine experiments. In L. Martini and W. F. Ganong (eds.). *Frontiers in Neuroendocrinology,* Vol. 2, Oxford University Press, New York, 3–44

Raynaud, J. P., Mercier-Bodard, C. and Baulieu, E. E. (1971). Rat estradiol-binding plasma protein (EBP). *Steroids,* **18**, 767–788

Reddy, V. V. R., Naftolin, F. and Ryan, K. J. (1973). Aromatisation in the central nervous system of rabbits: effects of castration and hormone treatment. *Endocrinology,* **92**, 589–594

Reddy, V. V. R., Naftolin, F. and Ryan, K. J. (1974*a*). Steroid 17-beta-oxydoreductase activity in the rabbit central nervous system and adenohypophysis. *J. Endocr.*, **62**, 401–402

Reddy, V. V. R., Naftolin, F. and Ryan, K. J. (1974*b*). Conversion of androstenedione to estrone by neural tissues from fetal and neonatal rats. *Endocrinology,* **94**, 117–121

Rommerts, F. F. G. and Van der Molen, H. J. (1971). Occurrence and localisation of 5-alpha-steroid reductase, 3-alpha- and 17-beta-hydroxysteroid dehydrogenases in hypothalamus and other brain tissues of the male rat. *Biochim. Biophys. Acta*, **248**, 489–502

Sanyal, M. K. and Todd, R. B. (1972). 5-alpha-dihydroxyprogesterone influence on ovulation of prepubertal rats. *Proc. Soc. exp. Biol. Med.*, **141**, 622–624

Sar, M. and Stumpf, W. E. (1973*a*). Cellular and subcellular localisation of radioactivity in the rat pituitary after injection of 1, 2-^{3}H-testosterone using dry-autoradiography. *Endocrinology,* **92**, 631–635

Sar, M. and Stumpf, W. E. (1973*b*). Autoradiographic localisation of radioactivity in the rat brain after the injection of 1, 2-^{3}H-testosterone. *Endocrinology,* **92**, 251–256

Sar, M. and Stumpf, W. E. (1973*c*). Neurons of the hypothalamus concentrate ^{3}H-progesterone or its metabolites. *Science,* **182**, 1266–1268

Sar, M. and Stumpf, W. E. (1975). Distribution of androgen-concentration neurons in rat brain. In W. E. Stumpf, and L. D. Grant (eds.). *Anatomical Neuroendocrinology,* Karger, Basel, 120–133

Seiki, K., Haruki, Y., Imanishi, Y. and Enomoto, T. (1977). Further evidence of presence of progesterone-binding proteins in female rat hypothalamus. *Endocr. Japon.*, **24**, 233–238

Seiki, K. and Hattori, M. (1971). A more extensive study on the uptake of labelled progesterone by the hypothalamus and pituitary gland of rats. *J. Endocr.*, **51**, 793–794

Seiki, K. and Hattori, M. (1973). *In vivo* uptake of progesterone by the hypothalamus and pituitary of the female ovariectomised rat and its relationship to cytoplasmic progesterone-binding protein. *Endocr. Japon.*, **20**, 111–119

Seiki, K., Miyamoto, M., Yamashita, A. and Kottani, M. (1969). Further studies on the uptake of labelled progesterone by the hypothalamus and pituitary of rats. *J. Endocr.*, **43**, 129–130

Shridharan, B. N., Meyer, R. K. and Karavolas, H. J. (1974). Effect of 5-alpha-dihydroprogesterone, pregn-5-ene-3, 20-dione, pregnelone and related progestins on ovulation in PMSG-treated immature rats. *J. Reprod. Fertil.*, **36**, 83–90

Sodersten, P. (1973). Estrogen-activated sexual behaviour in male rats. *Hormones Behav.*, **4**, 247–256

Stewart-Bentley, M., Odell, W. D. and Horton, R. (1974). The feedback control of luteinising hormone in normal adult men. *J. Clin. Endocr. Metab.*, **38**, 545–553.

Stumpf, W. E., Sar, M. and Keefer, D. A. (1975). Atlas of estrogen target cells in rat brain. In W. E. Stumpf and L. D. Grant (eds.). *Anatomical Neuroendocrinology*, Karger, Basel, 104–109

Stupnicka, E., Massa, R., Zanisi, M. and Martini, L. (1977). Role of anterior pituitary and hypothalamic metabolism of progesterone in the control of gonadotropin secretion. In P. O. Hubinont, M. l'Hermite and C. Robyn (eds.). *Clinical Reproductive Neuroendocrinology*, Karger, Basel, 88–95

Swerdloff, R. S., Grover, P. K., Jacobs, H. S. and Bain, J. (1973). Search for a substance which selectively inhibits FSH-effects of steroids and prostaglandins on serum FSH and LH levels. *Steroids*, **21**, 703–722

Swerdloff, R. S., Walsh, M. D. and Odell, W. D. (1972). Control of LH and FSH secretion in the male: evidence that aromatisation of androgens to estradiol is not required for inhibition of gonadotropin secretion. *Steroids*, **20**, 13–22

Tabei, T., Haga, H., Heinrichs, W. L. and Hermann, W. L. (1974). Metabolism of progesterone by rat brain, pituitary gland and other tissues. *Steroids*, **23**, 651–666

Tabei, T. and Heinrichs, W. L. (1974). Metabolism of progesterone by the brain and pituitary gland of sub-human primates. *Neuroendocrinology*, **15**, 281–289

Verjans, H. L. and Eik-Nes, K. B. (1977). Comparisin of effects of C19 (androstene or androstane) steroids on serum gonadotrophin concentrations and on accessory reproduction organ weights in gonadectomised adult male rats. *Acta Endocr.*, **84**, 829–841

Vertes, M. and King, R. J. B. (1971). The mechanism of oestradiol binding in rat hypothalamus: effect of androgenisation. *J. Endocr.*, **51**, 271–282

Wade, G. N. and Feder, H. H. (1972). Effects of several pregnane and pregnene steroids on estrous behavior in ovariectomised estrogen-primed guinea-pigs. *Physiol. Behav.*, **9**, 773–775

Wade, G. N., Harding, C. F. and Feder, H. H. (1973). Neural uptake of [1, 2-^{3}H] progesterone in ovariectomised rats, guinea-pigs and hamsters: correlation with species differences in behavioral responsiveness. *Brain Res.*, **61**, 357–367

Weiner, R. I. (1975). Role of brain catecholamines in the control of LH and prolactin secretion. In M. Motta, P. G. Crosignani and L. Martini (eds.). *Hypothalamic Hormones*, Academic Press, London, 249–253

Weisz, J. and Gibbs, C. (1974). Conversion of testosterone and androstenedione to estrogens *in vitro* by the brain of female rats. *Endocrinology*, **94**, 616–620

Whalen, R. E. and de Bold, J. F. (1974). Comparative effectiveness of testosterone, androstenedione and dihydrotestosterone in maintaining mating behavior in the castrated male hamster. *Endocrinology*, **95**, 1674–1679

Whalen, R. E. and Luttge, W. G. (1971). Differential localisation of progesterone uptake in brain. Role of sex, oestrogen pretreatment and adrenalectomy. *Brain Res.*, **33**, 147–155

Wilson, J. D. and Gloyna, R. E. (1970). The intranuclear metabolism of testosterone in the accessory organs of reproduction. *Recent Progr. Hormone Res.*, **26**, 309–336

Zanisi, M. and Martini, L. (1975). Effects of progesterone metabolites on gonadotrophin secretion. *J. Steroid Biochem.*, **6**, 1021–1023

Zanisi, M., Motta, M. and Martini, L. (1973). Inhibitory effect of 5-alpha-reduced metabolites of testosterone on gonadotrophin secretion. *J. Endocr.*, **56**, 315–316

Zanisi, M., Motta, M. and Martini, L. (1975). New findings on the feedback control of anterior pituitary function. In T. Charro (ed.). *Basic Applications and Clinical Use of Hypothalamic Hormones*, Excerpta Medica, Amsterdam, 178–191

Author index

Subject index